Groundwater Systems Management

Groundwater Systems Management

Edited by **Keith Wheatley**

SYRAWOOD
PUBLISHING HOUSE
New York

Published by Syrawood Publishing House,
750 Third Avenue, 9th Floor,
New York, NY 10017, USA
www.syrawoodpublishinghouse.com

Groundwater Systems Management
Edited by Keith Wheatley

International Standard Book Number: 978-1-68286-269-8 (Hardback)

Printed in China.

Contents

Preface

The world is advancing at a fast pace like never before. Therefore, the need is to keep up with the latest developments. This book was an idea that came to fruition when the specialists in the area realized the need to coordinate together and document essential themes in the subject. That's when I was requested to be the editor. Editing this book has been an honour as it brings together diverse authors researching on different streams of the field. The book collates essential materials contributed by veterans in the area which can be utilized by students and researchers alike.

Groundwater systems management is vital for adequate water consumption and administration. Water-soil interactions, assessment of groundwater quality, properties of aquifers, well hydraulics, and ground water flow are some of the topics included in the book that are of utmost significance and bound to provide incredible insights to readers. Different approaches, evaluations, methodologies and advanced case-studies on groundwater systems management have been included in this book. It is an excellent reference text for students, researchers, and scientists engaged in the field of water management, geology, etc.

Each chapter is a sole-standing publication that reflects each author´s interpretation. Thus, the book displays a multi-facetted picture of our current understanding of applications and diverse aspects of the field. I would like to thank the contributors of this book and my family for their endless support.

Editor

1

Grey Forecast Rainfall with Flow Updating Algorithm for Real-Time Flood Forecasting

Jui-Yi Ho [1] and Kwan Tun Lee [1,2,*]

[1] Hydrotech Division, Taiwan Typhoon and Flood Research Institute, National Applied Research Laboratories, 11F., No. 97, Sec. 1, Roosevelt Rd., Taipei City 10093, Taiwan; E-Mail: juiyiho@narlabs.org.tw

[2] Department of River and Harbor Engineering, National Taiwan Ocean University, No 2, Beining Rd., Keelung City 20224, Taiwan

* Author to whom correspondence should be addressed; E-Mail: ktlee@ntou.edu.tw;

Academic Editor: Kwok-wing Chau

Abstract: The dynamic relationship between watershed characteristics and rainfall-runoff has been widely studied in recent decades. Since watershed rainfall-runoff is a non-stationary process, most deterministic flood forecasting approaches are ineffective without the assistance of adaptive algorithms. The purpose of this paper is to propose an effective flow forecasting system that integrates a rainfall forecasting model, watershed runoff model, and real-time updating algorithm. This study adopted a grey rainfall forecasting technique, based on existing hourly rainfall data. A geomorphology-based runoff model can be used for simulating impacts of the changing geo-climatic conditions on the hydrologic response of unsteady and non-linear watershed system, and flow updating algorithm were combined to estimate watershed runoff according to measured flow data. The proposed flood forecasting system was applied to three watersheds; one in the United States and two in Northern Taiwan. Four sets of rainfall-runoff simulations were performed to test the accuracy of the proposed flow forecasting technique. The results indicated that the forecast and observed hydrographs are in good agreement for all three watersheds. The proposed flow forecasting system could assist authorities in minimizing loss of life and property during flood events.

Keywords: rainfall forecasting; flow forecasting; grey theory; geomorphology-based runoff model; flow updating algorithm

1. Introduction

Flood forecasting systems are nonstructural methods for reducing flood damage. An efficient forecasting system can assist with mitigating imminent disasters by providing information that can be disseminated rapidly to the flood-threatened areas. Standard practices in flood forecasting systems include hydrometeorological data transmission, database management, rainfall prediction, runoff estimation, and forecast information dissemination.

Brath *et al.* [1] indicated that quantitative rainfall forecasting plays a primary role in extending the lead time of river flow forecasting, which can improve the timeliness of flood control mechanisms. Because of improvements in the accuracy of weather radar systems, radar-based rainfall forecasting systems have superseded traditional rain gauges that provide measurements at only several locations for flood forecasting [2–4]. Toth *et al.* [5] indicated that radar detection is particularly difficult in mountainous regions because of the effect of ground occultation and altitude. Consequently, radar-based measurement techniques are limited under topographic conditions where radar reflectivity is poor [6,7]. Thus, hydrologists typically use probabilistic and stochastic methods for rainfall forecasting based on current and past rainfall measurements (e.g., [8–13]). Because these methods typically require large volumes of rainfall data for calibrating and training the model parameters, Deng [14] proposed a grey system theory-based model for future data prediction. Moreover, grey theory-based models require relatively few observations to predict outcomes [15–17]; thus, they are suitable for rainfall forecasting.

Rainfall forecasts are inputted into rainfall-runoff models to provide flood warning information for authorities. In recent decades, artificial neural networks (ANNs) have become a well-known tool for hydrologic forecasting [18–29]. However, ANNs require a large amount of hydrologic data to determine the adaptive weights, which are inadequate to be applied to data-sparse areas. Although fully distributed grid-based routing models can provide detailed information on flood wave transports, they may be unsuitable for real-time flood forecasting systems because the simulation process is typically time-consuming [30]. Hence, lumped and semi-distributed hydrological models are acceptable practical alternatives. For example, the Sacramento model was adopted by the National Weather Service River Forecast System in the United States [31,32], the tank model has been widely applied for runoff forecasting in Japan [33], the Hydrologiska Byråns Vattenbalansavdelning (HBV) model was adopted in Europe [34], and the Xinanjiang model was introduced in China [35]. To further minimize the requirement of observed flow data to develop semi-distributed models, hydrologists have adopted geomorphology-based runoff models [36–40].

Rodriguez-Iturbe and Valdes [35] proposed the geomorphologic instantaneous unit hydrograph (GIUH) model, which can be employed to derive the instantaneous unit hydrograph (IUH) of a watershed based on information from a topographic map or digital elevation data set. Subsequent studies have modified the GIUH model by incorporating kinematic-wave approximation, thereby providing reasonable estimations of flow velocity, which augments rainfall-runoff simulation used in both gauged and ungauged sites [41–43].

It is necessary to implement the effective data assimilation in the forecast process to bridge the immense gap between the theory and operational practice [44]. Comprehensive reviews of data assimilation approaches in operational hydrologic forecasting were presented by Liu *et al.* [45]. Since uncertainty is an inherent characteristic of watershed hydrodynamics, an ideal flood forecasting system should incorporate a real-time updating algorithm that revises the model state to improve the forecasting accuracy. Refsgaard [46] reported that techniques for updating real-time forecasting can be classified into the following four categories: (1) updating input variables; (2) updating state variables; (3) updating model parameters; and (4) updating output variables (*i.e.*, error prediction). Previous studies have developed updating techniques based on time-series analysis [47,48], statistical methods [17,49], multiple regression analysis [50], dimensional variational algorithms [51,52], and the filter approaches [53–58]. Selection of an appropriate updating algorithm depends on the availability of real-time feedback data and the structure of the rainfall-runoff model employed for flow forecasting.

The purpose of this study is to develop an effective flood forecasting system for midsize rural watersheds. We adopted the grey rainfall forecasting technique based on existing hourly rainfall data to avoid poor radar reflectivity in mountainous watersheds. In performing the watershed runoff simulation, a geomorphology-based runoff model which can account different geomorphologic and hydrological characteristics of the watershed was used in this study. Furthermore, a flow updating algorithm was linked to the runoff model to estimate watershed runoff in the next three hours. The proposed flow forecasting system can operate with high efficiency to meet the requirements of real-time flow forecasting. The system was applied to three watersheds; one in the United States (Goodwin Creek) and two in Northern Taiwan (Heng-Chi and San-Hsia). The results of the flood forecasting were compared with official records to confirm the validity of the proposed system. In the following sections, Section 2 describes the analytical methods including the short-term rainfall forecasting and the geomorphology-based runoff model with an updating algorithm. The application of the proposed methods and the forecast results are presented in Section 3. Section 4 summarizes the conclusions of this study.

2. Analytical Methods

The framework of the proposed flood forecasting system incorporates a grey rainfall forecasting model [14], kinematic-wave-based GIUH (KW-GIUH) model [36], and flow updating algorithm that is linked to the KW-GIUH model to improve the flow forecasting accuracy.

2.1. Short-Term Rainfall Forecasting

The grey rainfall forecasting is adopted herein for two reasons: (1) a short-term rainstorm system is too complex to be simulated by using deterministic approaches; (2) the grey system provides an efficient way for rainfall prediction using only small amount of past observed rainfall data. Consequently, the grey model proposed by Deng [14] is appropriate for the present system for rainfall forecasting.

Although hydrological time-series data typically exhibit random forms, a systematic trend can be observed after a repeatedly accumulated generating operations (AGO). According to the AGO concept,

Deng [14] developed an effective method for predicting future data based on a limited number of observations. A series of raw rainfall data can be expressed as follows:

$$R^{(0)}(t)=\left[r^{(0)}(1),\ r^{(0)}(2),\ \ldots,r^{(0)}(k)\ ,\ldots,\ r^{(0)}(n)\right] \tag{1}$$

where $R^{(0)}(t)$ denotes the raw time-series rainfall data set, and $r^{(0)}(k)$ is the kth observed rainfall. The first-order AGO series can be defined as [14]:

$$R^{(1)}(t)=\sum_{k=1}^{t}r^{(0)}(k)\ ;\ t=1,\ 2,\ 3,\ \ldots,n \tag{2}$$

where $R^{(1)}(t)$ is the first-order AGO rainfall time series, and $r^{(1)}(t)$ is the first-order AGO observed rainfall data at time *t*. A first-order differential equation is employed to fit the cumulative rainfall data, which can be expressed as:

$$\frac{dR^{(1)}(t)}{dt}+aR^{(1)}(t)=b \tag{3}$$

where a and b are the grey system model parameters. Deng [14] indicated that the whitening of the grey derivatives of discrete data with unit time intervals can be expressed as:

$$\left.\frac{dR^{(1)}(t)}{dt}\right|_{t=k}=r^{(1)}(k)-r^{(1)}(k-1)=R^{(0)}(k) \tag{4}$$

The whitening value of $R^{(1)}(t)\big|_{t=k}$ is defined as:

$$Z^{(1)}(k)=R^{(1)}(t)\big|_{t=k}\cong\frac{1}{2}\left(r^{(1)}(k)+r^{(1)}(k-1)\right),\quad \forall k=2,\ 3,\ \ldots\ ,\ t \tag{5}$$

where $Z^{(1)}(k)$ denotes the whitening value of $R^{(1)}(t)\big|_{t=k}$. Next, Equations (4) and (5) are substituted into Equation (3) to obtain a grey discrete differential form as:

$$R^{(0)}(k)+a\cdot Z^{(1)}(k)=b \tag{6}$$

The grey parameters a and b can be estimated using the least square method. Thus, the solution of Equation (6) is expressed as:

$$\hat{r}^{(1)}(k+\Delta t)=\left(r^{(0)}(k)-\frac{b}{a}\right)e^{-ak}+\frac{b}{a} \tag{7}$$

where $\hat{r}^{(1)}(k+1)$ is the forecast value of the first-order AGO series. Consequently, the rainfall depth at the subsequent time step can be obtained from the equation:

$$\hat{r}^{(0)}(k+\Delta t)=\hat{r}^{(1)}(k+\Delta t)-\hat{r}^{(1)}(k),\ \ \forall k=1,\ 2,\ 3,\ \ldots\ ,\ n\in N \tag{8}$$

where $\hat{r}^{(0)}(k+\Delta t)$ is the forecast rainfall depth at time $k+\Delta t$. The grey parameters in Equation (7) are updated when the new observed rainfall data are obtained. Yu *et al.* [17] indicated that the accuracy of rainfall forecasting decreases when the lead time is increased because the forecast error is cumulated from previous lead-time forecasting. Consequently, an algorithm called single-time-step forecasting proposed by Yu *et al.* [17] was used to overcome the shortcoming.

Several studies have successfully applied the first-order grey model for hydrological forecasting [14–17]. Because forecast errors are cumulative, grey forecasting models become less reliable as the forecast lead time increases. However, when a grey rainfall forecasting model is combined with a watershed rainfall-runoff model, the forecast lead time can be extended because of the time lag in the transporting of a flood wave from upstream to downstream.

This study adopted four criteria to evaluate the performance of the grey rainfall forecasting model. First, the error of total cumulative rainfall (*ETCR*) is defined as:

$$ETCR = \frac{\left|\sum_{t=1}^{n} r_t - \sum_{t=1}^{n} \hat{r}_t\right|}{\sum_{t=1}^{n} r_t} \tag{9}$$

where $\hat{r}_t$ is the forecast rainfall at time t; r_t denotes the observed rainfall at time t; and n represents the number of time steps to be estimated. A more accurate forecast can be obtained when *ETCR* is approximately zero. Second, the relative root mean square error (*RMSE*) is defined as:

$$RMSE = \sqrt{\frac{\sum_{t=1}^{n}\left[\left(R_t - \hat{R}_t\right)/R_t\right]^2}{n}} \tag{10}$$

where R_t denotes the observed cumulative rainfall at time t; $\hat{R}_t$ is the forecast cumulative rainfall at time t. A more accurate forecast can be obtained when *RMSE* is approximately zero. Third, the coefficient of efficiency (*CE*) is defined as [56]:

$$CE = 1 - \frac{\sum_{t=1}^{n}\left(R_t - \hat{R}_t\right)^2}{\sum_{t=1}^{n}\left(R_t - \overline{R}\right)^2} \tag{11}$$

$\overline{R}$ represents the mean of the observed cumulative rainfall. A more accurate fit between the forecast rainfall and the observed rainfall can be obtained when *CE* is approximately one. Moreover, the coefficient of correlation (*CC*) is defined as:

$$CC = \frac{\sum_{t=1}^{n}\left(R_t - \hat{R}_t\right)\left(\hat{R}_t - \hat{\overline{R}}\right)}{\sqrt{\sum_{t=1}^{n}\left(R_t - \overline{R}\right)^2 \cdot \sum_{t=1}^{n}\left(\hat{R}_t - \hat{\overline{R}}\right)^2}} \tag{12}$$

where $\hat{\overline{R}}$ is the mean of the forecast cumulative rainfall at time t. A more accurate forecast can be obtained when the value of *CC* is approximately one. The *ETCR* and *RMSE* represent a quantitative judgment of model performance. The *CE* is used to measure the similarity between the predicted and observed accumulated rainfall. The *CC* is used to measure the correlative relationship between the predicted and observed accumulative rainfall.

2.2. Geomorphology-Based Runoff Model with an Updating Algorithm

This study adopted an IUH model to provide an efficient method for estimating watershed runoff. The KW-GIUH model [36] was used because the IUH can be derived only by using watershed geomorphologic information obtained from a topographic map or digital elevation dataset. The hydrological response function of the watershed can be expressed analytically as follows [35]:

$$u(t)=\sum_{w\in W}\left[f_{x_{o_i}}(t)*f_{x_i}(t)*f_{x_j}(t)*\ldots*f_{x_\Omega}(t)\right]_w\cdot P(w) \tag{13}$$

where $u(t)$ is the IUH of the watershed; W is the flow path space, which is expressed as $W=\left\langle x_{o_i},\, x_i,\, x_j,\ldots,\, x_\Omega\right\rangle$; $f_{x_{oi}}(t)$ denotes the travel time probability density function in state x_j with a mean value of T_{x_j} ; $*$ denotes a convolution integral; and $P(w)$ represents the probability of a raindrop adopting a flow path w .

Kinematic-wave approximation can be employed to express the runoff travel time for the *i*th-order surface flow region as follows [59]:

$$T_{x_{o_i}}=\left(\frac{n_o\bar{L}_{o_i}}{\bar{S}_{o_i}^{1/2}\,i_e^{\,m-1}}\right)^{\frac{1}{m}} \tag{14}$$

where n_o is the overland-flow roughness coefficient; $\bar{L}_{o_i}$ denotes the mean *i*th-order overland length; $\bar{S}_{o_i}$ is the mean *i*th-order overland slope; i_e represents the effective rainfall intensity; and m is an exponent recognized as 5/3 in Manning's formula. The runoff travel time for the *i*th-other channel is expressed as follows [36]:

$$T_{x_i}=\frac{B_i}{2i_e\bar{L}_{o_i}}\left[\left(h_{co_i}^{m}+\frac{2i_en_c\bar{L}_{o_i}\bar{L}_{c_i}}{\bar{S}_{c_i}^{1/2}B_i}\right)^{\frac{1}{m}}-h_{co_i}\right] \tag{15}$$

where B_i is the *i*th-order channel width; n_c represents the channel roughness coefficient; $\bar{L}_{c_i}$ is the mean *i*th-order channel length; $\bar{S}_{c_i}$ denotes the mean *i*th-order channel slope; and h_{co_i} is the inflow depth of the *i*th-order channel caused by water transporting from upstream reaches. Hence, the runoff travel times for different orders of overland-flow paths and channels can be estimated, and the watershed IUH can then be derived by using Equation (13). Consequently, the watershed runoff simulated by using KW-GIUH model can be expressed as:

$$Q_{sim,t}=\int_0^{\tau}i(\tau)u(t-\tau)\,d\tau \tag{16}$$

where $Q_{sim,t}$ is the simulated direct runoff at time *t*; $i(\tau)$ is the rainfall intensity; and $u(t-\tau)$ is the unit impulse response function derived from the KW-GIUH model.

Uncertainty is an inherent hydrodynamic characteristic of watershed; therefore, this study adopted a real-time updating algorithm to improve the accuracy of flow forecasting. The change in measured discharge between time *t* and $t+\Delta t$ can be expressed as:

$$\Delta Q_{rec} = Q_{rec,\ t+\Delta t} - Q_{rec,\ t} \tag{17}$$

where $Q_{rec,\ t}$ and $Q_{rec,\ t+\Delta t}$ denote the measured discharges at time t and $t+\Delta t$, respectively; and ΔQ_{rec} represents the change in measured discharge between time t and $t+\Delta t$. Assuming that a reliable runoff model is used and short-term rainfall forecasting data (*i.e.*, with a lead time of several hours) are available, the value of ΔQ_{rec} in Equation (17) is assumed equal to the change in simulated discharge between time t and $t+\Delta t$; this is illustrated as follows:

$$\Delta Q_{rec} = \Delta Q_{sim} = Q_{sim,\ t+\Delta t} - Q_{sim,\ t} \tag{18}$$

where $Q_{sim,\ t}$ and $Q_{sim,\ t+\Delta t}$ denote the simulated discharges at time t and $t+\Delta t$, respectively; and ΔQ_{sim} represents the change in simulated discharge between time t and $t+\Delta t$. Consequently, the forecast discharge at time t can be approximated by:

$$\begin{aligned} Q_{fore,\ t+\Delta t} &= Q_{rec,\ t} + \Delta Q_{rec} \\ &\cong Q_{rec,\ t} + \Delta Q_{sim} \end{aligned} \tag{19}$$

where $Q_{fore,\ t+\Delta t}$ is the forecast discharge at time $t+\Delta t$ and ΔQ_{sim} is obtained from Equation (18). The schematic of the updating algorithm is shown in Figure 1. In the case that the real-time measured discharge $Q_{rec,t}$ cannot be transmitted successfully through the telemetric system during the rainstorm, the forecast discharge at time $t+\Delta t$ ($Q_{fore,\ t+\Delta t}$) is replaced by the model generated runoff discharge ($Q_{\text{sim},\ t+\Delta t}$) without using the updating techniques. Although the proposed updating algorithm is simple, it is an efficient method for watershed runoff forecasting.

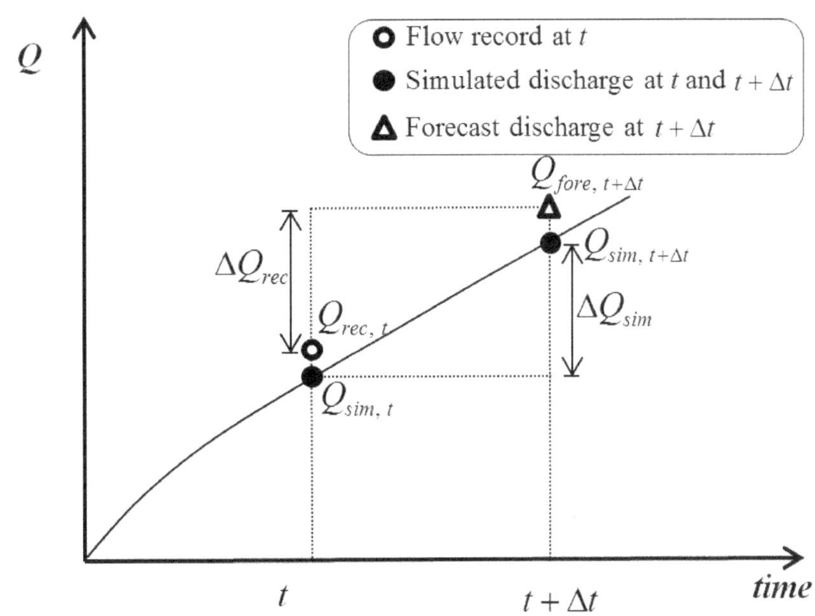

Figure 1. Schematic of the updating algorithm.

To evaluate the suitability of the KW-GIUH model, two criteria were chosen to determine the goodness-of-fit between the observed and simulated flow hydrographs. The coefficient of efficiency CE_Q is defined as follows [60]:

$$CE_Q = 1 - \frac{\sum_{t=1}^{n}\left(Q_{rec}(t) - Q_{fore}(t)\right)^2}{\sum_{t=1}^{n}\left(Q_{rec}(t) - \overline{Q}_{rec}\right)^2} \tag{20}$$

where $Q_{rec}(t)$ and $Q_{fore}(t)$ denote the recorded and forecast discharges at time t, respectively; $\overline{Q}_{rec}$ is the mean recorded discharge during a storm event, and n is the number of discharge records during the storm event. The CE_Q is used to measure the similarity between the predicted and observed discharge hydrographs. A more accurate fit is obtained when the value of CE_Q is approximately one. The peak discharge error is defined as follows:

$$EQ_P(\%) = \frac{\left|(Q_P)_{fore} - (Q_P)_{rec}\right|}{(Q_P)_{rec}} \times 100 \tag{21}$$

where $(Q_P)_{fore}$ is the forecast peak discharge, and $(Q_P)_{rec}$ denotes the recorded peak discharge. The EQ_P is used to measure the error of peak discharge directly. The error of time to peak discharge, ET_P, is defined as:

$$ET_P = (T_P)_{fore} - (T_P)_{rec} \tag{22}$$

where $(T_P)_{fore}$ and $(T_P)_{rec}$ are the forecast and recorded time to peak discharge, respectively.

3. Model Applications

3.1. Description of Study Watersheds

Three watersheds were selected to investigate the applicability of the proposed model; one in the United States (Goodwin Creek) and two in Northern Taiwan (Heng-Chi and San-Hsia). Goodwin Creek is a tributary of Long Creek that flows into the Yocona River, which is one of the main rivers of the Yazoo River Basin. Figure 2a shows the watershed stream network and locations of hydrological gauging stations. The terrain elevation of the Goodwin Creek watershed ranges from 71 to 128 m above sea level (mean). The land area is composed of cultivated land (13.79%), forests (26.00%), pastures (59.80%), and water (0.41%). The climate of the Goodwin Creek watershed is humid with hot temperatures during summer and mild temperatures during winter. The mean annual temperature and rainfall are approximately 17 °C and 1399 mm, respectively. Most of the rainfall occurs during winter and spring. Hydrological data were obtained from the Agricultural Research Service of the United States Department of Agriculture. Among the 32 rain-gauging stations in the area, this study obtained rainfall records from nine stations. The Thiessen polygons method [61] was employed to calculate the hourly spatial-average rainfall intensities. Fourteen flow gauging stations were set up in the Goodwin Creek watershed area. The control areas of the flow gauging stations ranged from 0.06 to 21.39 km^2. In this study, Flow-gauging Station No.1 (STA01), which has a drainage area of 21.39 km^2, was selected as the test site to verify the model.

The Heng-Chi and San-Hsia watersheds are subwatersheds in Ta-Han Creek, which is one of the main rivers of the Tam-Sui River Basin in Northern Taiwan. Figure 2b shows the watershed stream networks and locations of the hydrological gauging stations. The elevation of the Heng-Chi (San-Hsia) watershed ranges from 20 to 970 m (30 to 1770 m), and the land is composed of 70% (75%) forest, 25% (20%) cultivated land, and 5% (5%) buildings/road. The mean annual precipitation in these areas is approximately 3000 mm. Most of the severe storm events are from typhoon activity between May and October, and intense rainfall (>50 mm/h) occurs every year.

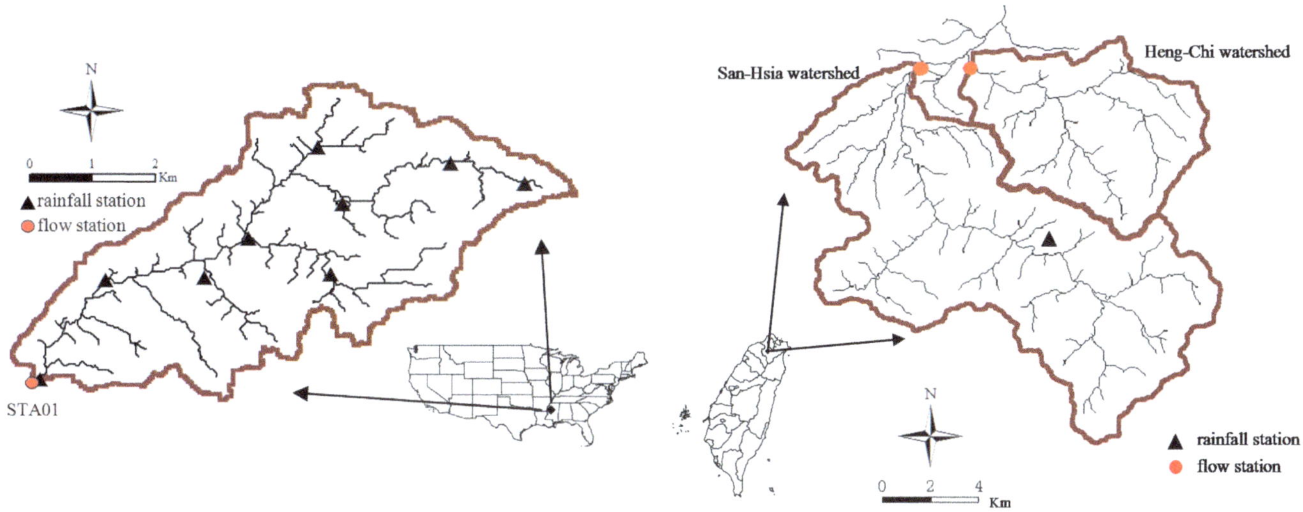

Figure 2. Watershed boundary and channel network of the study watersheds: (**a**) Goodwin Creek watershed; (**b**) Heng-Chi and San-Hsia watershed.

The geomorphologic factors were obtained from a digital elevation model [62] based on datasets of the Goodwin Creek watershed (30-m resolution) and the Heng-Chi and San-Hsia watersheds (40-m resolution). Table 1 shows the geomorphologic factors of the watersheds used in the KW-GIUH model.

Table 1. Geomorphologic factors of the study watersheds.

Watershed	i	N_i	$\overline{A}_i$ (km²)	$\overline{L}_{c_i}$ (km)	$\overline{S}_{c_i}$ (m/m)	$\overline{S}_{o_i}$ (m/m)
Goodwin (STA01)	1	76	0.18	0.40	0.0128	0.0228
	2	16	0.75	0.76	0.0090	0.0257
	3	4	3.08	1.56	0.0060	0.0260
	4	1	21.38	7.53	0.0019	0.0213
Heng-Chi	1	29	1.07	0.80	0.1304	0.3028
	2	6	6.91	3.13	0.0580	0.2957
	3	2	19.81	1.79	0.0105	0.2468
	4	1	53.15	4.98	0.0078	0.1977
San-Hsia	1	69	1.15	0.92	0.1613	0.3138
	2	16	4.99	2.08	0.0924	0.3016
	3	3	18.15	3.88	0.0372	0.3644
	4	1	125.88	17.83	0.0131	0.2918

Notes: N_i is the number of ith-order streams; $\overline{A}_i$ is the mean ith-order subwatershed area; $\overline{L}_{c_i}$ is the mean ith-order channel length; $\overline{S}_{c_i}$ is the mean ith-order channel slope; $\overline{S}_{o_i}$ is the mean ith-order hillslope slope.

3.2. Rainfall Forecasting

Table 2 shows the details of storm events that occurred in the study watersheds; these details were used for parameter calibration and model verification. In performing the grey rainfall model, parameters a and b (Equation (7)) can be estimated by using a least square method only based on small amount of past observed rainfall data. The watershed geomorphological factors in performing the KW-GIUH model are shown in Table 1, which can be obtained by applying a digital elevation model. The calibrated model parameters of the KW-GIUH model for the Heng-Chi and San-Hsia watersheds are $n_o = 0.6$ and $n_c = 0.05$, and $n_o = 0.2$ and $n_c = 0.02$ for the Goodwin watershed. The values of model parameters were stable for the test storms in the watersheds. Sensitivity analysis for the model parameters of KW-GIUH can be found in Lee and Yen [42].

Table 2. Storm records analyzed in this study.

Watershed (Rain Station)	Event Date	Rainfall Peak (mm/h)	Total Rainfall (mm)	Rainfall Duration (h)	Flow Peak (m^3/s)
Goodwin (STA01)	10/07/1989	11.13	92	48	29.7
	02/03/1991	16.80	62	18	21.1
	14/02/1992	4.66	30	11	9.1
	04/08/1995	13.29	113	28	16.7
	29/11/1996	6.95	44	29	10.2
	23/12/1997	7.17	45	13	19.6
	15/02/1998	10.20	62	48	27.3
	13/03/1999	11.92	95	52	31.0
	01/04/2000	24.09	152	63	32.9
	17/01/2001	7.26	78	60	10.3
Heng-Chi (Ta-Pao)	17/08/1984	36.00	372	51	169.0
	16/09/1985	69.00	348	25	620.0
	17/09/1986	46.00	420	61	457.0
	27/07/1987	32.00	114	18	164.0
	08/09/1987	59.00	261	45	329.0
	18/08/1990	48.00	342	48	492.0
	05/06/1993	54.00	146	18	179.0
	10/07/1994	22.00	150	31	58.2
	30/07/1996	31.00	450	42	243.0
	31/10/2000	33.00	508	38	317.0
12125San-Hsia (Ta-Pao)	17/08/1984	36.00	372	51	214.0
	16/09/1985	69.00	348	25	620.0
	17/09/1986	46.00	420	61	404.0
	27/07/1987	32.00	114	18	349.0
	08/09/1987	59.00	261	45	379.0
	18/08/1990	48.00	342	48	1060.0
	05/06/1993	54.00	146	18	339.0
	10/07/1994	22.00	150	31	257.0
	30/07/1996	31.00	450	42	720.0
	31/10/2000	33.00	508	38	435.0

Table 3 and Figure 3 show the performance of the grey rainfall forecasting model for the three watersheds. The *ETCR* and RMSE from Equations (9) and (10) represent the quantitative evaluation of the model performance, and *CE* from Equation (11) indicates the performance of the model based on cumulative rainfall. The performance of the model was assessed qualitatively based on the value of *CC* (Equation (12)) relative to the correlation between the forecast and observed cumulative rainfall. The results showed that *ETCR* is less than 0.24, *RMSE* is less than 0.38, *CE* is greater than 0.85, and *CC* is greater than 0.90, indicating that the forecast and recorded hyetographs are in good agreement. The forecast and recorded hyetographs in Figures 4 and 5 show the performance of the grey rainfall forecasting model based on lead times ranging from 1 to 3 h. Although the accuracy of the forecast rainfall decreases as the lead time increased, the results indicate that the proposed grey model is suitable for rainfall forecasting.

Table 3. Results of grey forecast rainfall.

		ETCR			*RMSE*			*CE*			*CC*		
Watershed	**Event Date**	**1-h Ahead**	**2-h Ahead**	**3-h Ahead**	**1-h Ahead**	**2-h Ahead**	**3-h Ahead**	**1-h Ahead**	**2-h Ahead**	**3-h Ahead**	**1-h Ahead**	**2-h Ahead**	**3-h Ahead**
Goodwin Creek	10/07/1989	0.03	0.18	0.22	0.03	0.08	0.17	0.99	0.98	0.92	0.99	0.95	0.88
	02/03/1991	0.05	0.16	0.20	0.08	0.17	0.29	0.96	0.93	0.88	0.99	0.95	0.88
	14/02/1992	0.01	0.03	0.08	0.02	0.14	0.27	1.00	0.98	0.95	0.99	0.96	0.90
	04/08/1995	0.06	0.11	0.16	0.04	0.17	0.24	0.97	0.90	0.87	0.99	0.98	0.92
	29/11/1996	0.03	0.09	0.10	0.08	0.24	0.31	0.97	0.90	0.85	0.99	0.98	0.96
	23/12/1997	0.06	0.15	0.21	0.03	0.19	0.25	0.98	0.91	0.88	0.99	0.99	0.97
	15/02/1998	0.04	0.07	0.16	0.06	0.11	0.19	1.00	0.94	0.88	1.00	0.96	0.97
	13/03/1999	0.09	0.14	0.20	0.06	0.18	0.21	0.99	0.95	0.89	1.00	0.98	0.90
	01/04/2000	0.03	0.08	0.15	0.03	0.12	0.16	0.99	0.96	0.91	1.00	1.00	0.98
	17/01/2001	0.07	0.11	0.19	0.07	0.11	0.16	0.98	0.91	0.88	0.98	0.94	0.91
Heng-Chi and San-Hsia	17/08/1984	0.05	0.08	0.16	0.11	0.18	0.27	0.99	0.98	0.94	1.00	0.98	0.96
	16/09/1985	0.06	0.12	0.24	0.09	0.22	0.31	0.96	0.93	0.86	0.99	0.95	0.88
	17/09/1986	0.01	0.03	0.05	0.05	0.18	0.21	1.00	0.98	0.95	1.00	1.00	0.99
	27/07/1987	0.09	0.18	0.22	0.03	0.11	0.18	0.97	0.91	0.88	0.98	0.93	0.91
	08/09/1987	0.01	0.09	0.10	0.14	0.26	0.38	0.89	0.83	0.80	1.00	0.98	0.96
	18/08/1990	0.03	0.09	0.12	0.08	0.17	0.24	0.95	0.85	0.81	1.00	0.99	0.97
	05/06/1993	0.05	0.08	0.11	0.01	0.04	0.19	1.00	0.94	0.87	0.99	0.96	0.97
	10/07/1994	0.09	0.11	0.18	0.02	0.13	0.20	0.98	0.99	0.96	0.99	0.98	0.90
	30/07/1996	0.04	0.07	0.11	0.05	0.15	0.22	0.99	0.95	0.83	1.00	1.00	0.98
	31/10/2000	0.04	0.05	0.17	0.08	0.09	0.13	0.99	0.97	0.93	1.00	1.00	0.99
Average		0.05	0.10	0.16	0.06	0.16	0.23	0.98	0.93	0.89	1.00	0.98	0.94

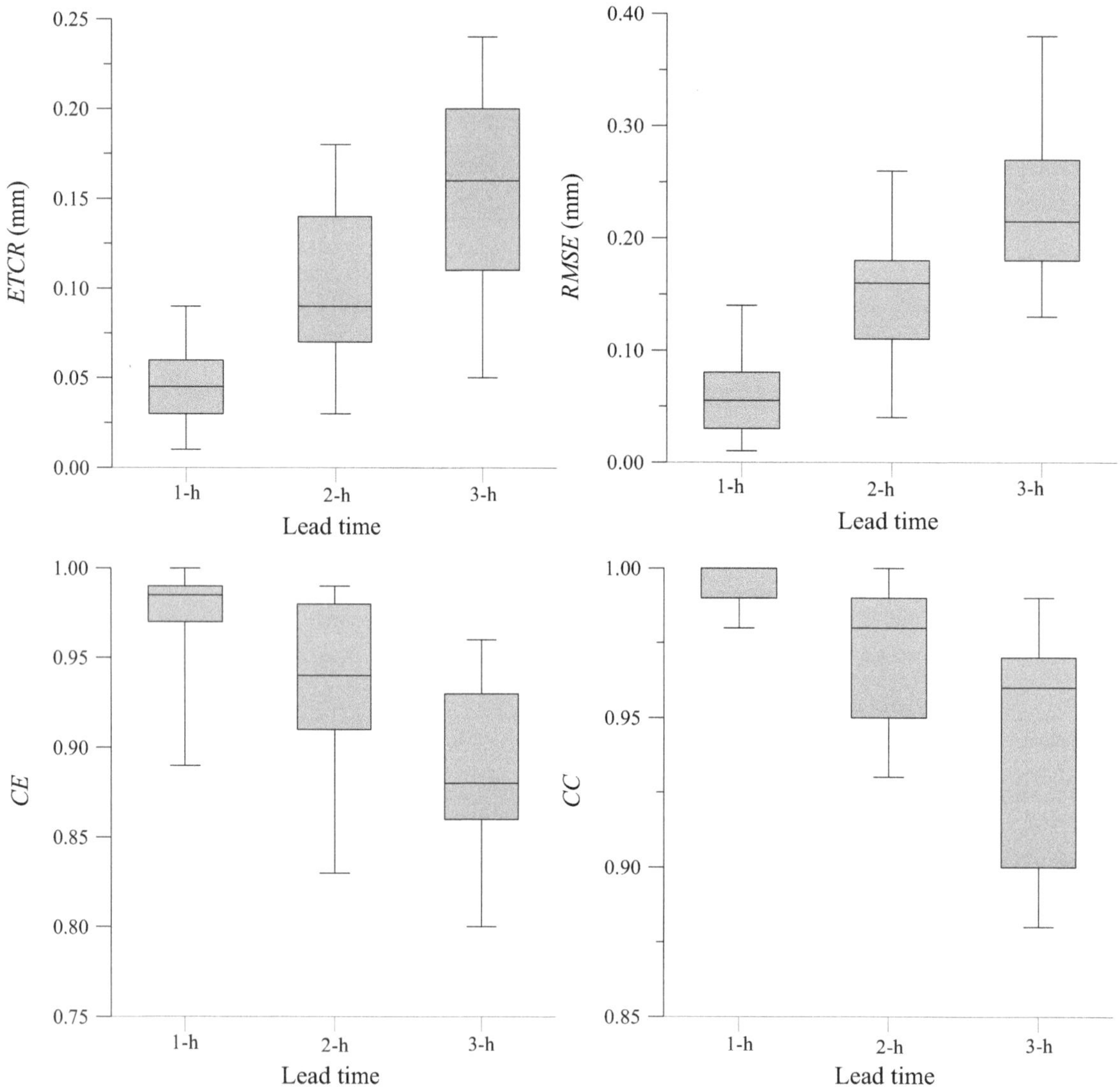

Figure 3. Results of evaluated criteria for grey forecast rainfall.

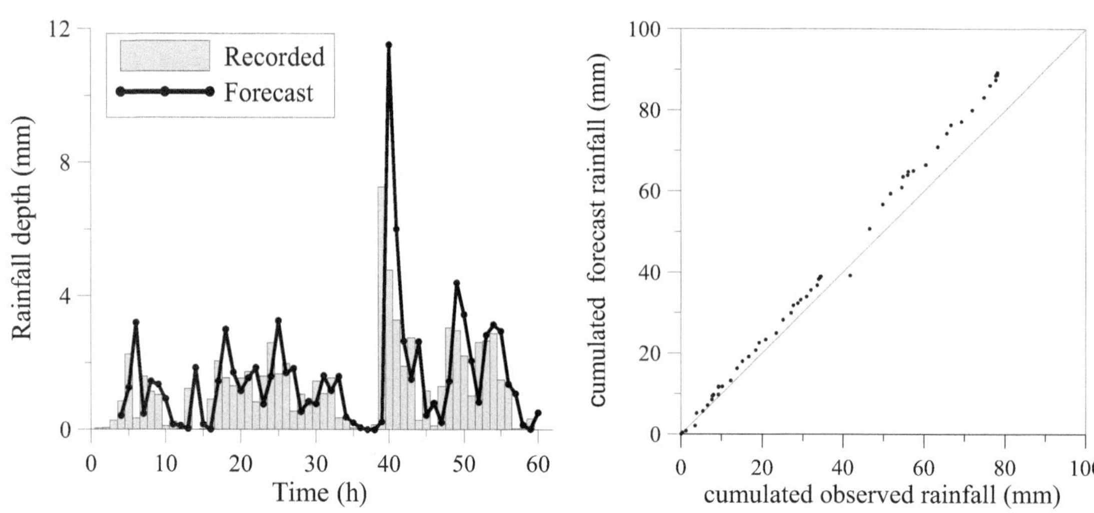

(a)

Figure 4. *Cont.*

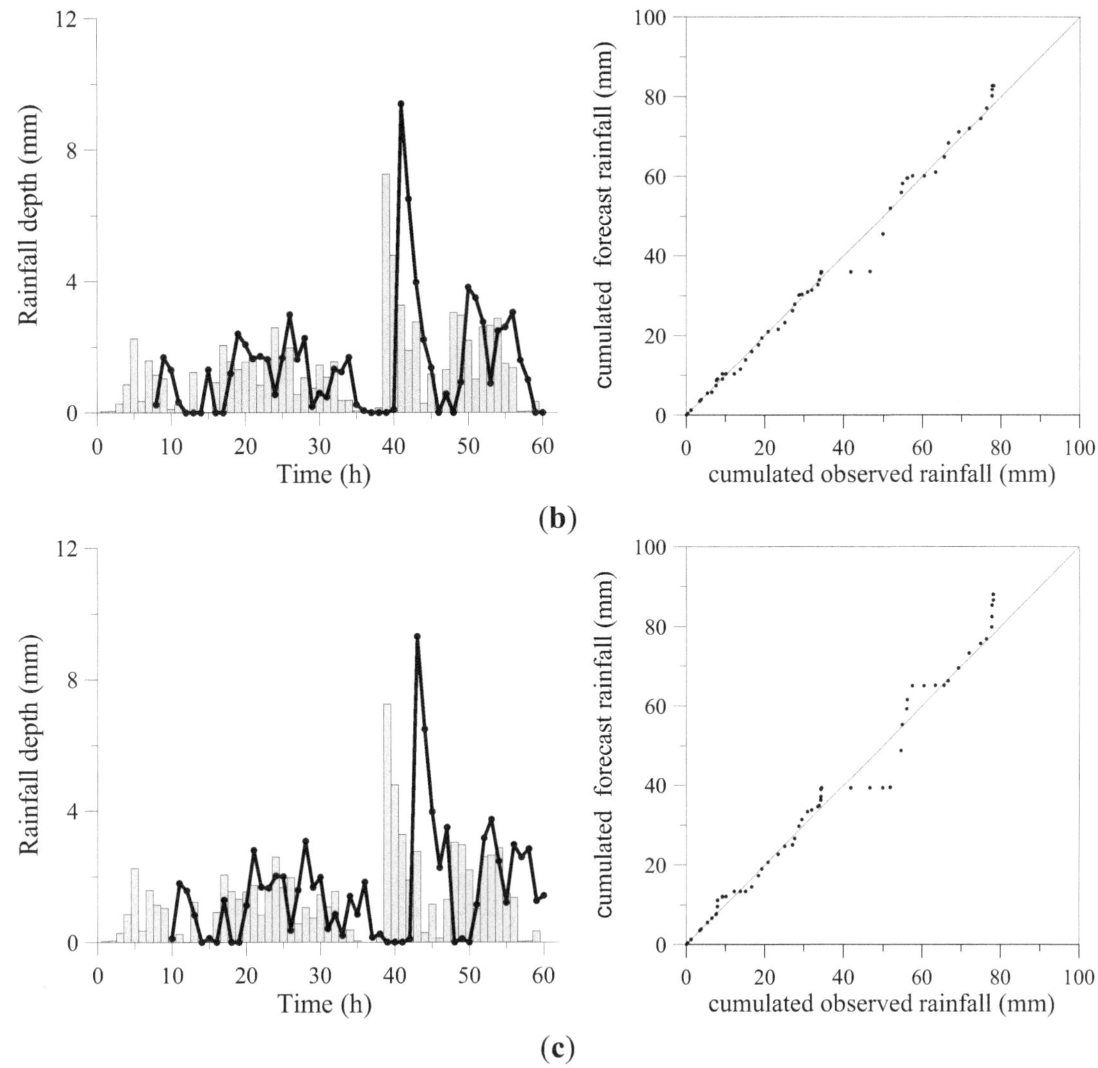

Figure 4. Grey model for rainfall forecasting in Goodwin Creek watershed: (**a**) 1-h ahead; (**b**) 2-h ahead; (**c**) 3-h ahead.

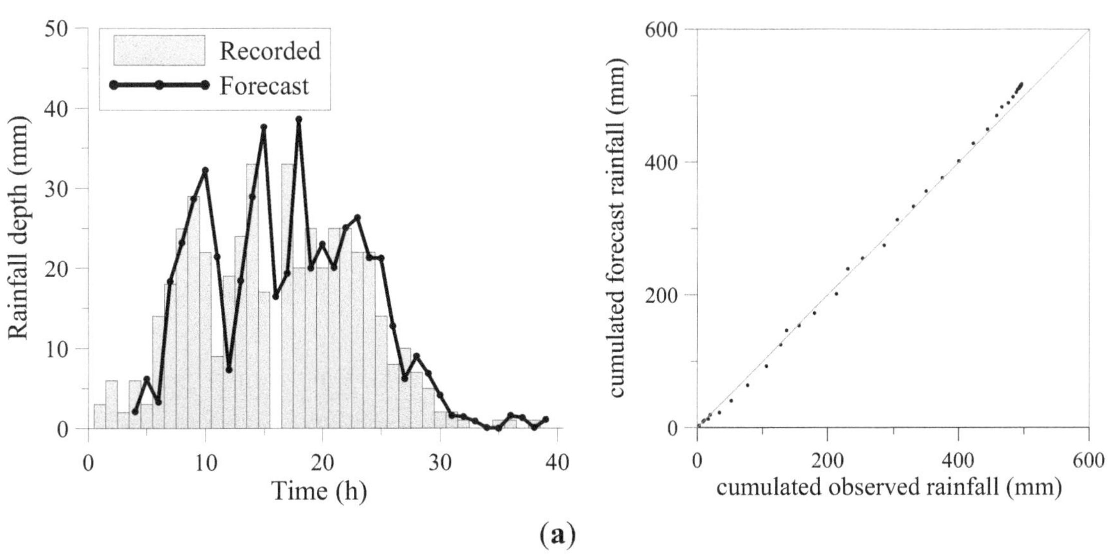

Figure 5. *Cont.*

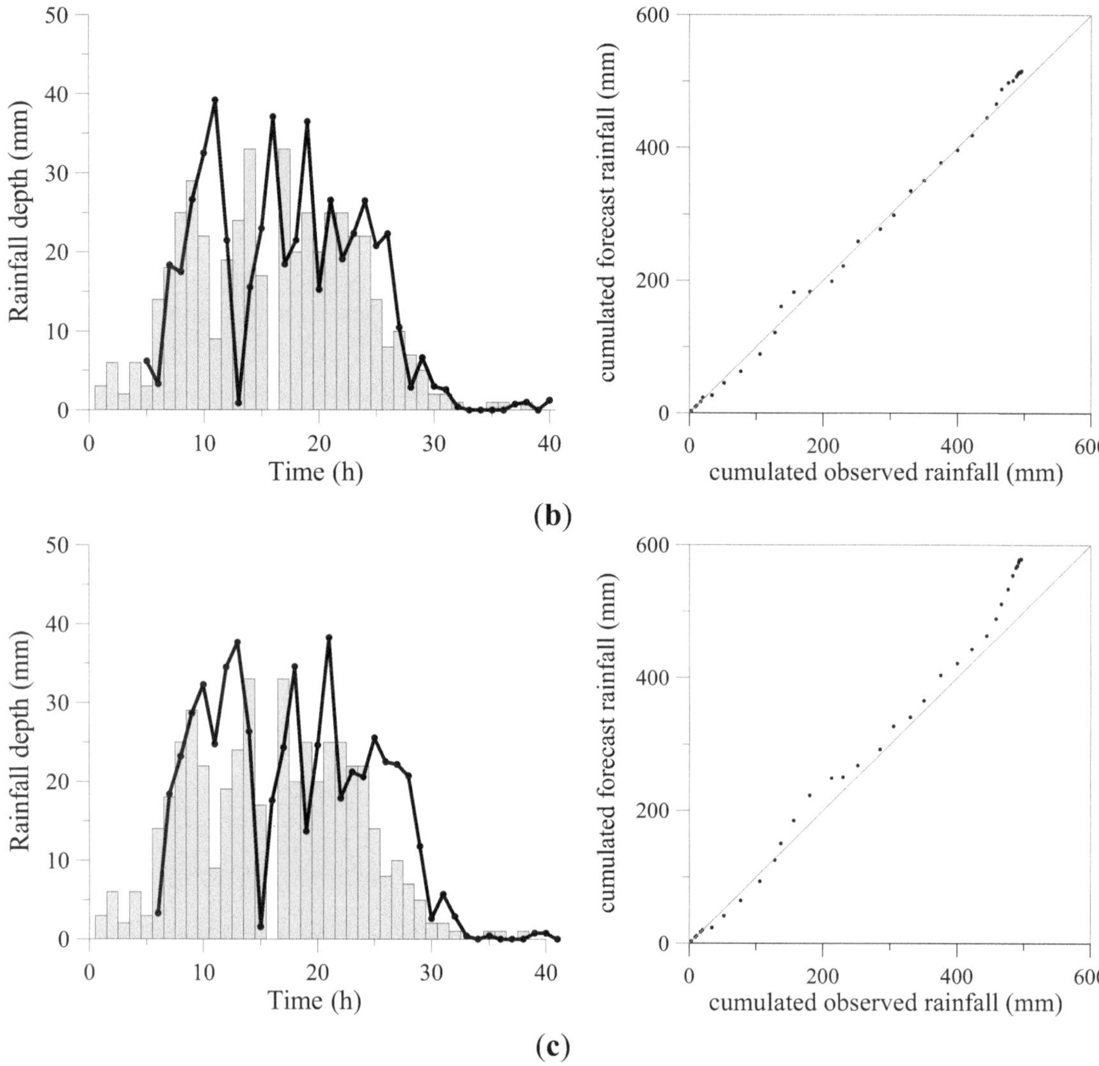

Figure 5. Grey model for rainfall forecasting in San-Hsia watershed: (**a**) 1-h ahead; (**b**) 2-h ahead; (**c**) 3-h ahead.

3.3. Flow Forecasting

Four sets of tests were performed to evaluate the applicability of the proposed system for real-time flood prediction. The simulation results are detailed shown as follows.

(1) Flow forecasting by using measured rainfall and without flow updating

This set of tests was conducted to evaluate the performance of the KW-GIUH model for simulating rainfall-runoff. Observed rainfall data were inputted into the KW-GIUH model and the flow updating algorithm was not used in the simulation. Figure 6 shows the results of runoff simulations for the Goodwin Creek and San-Hsia watersheds. As shown in Table 4, the simulated and observed hydrographs are in relatively good agreement in the study watersheds. The CE_Q values of the simulated hydrographs for all storm events are greater than 0.82, and most of the EQ_P and ET_P are lesser than 10% and 2 h, respectively. The results indicate that the KW-GIUH model is reliable for rainfall-runoff simulation in these two watersheds. Figure 6a shows that the temporal distributions of the observed rainfall hyetograph and flow hydrograph were inconsistent; specifically, the rainfall peak occurred at 45 h, whereas the flow peak occurred at 57 h. The reason for this inconsistency is unknown. However, this unusual hydrological record could be used to test the effectiveness of the proposed flow updating algorithm.

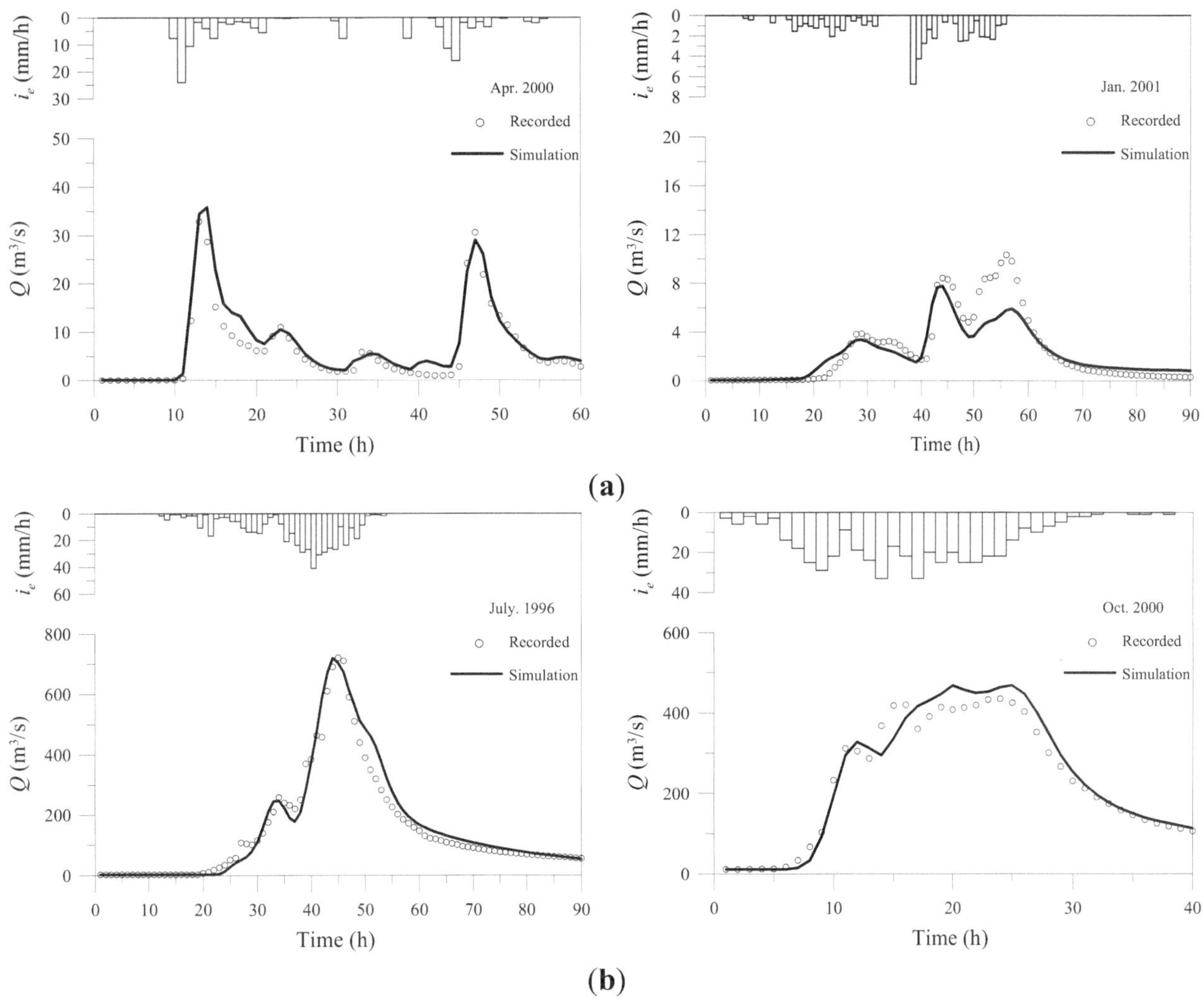

Figure 6. Flow forecasting using measured rainfall data and without flow updating in Goodwin Creek and San-Hsia watersheds: (**a**) Goodwin Creek watershed (STA 01); (**b**) San-Hsia watershed.

Table 4. Results of flow forecasting using measured rainfall and without flow updating technique.

Watershed	Event Date	CE_Q	EQ_p (%)	ET_p (h)
	10/07/1989	0.83	6.29	1
	02/03/1991	0.92	2.56	0
	14/02/1992	0.88	4.33	1
	04/08/1995	0.86	7.81	−1
Goodwin	29/11/1996	0.93	2.08	0
(SAT01)	23/12/1997	0.89	3.54	0
	15/02/1998	0.88	4.28	0
	13/03/1999	0.84	3.07	0
	01/04/2000	0.90	0.09	1
	17/01/2001	0.82	25.00	12

Table 4. *Cont.*

Watershed	Event Date	CE_Q	EQ_p (%)	ET_p (h)
Heng-Chi	17/08/1984	0.92	2.00	−1
	16/09/1985	0.88	0.24	−1
	17/09/1986	0.85	1.82	0
	27/07/1987	0.86	4.16	0
	08/09/1987	0.92	3.46	0
	18/08/1990	0.85	4.52	−1
	05/06/1993	0.97	2.01	−1
	10/07/1994	0.86	1.25	−1
	30/07/1996	0.94	3.54	−1
	31/10/2000	0.95	3.01	−2
San-Hsia	17/08/1984	0.90	6.54	0
	16/09/1985	0.89	3.33	0
	17/09/1986	0.86	6.29	1
	27/07/1987	0.84	5.54	1
	08/09/1987	0.88	5.19	0
	18/08/1990	0.83	2.24	0
	05/06/1993	0.87	5.64	−1
	10/07/1994	0.95	2.13	0
	30/07/1996	0.96	1.32	−1
	31/10/2000	0.95	7.65	−3
Average		0.89	4.36	0.10

(2) Flow forecasting by using forecast rainfall and without flow updating

For the second set of tests, flow forecasting was performed by inputting the forecast rainfall (obtained from the grey model) into the KW-GIUH model. Table 5 and Figure 7 show that the flow forecasting accuracy decreased as the lead time increased from 1 to 3 h. For the $t + 1$ forecast, the forecast flow is in good agreement with the observed flow. For the $t + 2$ and $t + 3$ forecasts, the temporal variation of the flow hydrograph is adequately represented in the simulation although the simulated flow peak is higher than the observed flow peak because the forecast peak rainfall was overestimated in the hyetograph. Regarding the storm event at the Goodwin Creek watershed on 17 January 2001, the results shown in Figure 7a indicate that the KW-GIUH model forecast the first flow peak accurately. However, the second flow peak is underestimated because of the inconsistency between the rainfall hyetograph and flow hydrograph as mentioned.

Table 5. Results of flow forecasting using forecast rainfall and without flow updating technique.

Watershed	Event Date	CE_Q			EQ_p (%)			ET_p (h)		
		1-h Ahead	**2-h Ahead**	**3-h Ahead**	**1-h Ahead**	**2-h Ahead**	**3-h Ahead**	**1-h Ahead**	**2-h Ahead**	**3-h Ahead**
Goodwin (SAT01)	10/07/1989	0.82	0.49	0.31	14.41	37.91	51.18	1	2	3
	02/03/1991	0.92	0.84	0.81	9.88	27.41	39.42	1	2	3
	14/02/1992	0.87	0.86	0.77	12.48	19.88	32.77	0	1	2
	04/08/1995	0.86	0.83	0.77	11.82	20.43	28.91	1	2	3
	29/11/1996	0.92	0.85	0.83	8.97	14.81	17.94	0	1	2
	23/12/1997	0.89	0.86	0.83	18.13	29.87	41.09	1	1	2
	15/02/1998	0.87	0.81	0.69	14.30	28.99	45.17	1	2	3
	13/03/1999	0.82	0.74	0.70	7.69	12.90	18.09	0	1	2
	01/04/2000	0.90	0.55	0.03	8.95	45.27	62.98	1	2	3
	17/01/2001	0.82	0.81	0.80	24.75	1.27	30.97	12	12	12
Heng-Chi	17/08/1984	0.90	0.86	0.81	4.81	11.85	19.28	1	2	2
	16/09/1985	0.87	0.81	0.74	3.29	8.74	20.32	1	1	2
	17/09/1986	0.83	0.79	0.71	10.93	24.31	31.88	1	2	3
	27/07/1987	0.85	0.83	0.78	8.49	18.41	24.31	1	2	2
	08/09/1987	0.92	0.90	0.86	7.96	16.19	19.22	1	2	2
	18/08/1990	0.75	0.43	0.09	14.41	19.84	31.03	1	2	3
	05/06/1993	0.95	0.94	0.88	2.09	7.31	9.08	0	1	2
	10/07/1994	0.85	0.81	0.70	1.09	8.54	11.72	0	1	1
	30/07/1996	0.93	0.91	0.82	9.75	14.32	18.97	−1	0	0
	31/10/2000	0.95	0.94	0.86	4.71	5.47	7.93	−3	−2	−1
San-Hsia	17/08/1984	0.90	0.81	0.79	8.49	14.55	21.09	1	2	2
	16/09/1985	0.88	0.80	0.74	6.39	11.52	17.92	1	1	2
	17/09/1986	0.86	0.83	0.52	9.31	18.45	37.01	1	2	3
	27/07/1987	0.84	0.83	0.67	4.09	11.12	14.17	1	2	2
	08/09/1987	0.87	0.81	0.63	4.41	9.18	22.97	1	2	2
	18/08/1990	0.81	0.62	0.31	6.31	11.48	18.02	1	2	3
	05/06/1993	0.85	0.81	0.76	7.31	9.52	16.55	0	1	2
	10/07/1994	0.95	0.91	0.85	4.86	7.59	16.31	0	1	1
	30/07/1996	0.95	0.94	0.89	1.96	4.48	12.05	−1	0	0
	31/10/2000	0.94	0.92	0.80	8.67	13.71	16.58	−3	−2	−1
Average		0.88	0.80	0.69	8.69	16.18	25.16	0.73	1.60	2.23

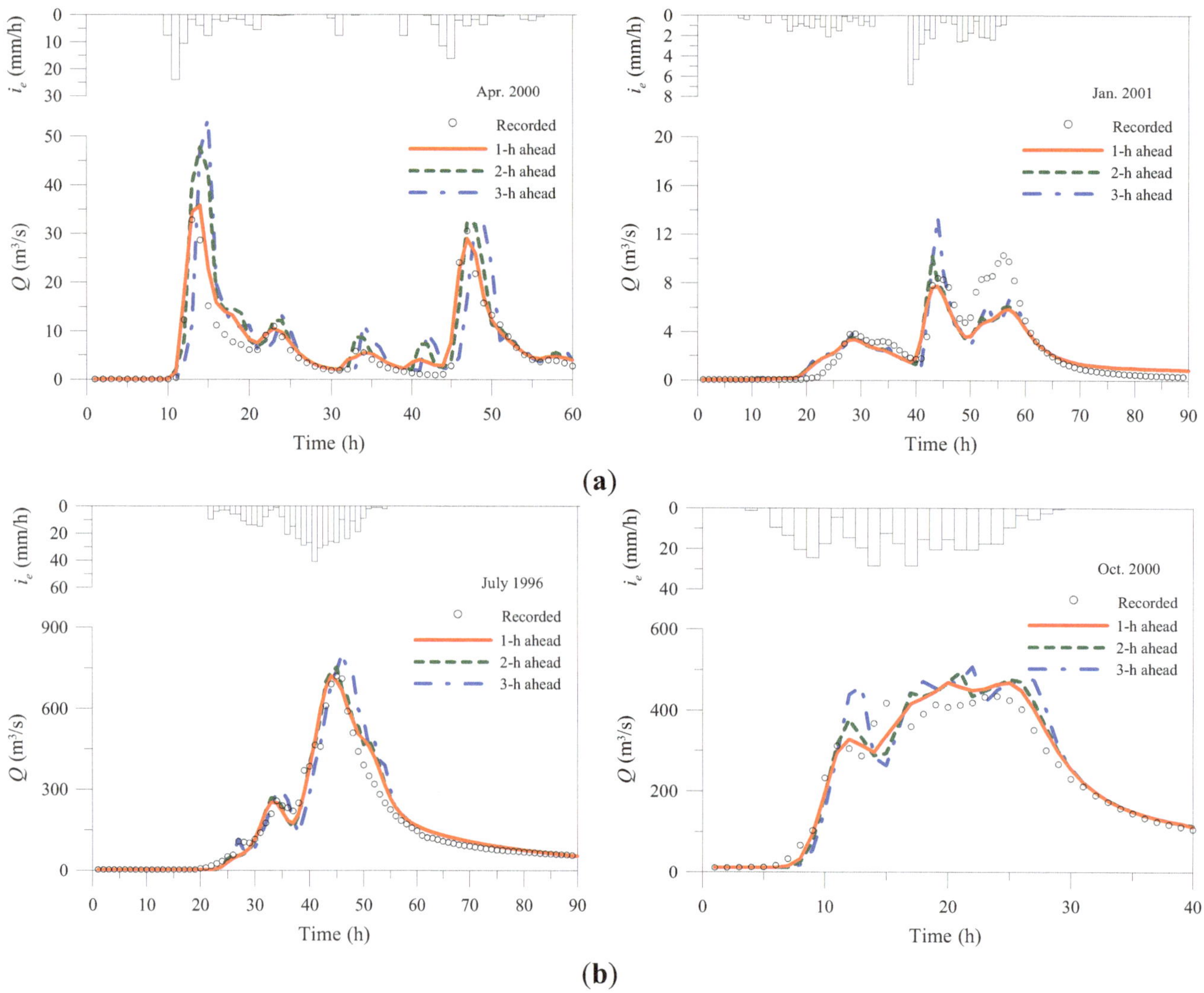

Figure 7. Flow forecasting using forecast rainfall and without flow updating in Goodwin Creek and San-Hsia watersheds: (**a**) Goodwin Creek watershed (STA 01); (**b**) San-Hsia watershed.

(3) Flow forecasting by using measured rainfall and flow updating technique

The third set of tests was conducted to evaluate the performance of the KW-GIUH model when the flow updating algorithm was used in the rainfall-runoff simulation, as shown in Equation (18). The measured rainfall at $t + 1$, $t + 2$, and $t + 3$ was inputted into the KW-GIUH model. Table 6 and Figure 8 show the simulation results, which were evaluated based on the coefficient of efficiency CE_Q, error of peak discharge EQ_P, and error of time to peak discharge ET_P. When the value CE_Q is approximately one and EQ_P and ET_P are approximately zero, good agreement between the recorded and simulated hydrographs is anticipated. The results in Figure 8 show that the CE_Q values are higher than 0.96, the mean EQ_P is 2.73%, and the mean ET_P is 0.17 h for the $t + 1$ simulation. For the $t + 2$ simulation, the CE_Q values are higher than 0.87, the mean EQ_P is 4.40%, and the mean ET_P is 0.67 h. Finally, for the $t + 3$ simulation, the CE_Q values are higher than 0.81, the mean EQ_P is 7.92%, and the mean ET_P is 1.23 h. Figure 9 shows that the forecast and recorded hydrographs are in good agreement for all storm events in this test, indicating that the proposed flow updating algorithm combined with the KW-GIUH model simulated the watershed runoff more accurately than do the KW-GIUH model alone. Moreover, regarding the storm event at the Goodwin Creek watershed on 17 January 2001, the flow hydrographs in Figures 7a and 9a show that the second peak was accurately forecasted when the flow updating algorithm is used, despite the recorded flow peak appearing to be unreasonable. The results show that using a purely deterministic approach to simulate watershed rainfall runoff is difficult without the assistance of a real-time adaptive algorithm.

Table 6. Results of flow forecasting using measured rainfall and flow updating technique.

Watershed	Event Date	CE_Q			EQ_p (%)			ET_p (h)		
		1-h Update	2-h Update	3-h Update	1-h Update	2-h Update	3-h Update	1-h Update	2-h Update	3-h Update
Goodwin (SAT01)	10/07/1989	0.97	0.93	0.89	2.54	5.77	8.43	1	1	1
	02/03/1991	0.98	0.94	0.90	0.14	3.43	4.95	1	1	2
	14/02/1992	0.96	0.91	0.84	3.83	5.46	5.57	0	0	1
	04/08/1995	0.96	0.92	0.83	2.53	4.59	7.33	0	0	1
	29/11/1996	0.96	0.91	0.85	3.64	4.22	8.36	1	2	2
	23/12/1997	0.97	0.92	0.85	4.81	5.65	8.78	−1	0	1
	15/02/1998	0.96	0.92	0.86	0.19	3.19	3.97	−1	−1	0
	13/03/1999	0.96	0.91	0.83	3.71	0.13	9.80	0	1	1
	01/04/2000	0.97	0.93	0.85	1.94	0.97	8.11	−1	−1	0
	17/01/2001	0.97	0.87	0.81	0.84	0.53	2.05	1	1	1
Heng-Chi	17/08/1984	0.96	0.92	0.88	4.75	5.69	10.54	0	0	1
	16/09/1985	0.97	0.93	0.89	4.55	6.29	14.12	0	0	1
	17/09/1986	0.97	0.92	0.90	2.30	1.89	4.92	0	1	2
	27/07/1987	0.96	0.91	0.88	0.76	2.69	5.50	0	1	1
	08/09/1987	0.96	0.92	0.89	1.62	5.86	8.80	1	2	2
	18/08/1990	0.97	0.93	0.91	0.50	3.40	4.14	−1	0	0
	05/06/1993	0.98	0.93	0.91	2.75	0.68	8.15	1	2	2
	10/07/1994	0.97	0.93	0.90	3.03	4.42	9.09	1	1	2
	30/07/1996	0.96	0.91	0.89	3.16	2.19	2.09	0	0	1
	31/10/2000	0.98	0.93	0.91	3.17	10.78	14.44	0	1	1
San-Hsia	17/08/1984	0.96	0.92	0.87	2.95	5.69	12.86	1	1	2
	16/09/1985	0.96	0.91	0.88	1.21	8.33	13.23	0	1	1
	17/09/1986	0.97	0.92	0.89	1.53	1.64	3.23	0	1	1
	27/07/1987	0.97	0.92	0.88	3.76	3.24	4.18	0	0	1
	08/09/1987	0.97	0.91	0.86	1.64	6.24	10.8	0	1	2
	18/08/1990	0.96	0.91	0.88	1.69	3.23	3.15	0	0	1
	05/06/1993	0.97	0.94	0.91	2.42	4.89	9.33	1	2	2
	10/07/1994	0.96	0.91	0.88	1.72	3.09	11.56	0	0	1
	30/07/1996	0.98	0.98	0.93	3.84	3.28	4.34	−1	0	1
	31/10/2000	0.99	0.96	0.95	6.22	7.35	9.42	1	2	2
Average		0.97	0.92	0.88	2.73	4.40	7.92	0.17	0.67	1.23

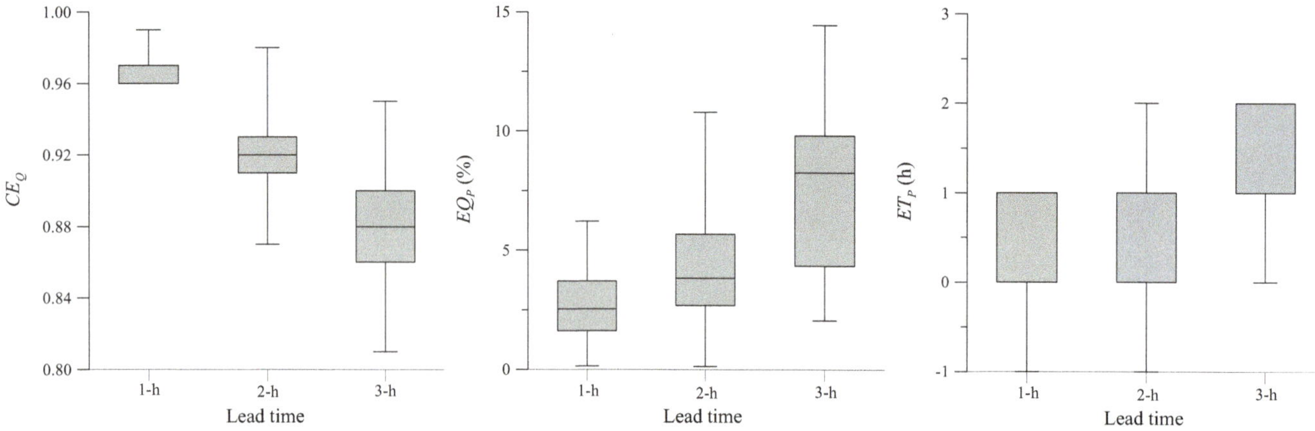

Figure 8. Results of evaluated criteria for flow forecasting using measured rainfall and flow updating technique.

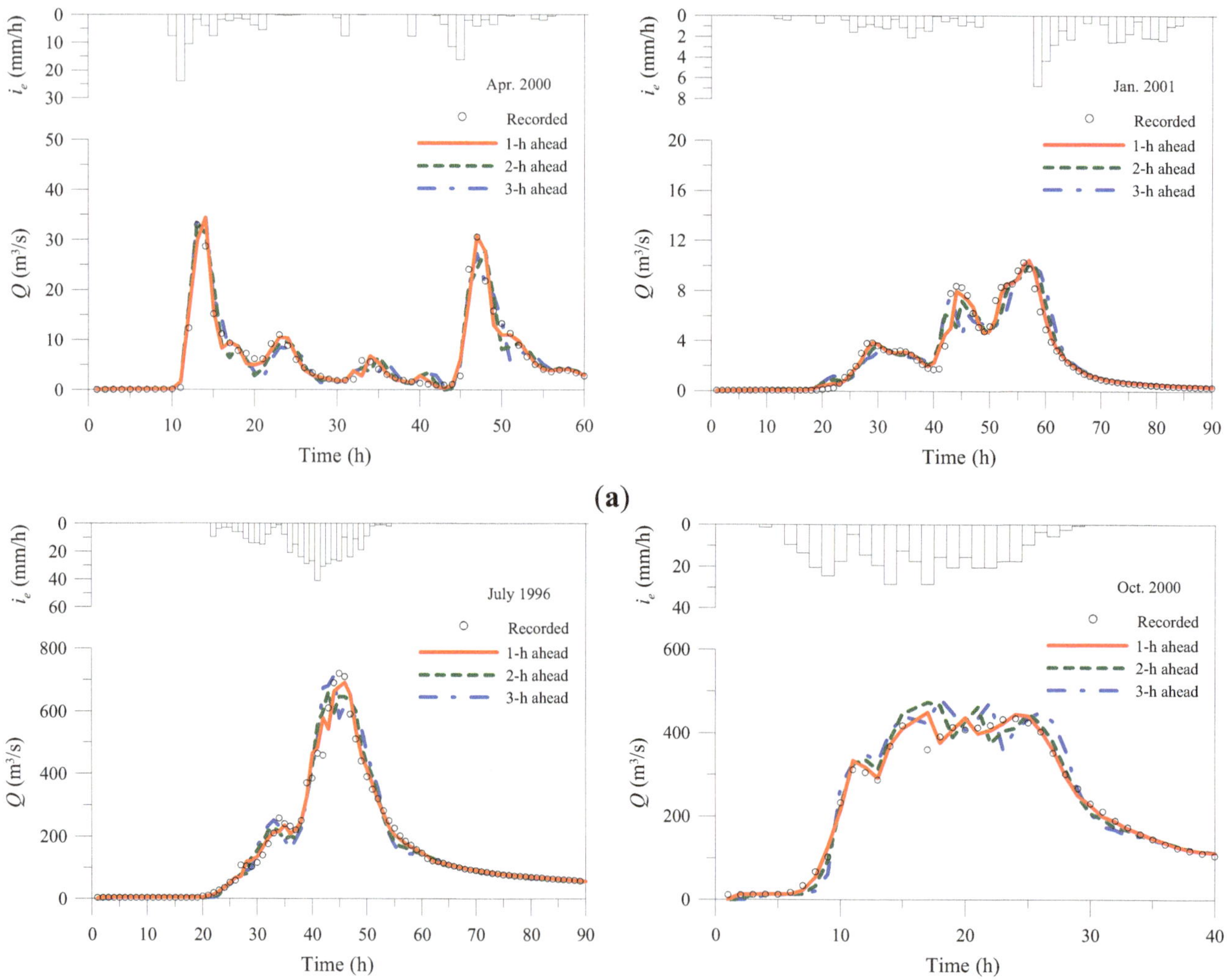

Figure 9. Flow forecasting using measured rainfall and flow updating algorithm in Goodwin Creek and San-Hsia watersheds: (**a**) Goodwin Creek watershed (STA 01); (**b**) San-Hsia watershed.

(4) Flow forecasting by using forecast rainfall and flow updating algorithm

The final set tests was conducted to confirm the performance of the proposed flood forecasting system. The forecast rainfall is generated by using the grey model, and the flow updating algorithm is included in the runoff simulation by using the KW-GIUH model to improve the forecasting accuracy. Table 7 and Figure 10 show that the mean CE_Q (EQ_P) values of the $t+1$, $t+2$, and $t+3$ forecasts are 0.92 (4.50%), 0.80 (9.12%), and 0.72 (13.57%). The mean ET_P values of the $t+1$, $t+2$, and $t+3$ forecasts are 0.70 h, 1.47 h, and 2.13 h, respectively. The results of the storm event simulations in Figure 11 shows that the recorded and simulated hydrographs are in good agreement for all the three watersheds under various geoclimate conditions, even as the lead time increases from 1 to 3 h.

Table 7. Results of flow forecasting using forecast rainfall and flow updating technique.

Watershed	Event Date	CE_Q			EQ_P (%)			ET_P (h)		
		1-h Ahead	2-h Ahead	3-h Ahead	1-h Ahead	2-h Ahead	3-h Ahead	1-h Ahead	2-h Ahead	3-h Ahead
Goodwin (SAT01)	10/07/1989	0.94	0.82	0.80	3.81	11.21	13.91	1	1	2
	02/03/1991	0.95	0.87	0.77	1.00	3.60	7.33	1	1	2
	14/02/1992	0.89	0.69	0.66	4.22	1.89	4.04	1	2	2
	04/08/1995	0.88	0.70	0.69	1.01	8.67	12.59	1	2	3
	29/11/1996	0.89	0.86	0.81	3.64	9.45	16.24	1	1	2
	23/12/1997	0.91	0.69	0.65	7.28	8.15	14.89	0	1	2
	15/02/1998	0.93	0.85	0.79	9.56	10.69	11.28	0	1	2
	13/03/1999	0.91	0.79	0.78	3.70	8.91	7.28	0	1	1
	01/04/2000	0.95	0.72	0.61	4.13	12.34	18.75	1	2	3
	17/01/2001	0.97	0.94	0.91	1.07	0.39	4.20	0	1	1
Heng-Chi	17/08/1984	0.95	0.76	0.62	8.42	11.21	3.71	1	1	2
	16/09/1985	0.89	0.71	0.61	4.88	13.77	5.66	1	1	2
	17/09/1986	0.90	0.84	0.79	1.52	8.15	19.10	1	2	3
	27/07/1987	0.96	0.76	0.61	6.79	9.97	11.52	1	2	2
	08/09/1987	0.89	0.75	0.62	0.04	7.44	16.60	1	2	3
	18/08/1990	0.91	0.78	0.66	3.21	0.22	9.67	0	1	1
	05/06/1993	0.88	0.77	0.61	7.93	12.73	19.02	1	2	3
	10/07/1994	0.92	0.74	0.64	3.28	11.82	23.11	1	1	2
	30/07/1996	0.96	0.90	0.83	1.70	3.62	6.61	1	2	2
	31/10/2000	0.95	0.89	0.82	3.68	5.03	10.02	1	2	2
San-Hsia	17/08/1984	0.93	0.76	0.63	9.31	11.59	17.42	1	2	3
	16/09/1985	0.89	0.74	0.61	5.81	14.60	19.88	1	2	2
	17/09/1986	0.91	0.91	0.78	8.44	18.00	22.30	0	1	2
	27/07/1987	0.89	0.72	0.59	7.32	11.52	15.39	1	2	3
	08/09/1987	0.96	0.83	0.74	1.75	5.47	14.50	0	1	1
	18/08/1990	0.91	0.75	0.65	1.68	3.13	9.50	1	2	2
	05/06/1993	0.88	0.71	0.63	6.03	22.14	27.56	1	2	3
	10/07/1994	0.92	0.78	0.69	1.32	13.36	28.36	0	1	2
	30/07/1996	0.98	0.97	0.92	4.60	4.83	5.03	1	1	2
	31/10/2000	0.98	0.95	0.94	7.75	9.65	11.62	0	1	2
Average		0.92	0.80	0.72	4.50	9.12	13.57	0.70	1.47	2.13

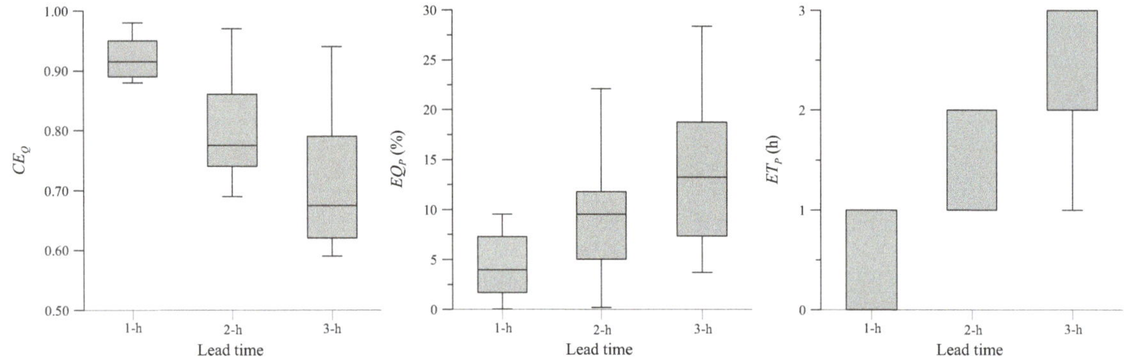

Figure 10. Results of evaluated criteria for flow forecasting using forecast rainfall and flow updating technique.

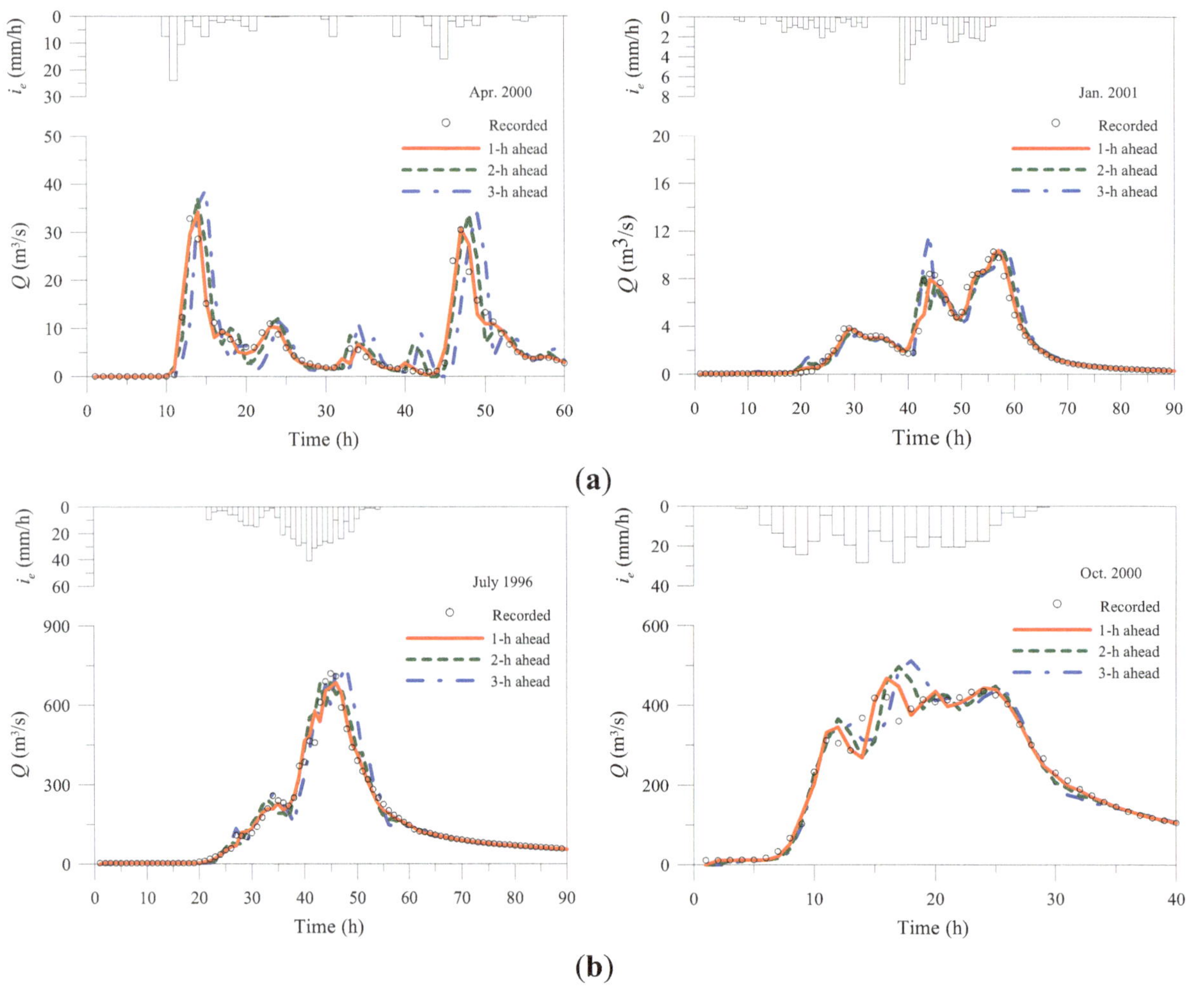

Figure 11. Flow forecasting using forecast rainfall and flow updating algorithm in Goodwin Creek and San-Hsia watersheds: (**a**) Goodwin Creek watershed (STA 01); (**b**) San-Hsia watershed.

4. Conclusions

This study developed an integrated framework for flood forecasting by using a rainfall forecasting model, watershed rainfall-runoff model, and real-time flow updating algorithm. Considering that current numerical meteorological models used in Taiwan cannot provide a 3-h prediction of the temporal distribution of rainfall, this study adopted a grey rainfall forecasting model. Using the KW-GIUH model for runoff simulation is advantageous because it can be developed based on only geomorphologic factors of the watershed. Moreover, a real-time flow updating algorithm was incorporated into the KW-GIUH structure to account for the uncertainty of watershed runoff processes. The proposed flood forecasting system was tested based on hydrological records from three watersheds under different geomorphological and hydrological conditions. For the 1-h, 2-h, and 3-h ahead forecast cases, the simulated mean coefficient of efficiency (error of peak discharge) is 0.92 (4.5%), 0.80 (9.12%), and 0.72 (13.57%). The mean ET_P values of the $t + 1$, $t + 2$, and $t + 3$ forecast cases are 0.70 h, 1.47 h, and 2.13 h, respectively. These results indicate that the proposed flood forecasting system can provide credible warning information for authorities. Furthermore, the proposed flow forecasting system can operate with high efficiency to meet the requirements of real-time flow forecasting. Nevertheless, in considering that the forecast rainfall is assumed to have the same tendency as previous rainfall; hence, only short-term prediction is applicable for this rainfall forecasting system. Since spatially-uniform rainfall is used in the KW-GIUH model, the proposed flow forecasting may be not used in a large watershed. Further validations to account for watersheds with various hydrological and geomorphologic characteristics are still required in future research.

Acknowledgments

This research was supported by the Ministry of Science and Technology, Taiwan, under grants MOST 103-2625-M-492-007 and 101-2218-E-006-003. The authors acknowledge the financial support provided by the Ministry of Science and Technology.

Author Contributions

Jui-Yi Ho was validated proposed model. Kwan Tun Lee was designed and performed the model. Both authors contributed about equally to compose this paper, discussed the results, and commented on the manuscript at all stages.

Conflicts of Interest

The authors declare no conflict of interest

References

1. Brath, A.; Burlando, P.; Rosso, R. Sensitivity analysis of real-time flood forecasting to on-line rainfall predictions. In *Selected Papers from the Workshop on Natural Disasters in European-Mediterranean Countries*; Siccardi, F., Bras, R.L., Eds.; National Research Council: Perugia, Italy, 1998; pp. 469–488.
2. Borga, M.; Anagnostou, E.N.; Frank, E. On the use of real-time radar rainfall estimates for flood prediction in mountainous basins. *J. Geophys. Res.* **2000**, *105*, 2269–2280.
3. Plessis, L.D. A review of effective flood forecasting, warning and response system for application in South Africa. *Water SA* **2002**, *28*, 129–137.

4. Walser, A.; Schär, C. Convection-resolving precipitation forecasting and its predictability in Alpine river catchments. *J. Hydrol.* **2004**, *288*, 57–73.
5. Toth, E.; Brath, A.; Montnari, A. Comparison of short-term rainfall prediction models for real-time flood forecasting. *J. Hydrol.* **2000**, *239*, 132–147.
6. Garbrecht, J.; Ogden, F.L.; DeBarry, P.A.; Maidment, D.R. GIS and distributed watershed models. I: Data converges and sources. *J. Hydrol. Eng.* **2001**, *6*, 506–514.
7. Moore, R.J.; Bell, V.A.; Jones, D.A. Forecasting for flood warning. *Comptes Rendus Geosci.* **2005**, *337*, 203–217.
8. French, M.N.; Krajewski, W.F.; Cuykendall, R.R. Rainfall forecasting in space and time using a neural network. *J. Hydrol.* **1992**, *137*, 1–31.
9. Burlando, P.; Rosso, R.; Gadavid, L.G.; Salas, J.D. Forecasting of short-term rainfall using ARMA models. *J. Hydrol.* **1993**, *144*, 193–211.
10. Lardet, P.; Obled, C. Real time flood forecasting using a stochastic rainfall generator. *J. Hydrol.* **1994**, *162*, 391–408.
11. Kelly, K.S.; Krzysztofowicz, R. Precipitation uncertainty processor for probabilistic river stage forecasting. *Water Resour. Res.* **2000**, *36*, 2643–2654.
12. Krzysztofowicz, R. Integrator of uncertainties for probabilistic river stage forecasting: Precipitation-dependent model. *J. Hydrol.* **2001**, *249*, 69–85.
13. Lin, G.F.; Wu, M.C. A hybrid neural network for typhoon-rainfall forecasting. *J. Hydrol.* **2009**, *375*, 450–458.
14. Deng, J.L. Introduction to grey system theory. *J. Grey Syst.* **1989**, *1*, 1–24.
15. Xia, J. Research and application of grey system theorem to hydrology. *J. Grey Syst.* **1989**, *1,* 43–52.
16. Huang, Y.P.; Huang, C.C. The integration and application of fuzzy and grey modeling methods. *Fuzzy Sets Syst.* **1996**, *78,* 107–119.
17. Yu, P.-S.; Chen, C.-J.; Chen, S.-C.; Lin, S.-C. Application of grey method toward runoff forecasting. *J. Am. Water Resour. Assoc.* **2001**, *37*, 151–166.
18. ASCE Task Committee on Application of Artificial Neural Networks in Hydrology. Artificial Neural Networks in Hydrology, part I: Preliminary concepts. *J. Hydrol. Eng.* **2000**, *5*, 115–123.
19. ASCE Task Committee on Application of Artificial Neural Networks in Hydrology. Artificial Neural Networks in Hydrology, part II: Hydrologic applications. *J. Hydrol. Eng.* **2000**, *5*, 124–137.
20. Xu, Z.X.; Li, J.Y. Short-term inflow forecasting using an artificial neural network model. *Hydrol. Process.* **2002**, *16*, 2423–2439.
21. Lin, G.F.; Chen, L.H. A non-linear rainfall-runoff model using radial basis function network. *J. Hydrol.* **2004**, *289*, 1–8.
22. Chetan, M.; Sudheer, K.P. A hybrid linear-neural model for river flow forecasting. *Water Resour. Res.* **2006**, *42*, doi:10.1029/2005WR004072.
23. Chen, W.; Chau, K.W. Intelligent manipulation and calibration of parameters for hydrological models. *Int. J. Environ. Pollut.* **2006**, *28*, 432–447.
24. Muttil, N.; Chau, K.-W. Neural network and genetic programming for modelling coastal algal blooms. *Int. J. Environ. Pollut.* **2006**, *28*, 223–238.
25. Chau, K.W. An ontology-based knowledge management system for flow and water quality modeling. *Adv. Eng. Softw.* **2007**, *38*, 172–181.

26. Toth, E.; Brath, A. Multistep ahead streamflow forecasting: Role of calibration data in conceptual and neural network modeling. *Water Resour. Res.* **2007**, *43*, W11405.
27. Taormina, R.; Chau, K.W. Neural network river forecasting with multi-objective fully informed particle swarm optimization. *J. Hydroinform.* **2015**, *17*, 99–113.
28. Wu, C.L.; Chau, K.W.; Li, Y.S. River stage prediction based on a distributed support vector regression. *J. Hydrol.* **2008**, *358*, 96–111.
29. Cheng, C.T.; Chau, K.W.; Sun, Y.; Lin, J. Long-term prediction of discharges in Manwan Reservoir using artificial neural network models. *Lecture Notes Comput. Sci.* **2005**, *3498*, 1040–1045.
30. Hapuarachchi, H.A.P.; Wang, Q.J. *A Review of Methods and Systems Available for Flash Flood Forecasting*; Report for the Bureau of Meteorology, Australia; Commonwealth Scientific and Industrial Research Organisation (CSIRO): Dickson, Austrilia, July 2008.
31. Burnash, R.J.C.; Ferral, R.L.; McGuire, R.A. *A Generalized Streamflow Simulation System: Conceptual Modeling for Digital Computers*; U.S. Department of Commerce: Washington, DC, USA, 1973.
32. Burnash, R.J.C. The NWS river forecast system-catchment modeling. In *Computer Models of Watershed Hydrology*; Singh, V.P., Ed.; Water Resources Publications: Highlands Ranch, CO, USA, 1995; pp. 311–366.
33. Sugawara, M. Automatic calibration of the tank model. *Hydrol. Sci. Bull.* **1979**, *24*, 375–388.
34. Bergstrom, S. The HBV model. In *Computer Models of Watershed Hydrology*; Singh, V.P., Ed.; Water Resources Publications: Highlands Ranch, CO, USA, 1995; pp. 443–476.
35. Zhao, R.J. The Xinanjiang model applied in China. *J. Hydrol.* **1992**, *135*, 371–381.
36. Jain, V.; Sinha, R. Derivation of unit hydrograph from GIUH analysis for a Himalayan river. *Water Resour. Manag.* **2003**, *17*, 355–375.
37. Zhang, B.; Govindaraju, R.S. Geomorphology-based artificial neural networks (GANNs) for estimation of direct runoff over watersheds. *J. Hydrol.* **2003**, *273*, 18–34.
38. Shadeed, S.; Shaheen, H.; Jayyousi, A. GIS-based KW-GIUH hydrological model of semiarid catchments: the case of Faria Catchment, Palestine. *Arabian J. Sci. Eng.* **2007**, *32*, 3–16.
39. Chiang, S.; Tachikawa, Y.; Takara, K. Hydrological model performance comparison through uncertainty recognition and quantification. *Hydrol. Process.* **2007**, *21*, 1179–1195.
40. Kumar, A.; Kumar, D. Predicting direct runoff from hilly watershed using geomorphology and stream-order-law ratios: Case Study. *J. Hydrol. Eng.* **2008**, *13*, 570–576.
41. Rodriguez-Iturbe, I.; Valdes, J.B. The geomorphologic structure of hydrologic response. *Water Resour. Res.* **1979**, *15*, 1409–1420.
42. Lee, K.T.; Yen, B.C. Geomorphology and kinematic-wave based hydrograph deviation. *J. Hydraul. Eng.* **1997**, *123*, 73–80.
43. Yen, B.C.; Lee, K.T. Unit hydrograph derivation for ungauged watersheds by stream order laws. *J. Hydrol. Eng.* **1997**, *2*, 1–9.
44. Clark, M.P.; Rupp, D.E.; Woods, R.A.; Zheng, X.; Ibbit, R. P.; Slater, A.G.; Schmidt, J.; Uddstrom, M.L. Hydrological data assimilation with the ensemble Kalman filter: Use of streamflow observation to update states in a distributed hydrological model. *Adv. Water Resour.* **2008**, *31*, 1309–1324.
45. Liu, Y.Q.; Weerts, A.H.; Clark, M.; Hendricks Franssen, H.J.; Kumar, S.; Moradkhani, H.; Seo, D.-J.; Schwanenberg, D.; Smith, P.; van Dijk, A.I.J.M.; *et al.* Advancing data assimilation in operational

hydrologic forecasting: Progresses, challenges, and emerging opportunities. *Hydrol. Earth Syst. Sci.* **2012**, *16*, 3863–3887.
46. Refsgaard, J.C. Validation and intercomparison of different updating procedures for real-time forecasting. *Nordic Hydrol.* **1997**, *28*, 65–84.
47. Reed, D.W. *A Review of British Flood Forecasting Practice*; Report 90; Institute of Hydrology: Wellington, UK, 1984.
48. Bobinski, E.; Mierkiewicz, M. Recent developments in simple adaptive flow forecasting models in Poland. *Hydrol. Sci. J.* **1986**, *31*, 297–320.
49. Nayak, P.C.; Sudheer, K.P.; Ramasastri, K.S. Fuzzy computing based rainfall-runoff model for real time flood forecasting. *Hydrol. Process.* **2005**, *19*, 955–968.
50. Chatfield, C. *The Analysis of Time Series an Introduction*; Chapman and Hall: London, UK, 1989.
51. Seo, D.J.; Koren, V.; Cajina, N. Real-time variational assimilation of hydrologic and hydrometeorological data into operational hydrologic forecasting. *J. Hydrometeorol.* **2003**, *4*, 627–641.
52. Valstar, J.R.; McLaughlin, D.B.; te Stroet, C.B.M.; van Geer, F.C. A representer-based inverse method for groundwater flow and transport applications. *Water Resour. Res.* **2004**, *40*, doi:10.1029/2003WR002922.
53. Da Ros, D.; Borga, M. Adaptive use of a conceptual model for real time flood forecasting. *Nordic Hydrol.* **1997**, *28*, 169–188.
54. Madsen, H.; Skotner, C. Adaptive state updating in real-time river flow forecasting-a combined filtering and error forecasting procedure. *J. Hydrol.* **2005**, *308*, 302–312.
55. Weerts, A.H.; el Serafy, G.Y.E. Particle filtering and ensemble Kalman filtering for state updating with hydrological conceptual rainfall-runoff models. *Water Resour. Res.* **2006**, *42*, doi:10.1029/2005WR004093.
56. Wang, D.; Cai, X. Robust data assimilation in hydrological modeling-a comparison of Kalman and H-infinity filters. *Adv. Water Resour.* **2008**, *31*, 455–472.
57. Matgen, P.; Montanari, M.; Hostache, R.; Pfister, L.; Hoffmann, L.; Plaza, D.; Pauwels, V.R.N.; de Lannoy, G.J.M.; de Keyser, R.; Savenije, H.H.G. Towards the sequential assimilation of SAR-derived water stages into hydraulic models using the Particle Filter: proof of concept. *Hydrol. Earth Syst. Sci.* **2010**, *14*, 1773–1785.
58. DeChant, C.M.; Moradkhani, H. Examining the effectiveness and robustness of data assimilation methods for calibration and quantification of uncertainty in hydrologic forecasting. *Water Resour. Res.* **2012**, *48*, doi:10.1029/2011WR011011.
59. Nash, J.E.; Sutcliffe, J.V. River flow forecasting through conceptual models part I—A discussion of principles. *J. Hydrol.* **1970**, *10*, 282–290.
60. Henderson, F.M.; Wooding, R.A. Overland flow and groundwater flow from a steady rainfall of finite duration. *J. Geophys. Res.* **1964**, *69*, 1531–1540.
61. Thiessen, A.H. Precipitation averages for large areas. *Mon. Weather Rev.* **1911**, *39*, 1082–1084.
62. Lee, K.T. Generating design hydrographs by DEM assisted geomorphic runoff simulation: A case study. *J. Am. Water Resour. Assoc.* **1998**, *34*, 375–384.

2

Investigation of Residence and Travel Times in a Large Floodplain Lake with Complex Lake–River Interactions: Poyang Lake (China)

Yunliang Li [1,*], Qi Zhang [1,2] and Jing Yao [1]

[1] Key Laboratory of Watershed Geographic Sciences, Nanjing Institute of Geography and Limnology, Chinese Academy of Sciences, 73 East Beijing Road, Nanjing 210008, China; E-Mails: qzhang@niglas.ac.cn (Q.Z.); jyao@niglas.ac.cn (J.Y.)

[2] Key Laboratory of Poyang Lake Wetland and Watershed Research, Ministry of Education, Jiangxi Normal University, 99 Ziyang Road, Nanchang 330022, China

* Author to whom correspondence should be addressed; E-Mail: yunliangli@niglas.ac.cn;

Academic Editor: Miklas Scholz

Abstract: Most biochemical processes and associated water quality in lakes depends on their flushing abilities. The main objective of this study was to investigate the transport time scale in a large floodplain lake, Poyang Lake (China). A 2D hydrodynamic model (MIKE 21) was combined with dye tracer simulations to determine residence and travel times of the lake for various water level variation periods. The results indicate that Poyang Lake exhibits strong but spatially heterogeneous residence times that vary with its highly seasonal water level dynamics. Generally, the average residence times are less than 10 days along the lake's main flow channels due to the prevailing northward flow pattern; whereas approximately 30 days were estimated during high water level conditions in the summer. The local topographically controlled flow patterns substantially increase the residence time in some bays with high spatial values of six months to one year during all water level variation periods. Depending on changes in the water level regime, the travel times from the pollution sources to the lake outlet during the high and falling water level periods (up to 32 days) are four times greater than those under the rising and low water level periods (approximately seven days).

Keywords: 2D hydrodynamic model; residence time; travel time; dye tracer; water level dynamics; Poyang Lake

1. Introduction

Lakes are vitally important components of the hydrosphere with indispensable ecological value and a large inventory of surface freshwater that is readily accessible for human consumption [1]. Many physical, chemical and biological processes that occur within particular environments, including the transport of pollutants, sediments and other materials, are influenced by lake hydrodynamics [2].

One important physical attribute of lakes is the transport time scale (e.g., residence time, water age, flushing time, and travel time) that describes the lake's ability to renew the water contained within it via water level regime dynamics [3,4]. Although the time scale concept was designed to characterize the transport processes of lakes, this scale does not have a universally accepted definition. The terms 'residence time' and 'travel time' are most commonly used to represent the time scale of the physical transport processes and to elucidate associated water quality problems in lakes and other hydrological systems [5]. In the literature, a widely used concept for quantifying the residence times for lakes is known as the *e*-folding time [2,4,5]. The *e*-folding time corresponds to the time when the average concentration of contaminants in the lake is reduced to $1/e$ or 37% of the initial concentration [6–8]. Travel time is defined as the time that elapses between the introduction of a conservative and passive tracer to the water column and the time required to reach its peak concentration at a given location [9–11]. When two or more peaks are observable, the representative value of the travel time is determined by using the peak that corresponds to the higher concentration [12,13]. Water circulation in a given lake is the end result of the complex interplay of various forces (e.g., inflows, outflows, winds, and tides), and therefore, knowledge of the transport time scale can substantially aid in the assessment of the environmental state of lakes and their sustainable management for engineers, scientists, managers and policy makers.

Poyang Lake, the largest freshwater lake in China, is one of the few lakes in China that remains naturally connected to surrounding rivers. The lake predominantly receives inflows from catchment rivers, which subsequently exchange flows with the Yangtze River [14]. Poyang Lake is an example of a highly valued water resource with particularly complex hydrological forcing and thereby exhibits a highly dynamic hydrological regime [15]. The water level in the lake is primarily controlled by seasonal variations in catchment river inflows during the water level rising period [14,16]. However, the flow regime changes in the Yangtze River largely influence the lake's behavior, mainly due to the 'blocking effect' of the Yangtze River during the high water level period [17,18] and to the 'empty effect' during the water level recession period [19]. Therefore, the lake exhibits varying interactions with the lake's catchment and the Yangtze River on a seasonal scale.

Poyang Lake serves multiple functions relating to industrial production, drinking water, irrigation, ecosystem maintenance, and the regulation of environmental functions as well as supporting cultural and recreational functions for approximately 12.4 million residents [20]. However, environmental degradation has begun due to increasingly intensive economic activities. Steadily deteriorating water quality and the associated rapidly changing environmental conditions have intensified due to wastewater discharge from industrial and domestic sources as well as agricultural chemical inputs [20]. For example, the average total nitrogen (TN) and total phosphorus (TP) concentrations in the lake water were 1.1 mg/L and 0.07 mg/L, respectively, indicating severe eutrophication [21]. The water quality of Poyang Lake was very good prior to year 2000 (Grade I); however, it has deteriorated to Grades III and IV in the last 10 years (according to the Chinese Water Quality Classification System, Grade I is the best and Grade V is the worst) [22]. These urgent problems concerning Poyang Lake have recently gained considerable attention [23,24] because they have resulted in a serious threat to inhabitants that rely on this freshwater resource [25]. The lake water quality is known to exhibit significant temporal and spatial variations that are closely related to lake-dependent flushing [21,22,26]. Therefore, it is important to determine the residence and travel times within Poyang Lake, particularly with respect to its seasonal water level dynamics. However, no references or estimates that consider these specific water level regime changes are available in the literature.

Previous efforts have focused on estimating the average residence times of Poyang Lake by using a single value based on the ratio of the lake volume to the outflow rate [26–29]. In its simplest form, this method may not be appropriate for systems that are similar to Poyang Lake, which have highly variable volumes (varying from 4.5×10^8 to 150×10^8 m^3) and outflows (the lake discharges to the Yangtze River but with frequent backflow) due to complex water level dynamics [17]. In this case, it is unclear which variable values should be employed and which temporal averaging scales (e.g., annual, monthly, or bi-weekly) are reasonable [30]. In addition, a single ratio for a system is not valid for all periods, locations or constituents; thus, a single ratio cannot represent all transport processes [5]. More importantly, the single ratio provides no information regarding the connections between the transport and the spatial heterogeneity of water environmental factors within the lake, such as specific conductivity, temperature and chlorophyll *a* [31]. However, field experiments that infer the residence time are expensive; thus, the overall spatial and temporal data cannot be surveyed [32]. Due to large variations in the published values of the residence times (varied from 10 to 59 days) [26–29], approaches for more accurate and realistic expressions that identify and quantify the spatially heterogeneous phenomena of residence time are required for this highly dynamic lake. Therefore, this particular case can be considered to be a first study.

Most existing methodologies for investigating transport time scale are based on physically based hydrodynamic modeling. Hydrodynamic models combined with dye-tracer simulations are used to explore pollutant transport processes and associated residence time (or travel time) in different water bodies, such as the tidal-driven York River estuary (USA) [33], the Venice Lagoon (Italy) [8], Lake Tana (Ethiopia) [34], the Irish Sea (Irish) [3] and Mediterranean lagoons (Italy and Spain) [35]. These previous studies indicate that hydrodynamic models are efficient for investigating the transport time scale (e.g., residence time and travel time) because both advection and dispersion processes are considered using a combined dye tracer-based approach. An additional advantage of using a hydrodynamic model is that it can be used to test hypotheses and to simulate various scenarios that can affect the derived transport time scale, which strengthens confidence in the approach.

In this study, a 2D hydrodynamic model (MIKE 21) linked to a dye tracer approach was used in the first attempt to explore the detailed transport time scales of a large floodplain lake (Poyang Lake) using the concepts of residence and travel time. Although the ultimate purpose of this study was to provide transport time scales for future assessments of the ecological and biochemical changes within Poyang Lake, this study was limited to three main objectives: (1) to characterize the spatial distribution of water residence times; (2) to investigate the seasonal behaviors of travel times for various water regime dynamics of the lake; and (3) to provide physical background information relating to Poyang Lake hydrodynamics and associated transport time scales. The spatial and temporal variability of the residence and travel times were investigated using a series of hydrodynamic model simulations.

2. Study Site

Poyang Lake (28°4'–29°46' N, 115°49'–116°46' E) is located in the northern region of Jiangxi Province and on the south bank of the middle-lower reaches of the Yangtze River (Figure 1). Poyang Lake is internationally recognized as an important wetland system for its large numbers of wild water birds, including many endangered species [36]. Poyang Lake predominantly receives inflows from the Ganjiang, Fuhe, Xinjiang, Raohe and Xiushui Rivers within its 16.22×10^4 km^2 catchment [14], which subsequently flows into the Yangtze River through a narrow channel at Hukou (the junction of the Yangtze River and Poyang Lake) and at its northern outlet (see Figure 1). The Ganjiang, Fuhe and Xinjiang Rivers contribute approximately 89% of the lake's inflow, and the remaining 11% of the lake's inflow is from the Raohe and Xiushui Rivers [37]. In addition, these five rivers are the greatest contributors of industrial wastewater and domestic sewage to the lake [24,25].

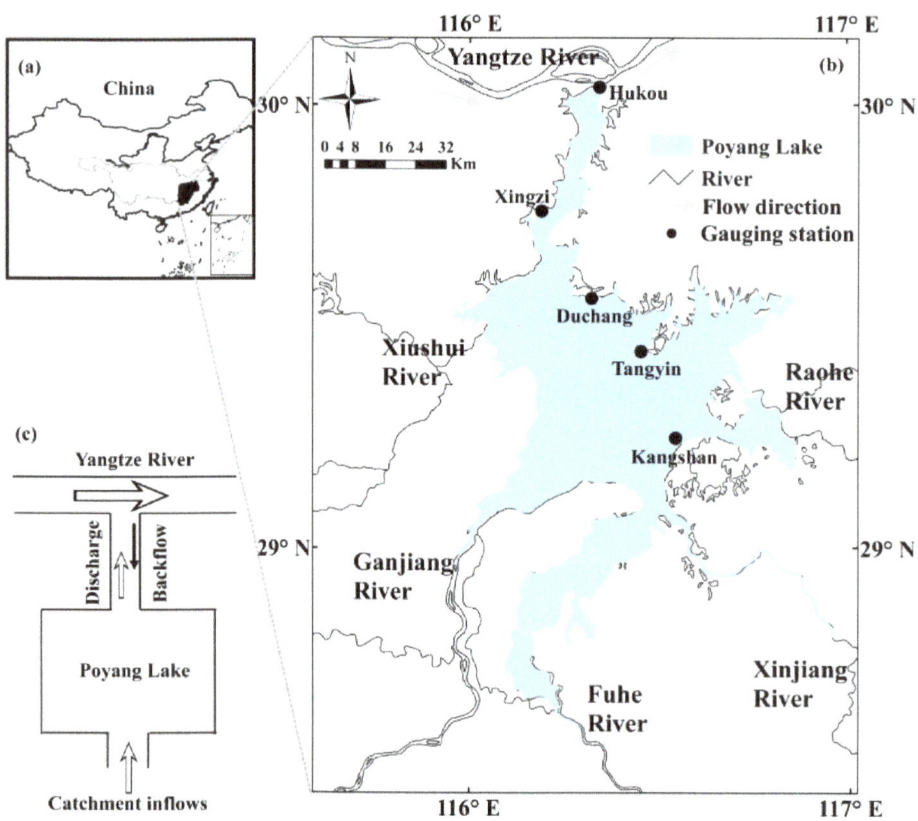

Figure 1. (**a**) Location of the Poyang Lake catchment in China; (**b**) the lake gauging stations and the major rivers surrounding the lake and (**c**) a schematic diagram of the river-dominated lake system.

Poyang Lake is a geometrically complex lake with tortuous shorelines and an incised bottom morphology [15]. The bottom of the lake is shaped by a combination of lacustrine and riverine morphological processes. The lake bed elevation decreases from the south to the north, with a difference of approximately 6.5 m [37]. Flow throughout the lake follows a complex circulation pattern that is defined by topographical features, including deep channels, floodplains and islands. Poyang Lake is a river-dominated lake with dramatic seasonal water level fluctuations that vary from 8 to 18 m each year. During the late spring and early summer, the water level dramatically increases and the water surface area expands to >3000 km^2 [38]. In contrast, during the winter, the water level decreases sharply, reaching a surface area of <1000 km^2 [38]. The differences in the water surface elevation can reach up to 5 m in the lake during the low water level period (Figure 2), whereas the surface of the lake is nearly horizontal during the high water level period (Figure 2).

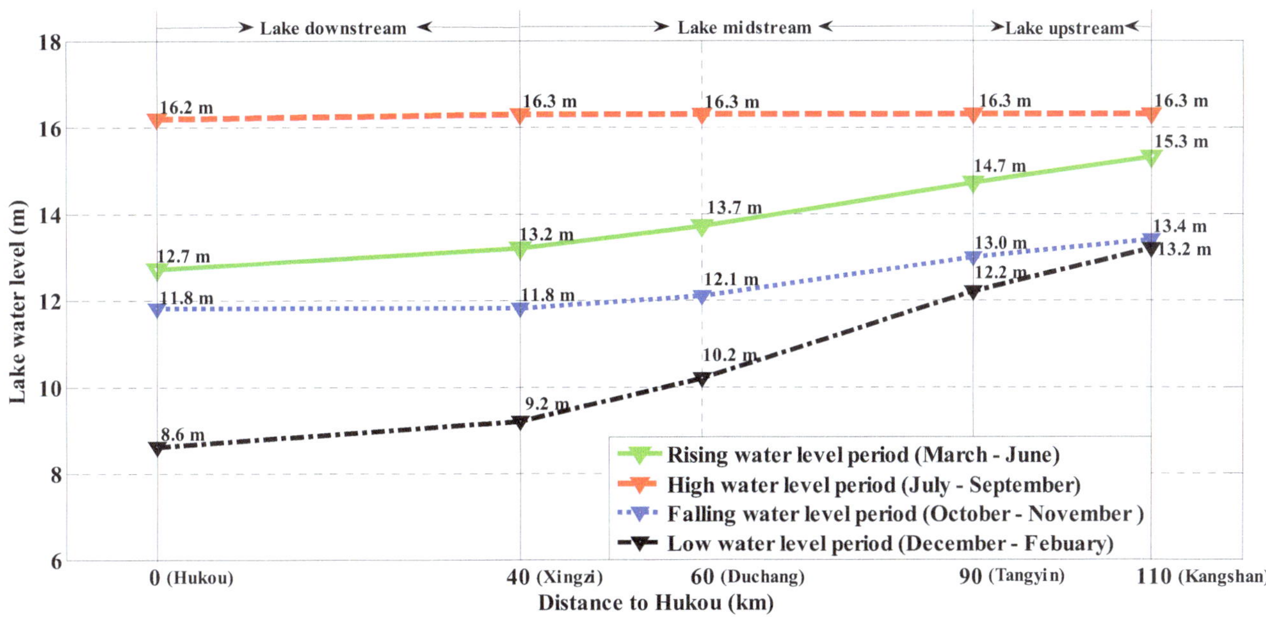

Figure 2. Observed lake water surface profiles for different water level periods based on the average water levels of the Hukou, Xingzi, Duchang, Tangyin and Kangshan gauging stations for the years 2000–2010.

3. Methodology

3.1. Hydrodynamic Model

As a river-dominated lake, Poyang Lake exhibits an extensive surface area (>3000 km^2) [38], is shallow (the depth is 8.4 m averaged over the whole lake) and is vertically well mixed [28]. Therefore, the vertically averaged currents (or two-dimensional currents) may be suitable to represent the majority of the flow in the lake. In response to this, a 2D depth-averaged hydrodynamic model for the lake has been implemented [37] using the MIKE 21 code [39]. The MIKE 21 code has been extensively applied in studying various water bodies, including the Alberni Inlet (Canada) [40], the Emilia Romagna coastal area (Italy) [41], the Arabian Sea (India) [42] and Lake St Lucia (South Africa) [43]. Therefore, only a brief description of the code is given here.

For the large floodplain of Poyang Lake, the irregular shorelines for the lake model based on historical flood inundation records was determined using GIS (Geographic Information System) technique [37]. The wet-dry point treatment method of MIKE 21 [39] was suitable for simulating the wetting and drying processes associated with the considerable variations in the lake area. A 2D grid system with variable mesh resolutions (unstructured triangular grid) was adopted to capture the complex lake bathymetry (in year 2000; Figure 3a) and to provide a reliable simulation of lake hydrodynamics by using grid sensitivity test [37]. This approach provided computational efficiency and allowed sharper gradients to be maintained with coarse mesh resolution. A higher mesh resolution was applied to the relatively deep and narrow flow channels across the lake, whereas a coarse mesh covered the relatively flat plain area of the adjacent lake floodplains (Figure 3b). Thus, the grid has 11,251 nodes and 20,450 triangular elements with approximate minimum and maximum element sizes of 70 and 1500 m [37]. Daily catchment inflows from the five main rivers were specified as upstream boundary conditions in the hydrodynamic model. The connection of the lake to the Yangtze River was simulated by specifying the downstream boundary water levels as those observed at Hukou gauging station. Spatially uniform but time-varying precipitation and evaporation rates were averaged from five lake gauging stations (locations shown in Figure 1b) and used for the Poyang Lake surface in the hydrodynamic model [37]. The minimum time step of the numerical integration was set to 5 s to maintain the Courant–Friedrich–Levy (CFL) condition for a stable solution [37].

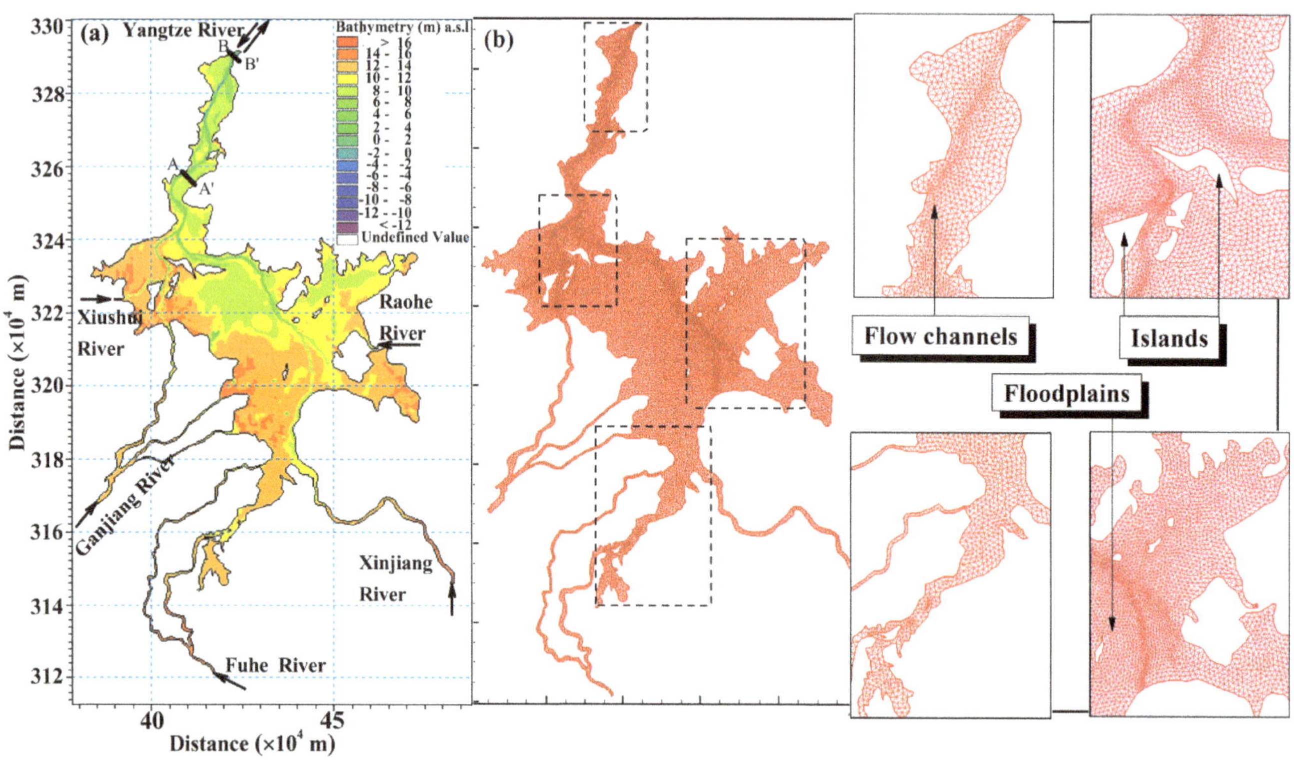

Figure 3. (**a**) Poyang Lake bathymetry, cross-sections of velocity observations (A-A' and B-B') and inflow/outflow sources of the lake (black arrows); (**b**) the unstructured grid for the MIKE 21 model and a close-up of the variable mesh resolution.

The performance of MIKE 21 was successfully validated against a wide variety of field observations conducted over 9 consecutive years (2000–2008) [37], including (1) lake water-level records at the Xingzi,

Duchang, Tangyin and Kangshan gauging stations (locations shown in Figure 1b); (2) flow rates at the Hukou gauging station (location shown in Figure 1b); and (3) lake water-surface areas and 2D inundation maps obtained from 14 MODIS (Moderate Resolution Imaging Spectroradiometer) images from year 2004. In an effort to further validate the model's ability to predict the lake hydrodynamics, a particular focus is placed on the predicted velocity variations at two typical cross-sections (see A-A' and B-B' in Figure 3a). In the present study, a comparison between depth-averaged velocity magnitudes from the ADCP (Acoustic Doppler Current Profile) velocity profile observations and model simulations at the cross-sections of Hukou and Xingzi were conducted over the wet (May 2008) and dry (December 2008) seasons (Figure 4). Although the model appears to be in agreement with the observed flow velocities for different water level periods (Figure 4), some model components proved somewhat difficult to accurately reproduce, such as the under- or over-estimated flow velocities in magnitude and phase (Figure 4), because the complex bathymetry exerted more influence on the velocity than the water level [37]. However, as the model is able to capture most of the important hydrodynamic characteristics and has an acceptable accuracy (the determination coefficient R^2 varies from 0.79 to 0.84; see Figure 4), it meets the requirements for using the 2D depth-averaged hydrodynamic model (see the insert chart in Figure 4) for transport time scale investigations.

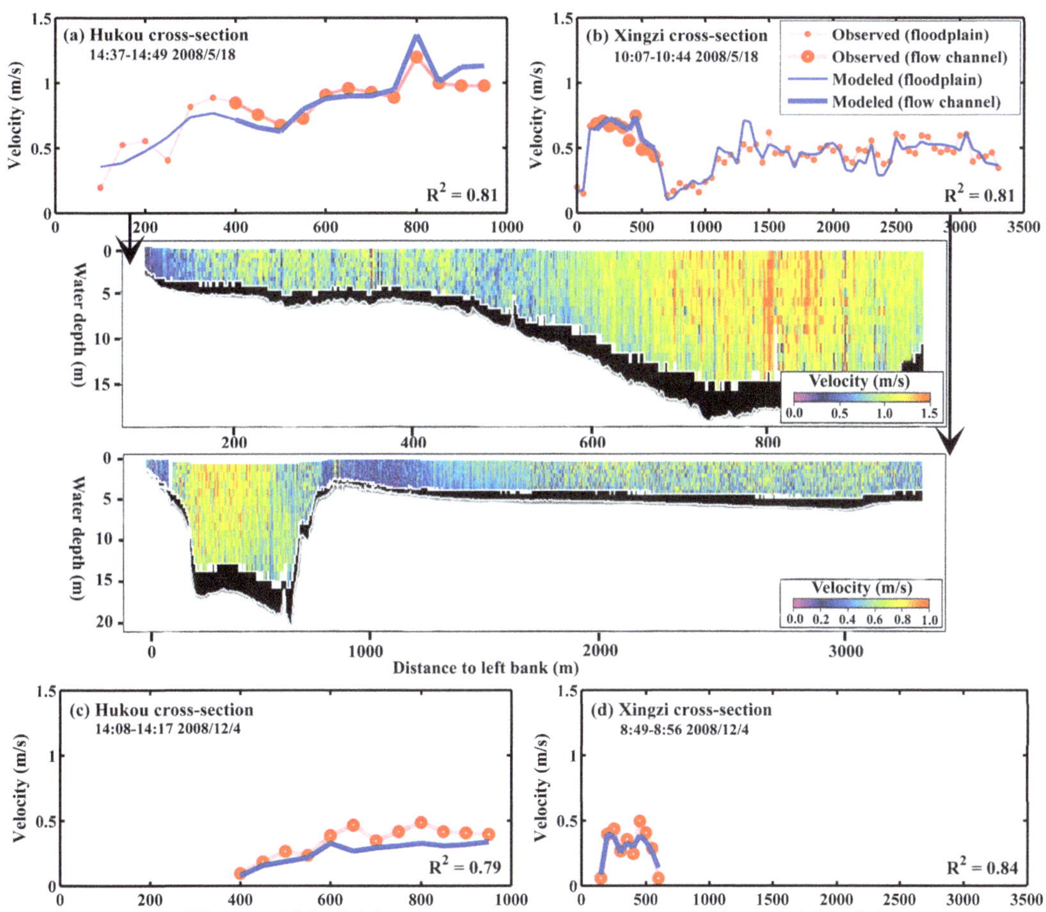

Figure 4. Comparison between depth-averaged velocity magnitudes from ADCP observations and simulations at two cross-sections of Hukou and Xingzi for periods of rising (**a**,**b**) and low water levels (**c**,**d**). Note that the insert chart represents the ADCP velocity profile observations sampling every 1 s.

3.2. Physical Setting of the Simulation Scenarios

To investigate the residence and travel times in Poyang Lake, the hydrodynamic model MIKE 21 [37] was used to generate the relevant hydrodynamic fields that drive the dye tracer simulations (2D advection-dispersion) [39].

Residence time (*e*-folding time in the present study) was measured to track the spatial and temporal distributions of a conservative tracer following its uniform distribution throughout the lake at an initial concentration (instantaneous release) corresponding to 100% (arbitrary unit; kg/m^3 used in this study) and a concentration of zero at the boundaries. Consequently, the *e*-folding times for each computational cell of the hydrodynamic model MIKE 21 were obtained to show the spatial distributions of the residence times. Seasonal variability in water circulation implies a need to perform separate simulations for various water level regimes of Poyang Lake. Hence, the releases were assumed to take place in the spring (on the 1st of April; water level rising period), summer (on the 1st of July; high water level period), autumn (on the 1st of October; water level recession period) and winter (on the 1st of January; low water level period). Subsequently, the travel times were calculated for these four seasons by assuming continuous injection of a conservative tracer at three different locations (see Section 4.3 for more detail). Three locations were chosen because they best represented the pollution sources frequently occurring at the mouths of the rivers entering Poyang Lake [23,24]. In these locations, the tracer was released continuously (1.0 kg/m^3 used in this study) over a period of 24 h (*i.e.*, the first day of the simulations), and a concentration of zero was imposed for the boundaries and for any other areas in the lake domain. Finally, travel times were determined based on the simulated peak concentrations at the lake outlet (Hukou).

Before running the model, the length of the residence and travel times were unknown; therefore, a reasonable simulation period had to be selected. In each case, starting from the date of release, the simulation covered a period of 365 days. The initial condition for each case was specified according to the previous lake water surface of the release date. To define a representative condition for hydrological regimes, the major forcings such as the observed five river inflows and Hukou water levels were averaged over a period of 10 years (2001–2010) and were used as model inputs rather than using an arbitrary year. Additionally, uniform precipitation and evaporation to/from the Poyang Lake surface and a constant wind field (wind direction = 45° NE, wind speed = 3.03 m/s) were considered. It should be noted that these meteorological conditions across the lake, though variable, were proven to have a relatively minor effect on the hydrodynamics of this unique lake system [28]. Furthermore, the values of the validation coefficients and other settings of the MIKE 21 model are the same as previous work [37]. The simulations differed in the tracer release dates and forcing functions applied.

4. Results and Discussion

4.1. Spatial Distribution of the Residence Time and Physical Interpretation

Figure 5 shows the spatial distribution of the residence times for the tracer release in the spring (1st of April) during the water level rising period. The results indicated that the water residence times at magnitudes of 5–10 days were observed in the main flow channels of the lake, which is mainly governed by along-channel advection during this wet season [27]. This result is consistent with previous study [44]. The study concluded that the monthly mean residence time varies from 5 to approximately 15 days for the water level rising period of Poyang Lake. In addition, the downstream values of the residence times appeared to increase upstream (e.g., 8 days for Hukou gauging station and 5 days for Duchang gauging station; see Figure 5b), which was partly attributed to hydraulic gradient differences for the Hukou–Xingzi and Duchang–Kangshan sections during this water level rising period (Figure 2). The residence times were shorter than 2 days for most upstream areas with high bed elevations (Figure 5a), depending on the frequent wetting and drying processes of the lake as a result of the changing catchment inflows [16]. The residence times in the eastern bay of the lake reached approximately 150 days (Figure 5c) due to the regional clock-wise water recirculation (Figure 6) caused by the combined effects of the islands and the relatively sharp bathymetric features in this region (Figure 3a). This type of water recirculation has been confirmed by the previous field observations [27] and is most likely responsible for the slower tracer removal in this area. Overall, the residence times in the main flow channels are significantly shorter than the values in other areas due to the velocity differences and associated water flows (Figure 6).

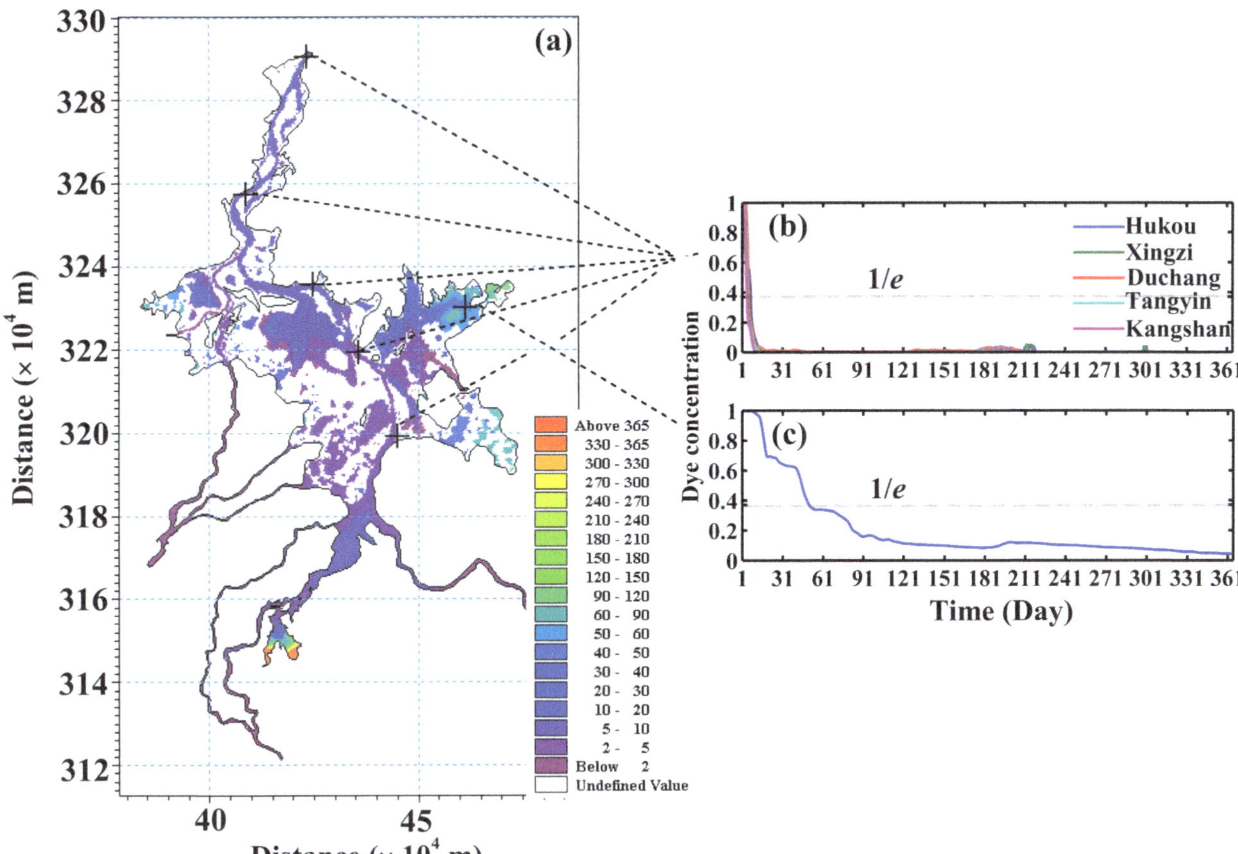

Figure 5. (**a**) Spatial distribution of the depth-averaged residence times (in days) for the water level rising period; (**b**,**c**) the computed dye concentrations (in kg/m^3) with time for the areas of interest. Releases were performed on the 1st of April during the spring. Note that the "Undefined Value" represents dry land.

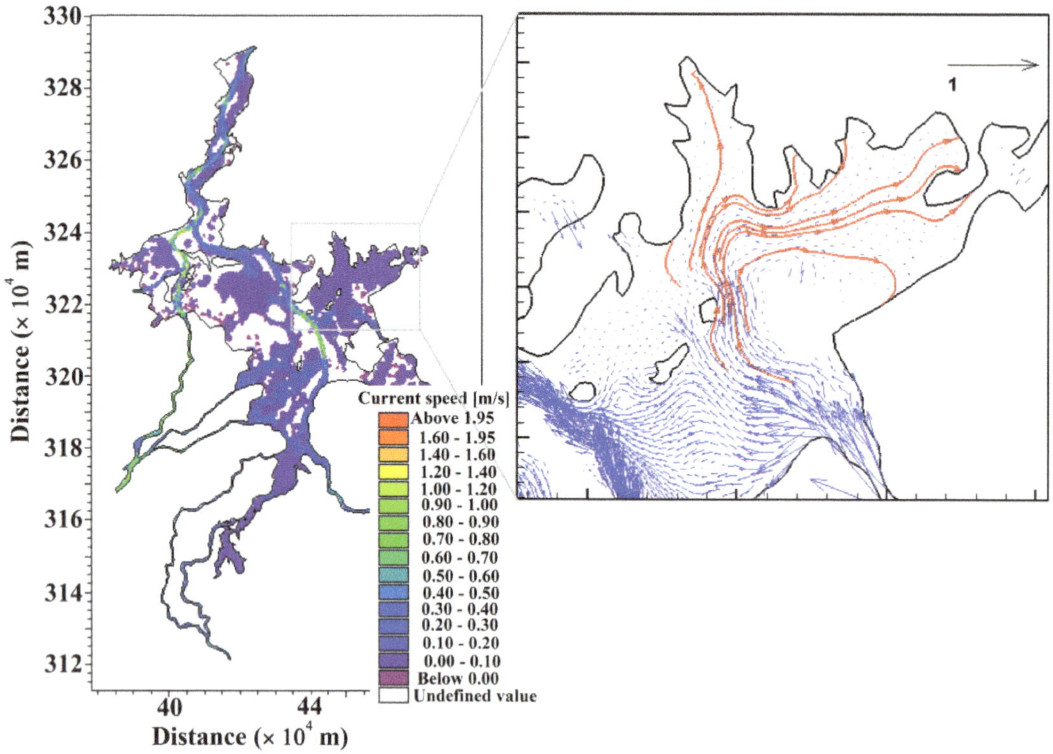

Figure 6. Map of the depth-averaged velocity (m/s) averaged over the water level rising period of MIKE 21 runs and a close-up of the velocity field for the region of interest. Red arrows represent the streamlines and associated flow directions. Note that the "Undefined Value" represents dry land.

The summer release (1st of July) with high lake water levels is illustrated in Figure 7. It is evident that the residence times of this "lake behavior" are significantly higher than the residence times of the "river behavior" in the spring during the water level rising period. This expected result coincides with evidence that suggests a negligible hydraulic gradient in the water surface (Figure 2) with significantly lower velocities (<0.1 m/s; Figure 8) across the entire lake domain, and a higher lake volume during the summer seasons. The waters adjacent to the lake's catchment river mouths are renewed considerably more quickly, with residence times of shorter than 2 days (Figure 7a,d), due to the stronger flushing of the catchment river inflows (Figure 8). From the upstream gauging station of Kangshan to the lake outlet of Hukou, the tracer concentrations along this main flow channel and its floodplain drop below the *e*-folding time after approximately 20–30 days following its introduction (Figure 7b), which is consistent with the average value of 26 days (in July and August) found by previous study [26]. However, areas characterized by slightly higher residence times are located in the downstream flow channels of the lake (Figure 7a), which is mainly attributed to the strong effects of backflow from the Yangtze River on northern Poyang lake (Figure 8) [17,27]. In addition, this complex water exchange process causes some perturbation in the dye concentrations from the downstream extremity to the most upstream ends of the lake (Figure 7b). The effect of this type of lake-river interactions on the hydrodynamics can reach the upstream ends of the lake [19]. The strong gradient in residence times from the catchment river mouths to the main flow channels in the lake were clearly reproduced (Figure 7a). This finding indicates that the influences of the river catchments (on residence times) gradually decrease in the center of the lake from the surrounding river mouths. Furthermore, the significantly lower velocities resulted in the highest residence time (up to 365 days) in some bays relative to any other areas in the lake domain (Figure 7a). Previous study found that the residence times in the bays of the lake reached a peak of approximately 270 days during this high water level period (in year 2008) [44,45]. In general, this study delivers the estimates of the same order for this high water level period.

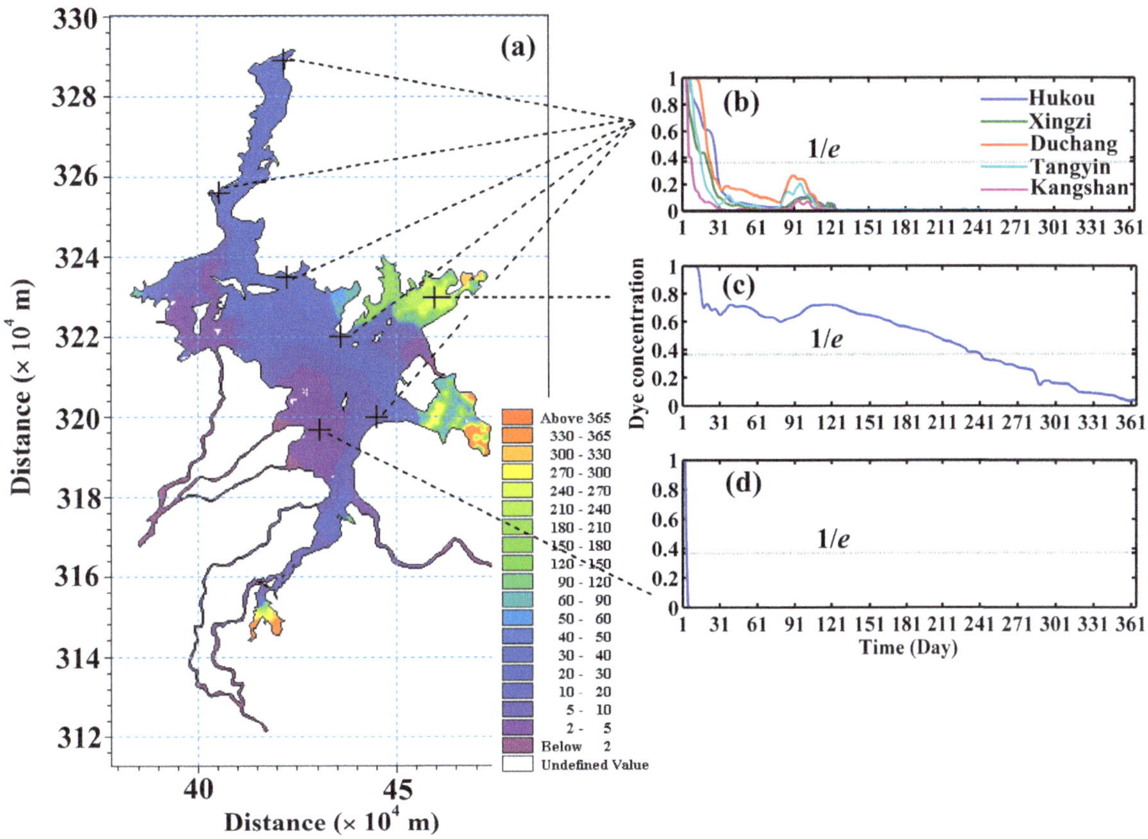

Figure 7. (**a**) Spatial distribution of the depth-averaged residence times (in days) for the high lake level period; (**b**–**d**) the computed dye concentrations (in kg/m^3) with time for the areas of interest. Releases were performed on the 1st of July during the summer. Note that the "Undefined Value" represents dry land.

Although no observation was made to confirm the spatial pattern of the residence times, the locations and magnitudes of the concentrations of water environmental factors can be explained, in part, by the "residence time distribution" phenomenon [5,26,32,46]. Figure 9 shows the spatial distribution of algae biomass averaged over 83 field samplings during the high water level period (in July 2012) of Poyang Lake. Notably, the maximum accumulation of algae biomass (e.g., the eastern and southern bays of the lake) appears to coincide with the maximum residence time shown in Figure 7. The low concentrations are in agreement with the short residence times along the main flow channels of the lake (Figure 9). Moreover, the spatial distributions of chlorophyll *a*, TN, TP and COD_{Mn} (chemical oxygen demand, measured by potassium permanganate) concentrations (averaged over July-September from year 2009 to 2012) are also consistent with the residence time distribution of the present study [46], indicating that the spatial pattern of the residence times has important environmental implications for Poyang Lake [26,45].

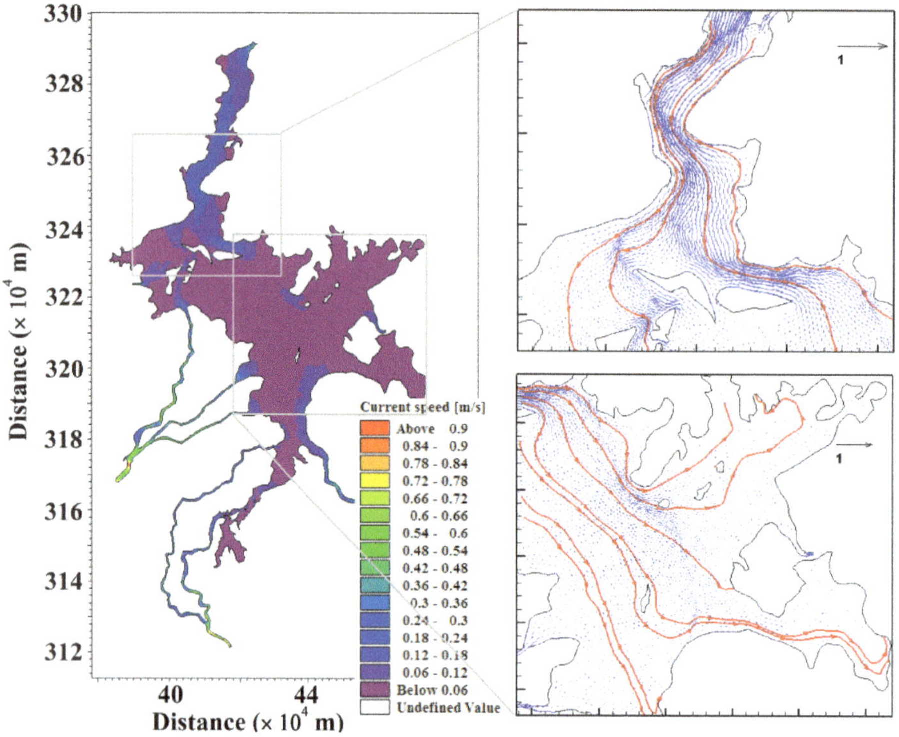

Figure 8. Map of the depth-averaged velocity (m/s) averaged over the high water level period of MIKE 21 runs and a close-up of the velocity field for the region of interest. Red arrows represent the streamlines and associated flow directions. Note that the "Undefined Value" represents dry land.

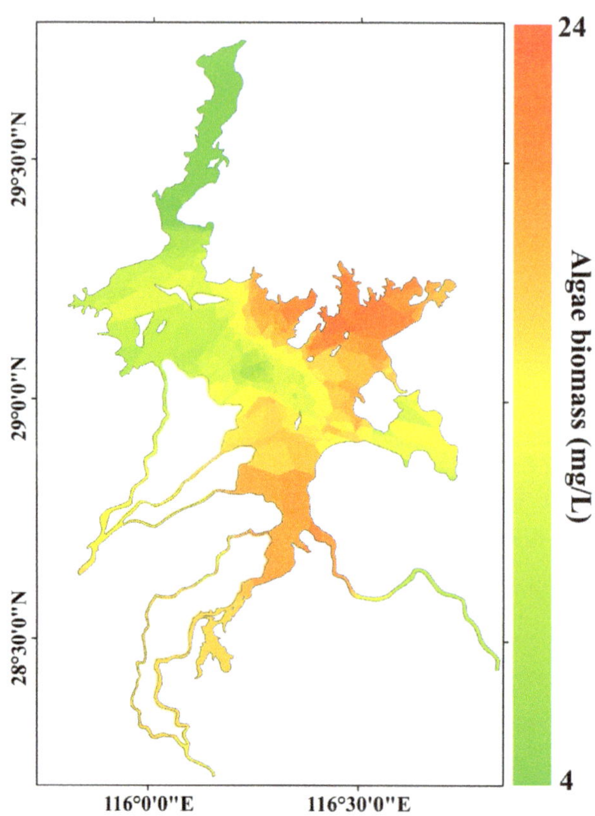

Figure 9. Spatial distribution of algae biomass (in mg/L) in Poyang Lake. Algae biomass data were obtained from 83 field samplings in July 2012.

The release case in the autumn (1st of October) represented the water level recession period and is shown in Figure 10. The residence times significantly decreased relative to the period with high water levels in the summer (compare Figure 10 to Figure 7), mainly due to the gradually decreasing water level of the Yangtze River, which increased the hydraulic gradient between the lake and the river (Figure 2) and intensified the lake's discharges (Figure 11). It should be noted that similar distribution patterns of the residence times are represented between the autumn and summer releases. This similarity can be partly attributed to the contraflow of the Yangtze River, although the water levels of the Yangtze River for the recession period were relatively lower than the water levels in the summer [28]. The residence time of the main flow channels of the lake is approximately 20 days (Figure 10a,b), which is slightly shorter than the residence times in the adjacent floodplain due to the relatively faster water velocities (0.3–0.4 m/s) in the main flow channels (Figure 11). The highest values (approximately 200 days or higher) were predicted for the eastern bays that are adjacent the lake shorelines (Figure 10a,c), indicating a poor flushing for local regions (the velocity ranges in magnitude from 0.1 to 0.2 m/s; Figure 11).

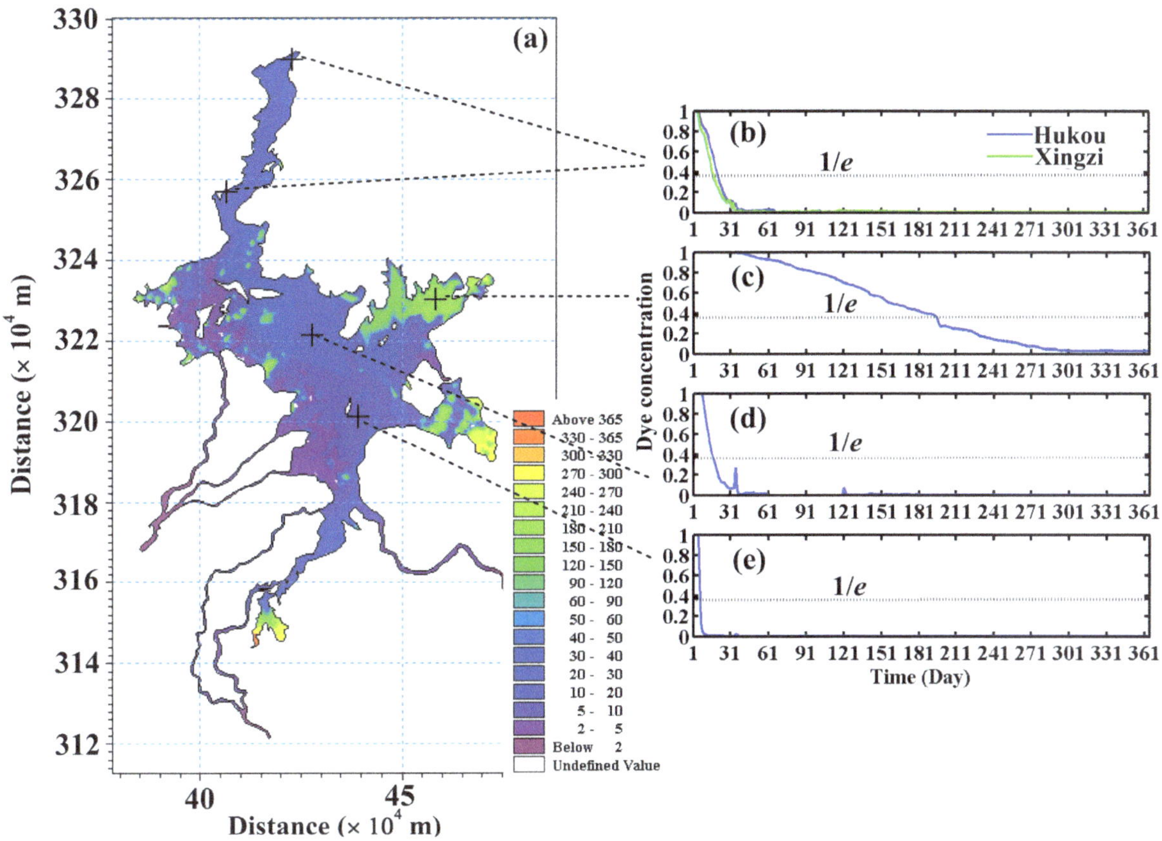

Figure 10. (**a**) Spatial distribution of the depth-averaged residence times (in days) for the water level recession period; (**b–e**) the computed dye concentrations (in kg/m^3) with time for the areas of interest. Releases were performed on the 1st of October during the autumn. Note that the "Undefined Value" represents dry land.

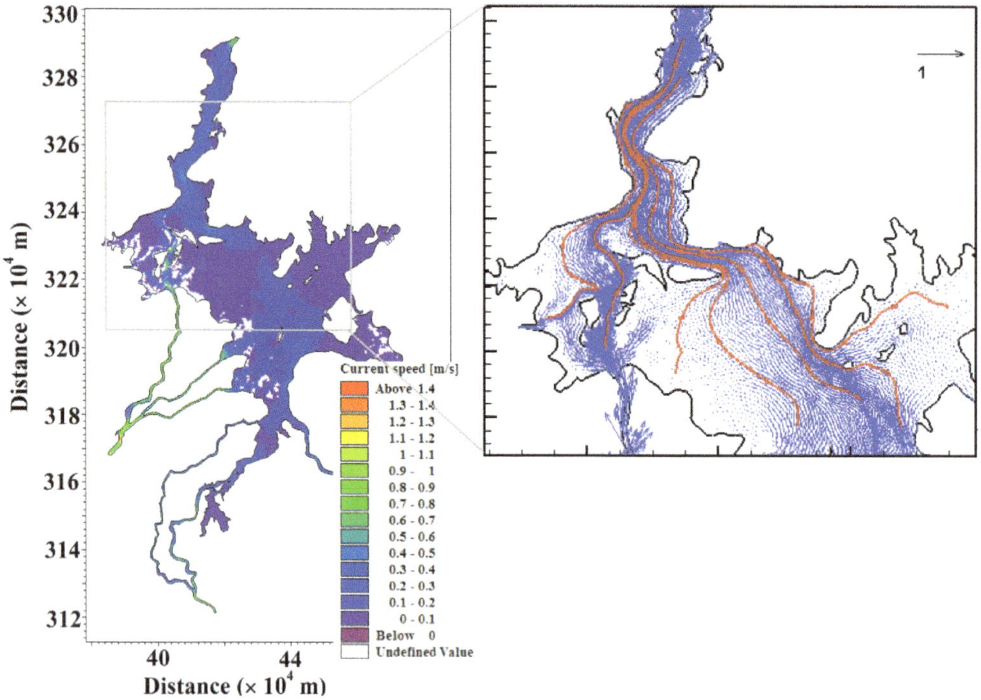

Figure 11. Map of the depth-averaged velocity (m/s) averaged over the water level recession period of MIKE 21 runs and a close-up of the velocity field for the region of interest. Red arrows represent the streamlines and associated flow directions. Note that the "Undefined Value" represents dry land.

Figure 12 demonstrates the high spatial variability of the residence times based on the winter release simulation (1st of January) during the low water level period. For this low lake water level regime with dominated gravity flow (the considerably greater water level gradient increases the horizontal component of gravity) [28], a relatively shorter residence time of approximately 10 days was observed for the main flow channels (Figure 12a,b), which agrees with the previous result for mean conditions in winter [44]. This finding is not unexpected because during the period with extremely low water levels, the along-channel water movement is northward and moves quickly from the shallower (upstream) to the deeper regions (downstream) of the lake channels (the velocity is up to 1.2 m/s; Figure 13) [26,37]. In addition, the water surface area of the lake decreases and the tracer removal is intensified in this area (Figure 12). It is interesting to note that the sublakes are separated from the main flow channels and exhibit significantly different residence times at magnitudes of 30–100 days (Figure 12); this pattern strictly coincides with the slower water movements in these relatively closed water bodies than in the main flow channels (Figure 13), in which tracer drift depends primarily on meteorological conditions (e.g., precipitation, evaporation, and winds). The highest residence times occurred in some bays that are adjacent to the lake shorelines, which indicated weak flushing (Figure 12). These results can be further explained with the water environmental factors [47], *i.e.*, the TN, TP and COD_{Cr} (chemical oxygen demand, measured by potassium dichromate) concentrations in the eastern and southern bays of the lake are higher than the concentrations in the main flow channels during the low water level period (in February 2008), which is consistent with the spatial pattern of the residence times shown in Figure 12. As noted previously, the concentration distribution can be partly supported the spatial patterns of the residence times [26,32].

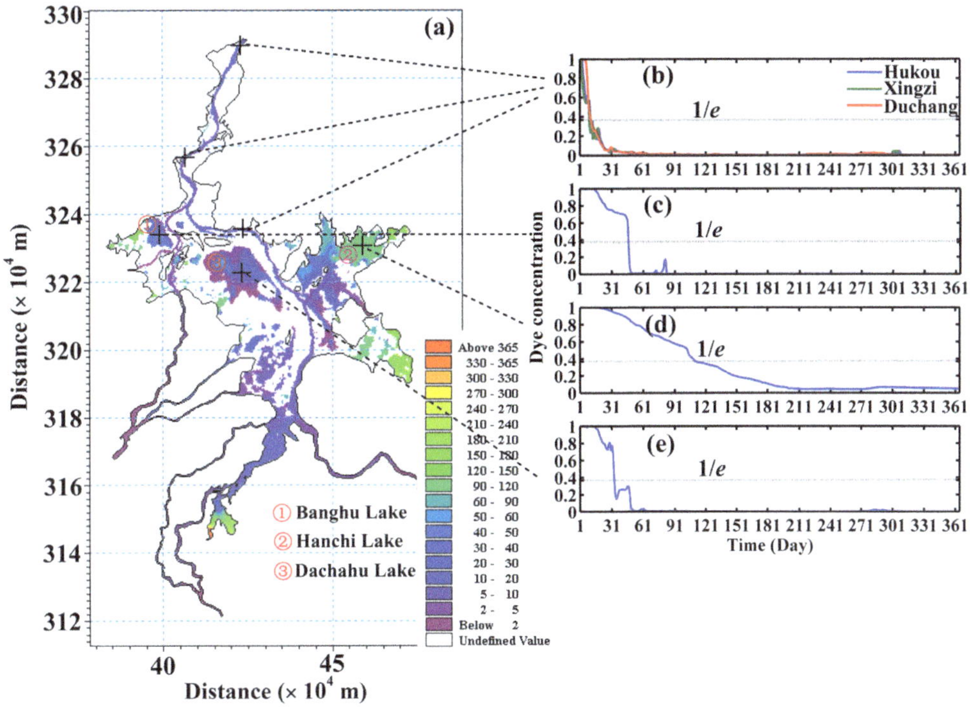

Figure 12. (**a**) Spatial distribution of the depth-averaged residence times (in days) for the low water level period; (**b–e**) the computed dye concentrations with time (in kg/m^3) for areas of interest. Releases were performed on the 1st of January during the winter. Note that the "Undefined Value" represents dry land.

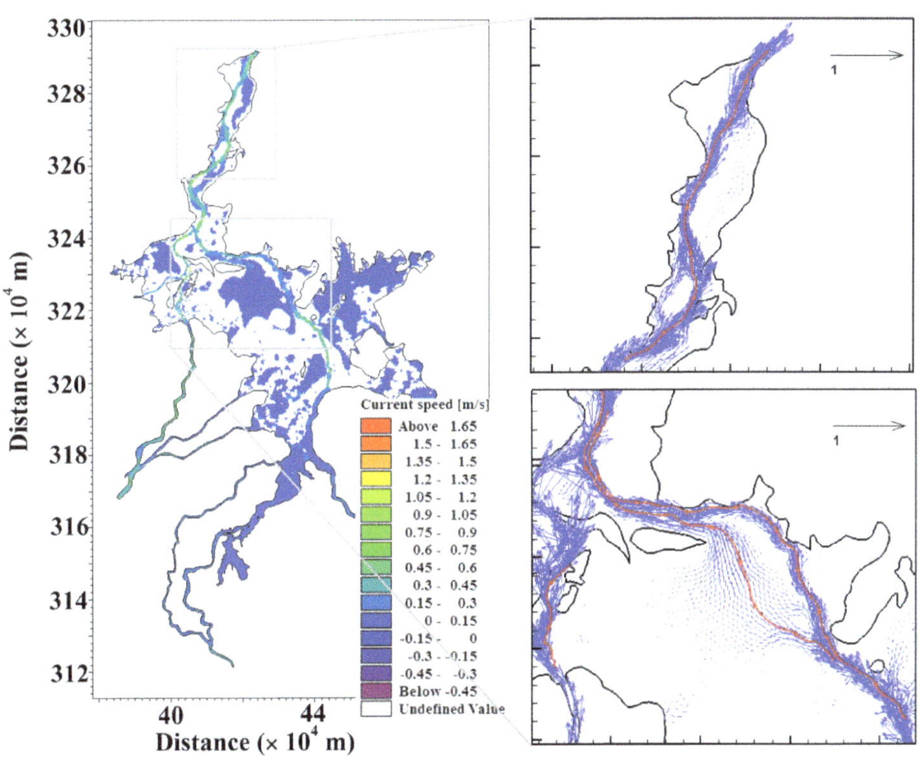

Figure 13. Map of the depth-averaged velocity (m/s) averaged over the low water level period of MIKE 21 runs and a close-up of the velocity field for the region of interest. Red arrows represent the streamlines and associated flow directions. Note that the "Undefined Value" represents dry land.

4.2. Frequency Distribution Curves

Figure 14 shows the frequency distribution curves of the four water level regimes calculated from the spatial distributions of the residence times. Generally, the steeper the frequency distribution curve, the less variation in the residence time values for any hydrological system [3]. The characteristic shapes of the curves indicate that although large portions of the lake are characterized by similar residence times, places with sharp τ_e gradients can be expected (Figure 14). This finding is represented as a characteristic 'step' in the frequency distribution curves. Figure 14 shows that a large percentage of the area (approximately 80%) features τ_e values of between a few days and approximately 1 month. For τ_e values of more than 1 month, the gradient of the curves is significantly shallower. The characteristic shapes of the frequency distribution curves with apparent contraflexure points were similar for all of the tracer release cases, which indicated the transition of the flushing characteristic (Figure 14). The most pronounced feature of the frequency distribution curves was the characteristic 'step' that indicated that Poyang Lake is a mixing system, which can be divided into fast (the main flow channels of the lake) and slow flushing (some bays adjacent to the lake shorelines) systems during different water level regimes (Figure 14).

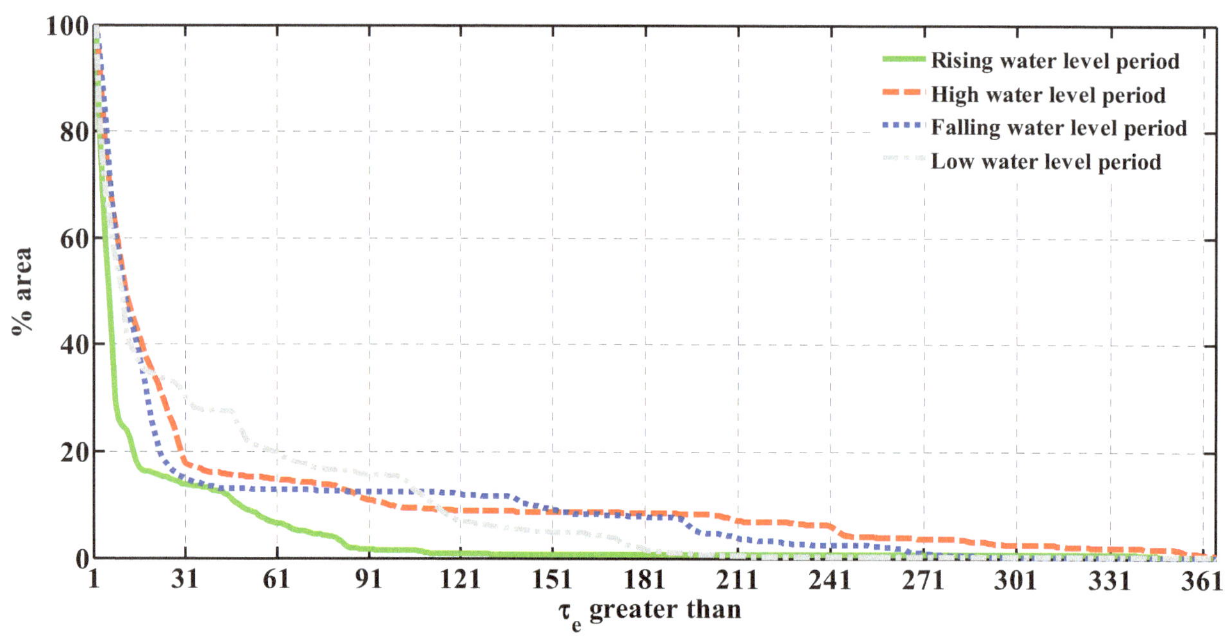

Figure 14. Residence time frequency distribution curves of the four water level regimes obtained from the MIKE 21 hydrodynamic modeling runs.

4.3. Travel Time for the Potential Pollution Sources and Physical Interpretation

A travel time analysis was conducted to further investigate the flushing ability of Poyang Lake. For the transport from the C1 release point to the Hukou outlet, the travel times fell within a range of 4–17 days, with higher values occurring during the periods of high and falling water levels (see Figure 15i). It took nearly four times longer for the tracer to arrive at the lake outlet during these periods relative to the periods with rising and low water levels (Figure 15i). In addition, this phenomenon can be observed in Figure 15ii (C2 release point; travel time ranging from 7 to 32 days) and Figure 15iii (C3 release point; travel time ranging from 6 to 30 days). The seasonal behaviors of the travel times are consistent with previous results [45], *i.e.*, the travel times (in year 2008) were approximately 15 days for periods of rising and low water levels and approximately 40 days for high and falling water levels.

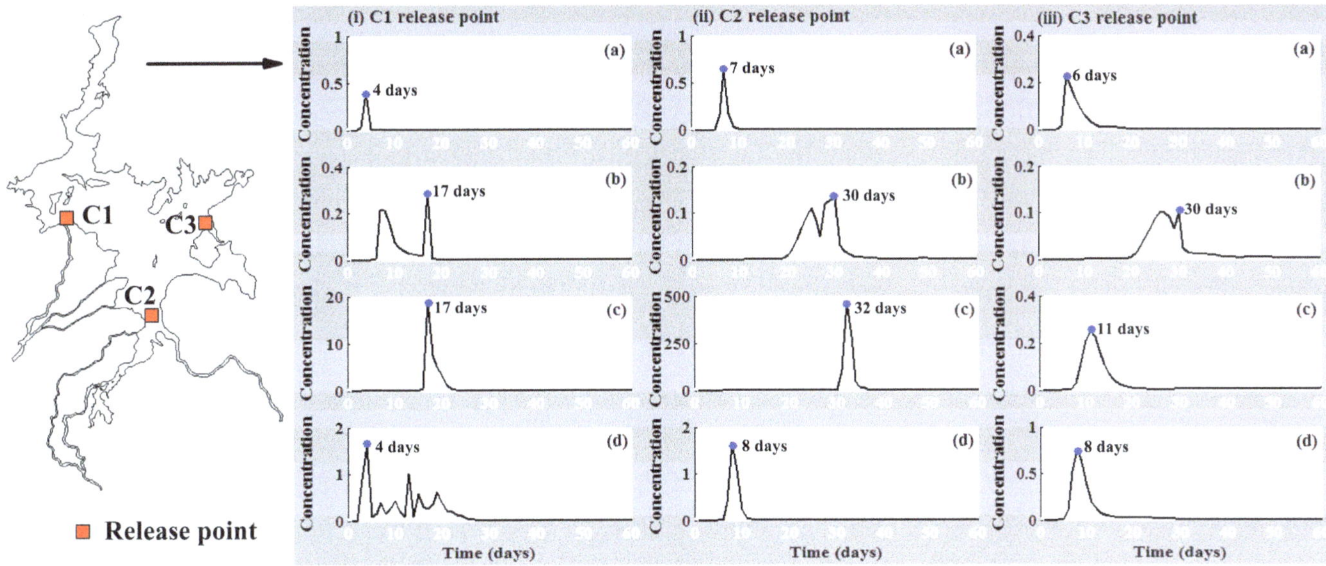

Figure 15. Time series of tracer concentrations at the Hukou lake outlet (concentrations after 60 days are not shown due to low values). Releases were performed during (**a**) the water level rising period on the 1st of April during the spring; (**b**) the high water level period on the 1st of July during the summer; (**c**) the water level recession period on the 1st of October during the autumn; and (**d**) the low water level period on the 1st of January during the winter. Here, time represents the time elapsed since the release. The blue dot represents the peak concentration and the associated travel times that are defined in this study. A unit dye concentration of 10^{-4} kg/m^3 was used.

The most notable finding obtained from the modelling results is that the travel times were nearly the same during the periods of rising and low water levels and during the periods of high and falling water levels (Figure 15). This phenomenon can be explained by the advection-dispersion processes of the lake. Due to the substantial influences of the Yangtze River, the lake water surface is almost horizontal with relatively small water level differences for the high and falling water level periods (Figure 2), and the resulting small flow velocities across the lake (Figures 8 and 11) increase the dispersion process of the dye tracer. Although the average travel time is generally determined by the advection process of the horizontal transport, the dispersion process plays a considerable role during the high and falling water level periods of the lake, which is similar to previous reports [48,49]. However, the relatively small dispersion in the lake can be neglected for the rising and low water level releases. The dye quickly arrived at the northward outlet, attributed to the along-channel advection-dominated regime, which is mostly due to the rapid along-channel water flows from shallower to deeper regions of the lake (Figures 6 and 13) as a result of the large water level gradients (Figure 2). It should be noted that although the dispersion coefficient changes with time and location (this is the same as the flow velocity), flow velocity changes may play a larger role than the variable dispersion coefficient of the hydrodynamic model [49,50].

Overall, our results demonstrate that the travel times are mainly determined by the highly dynamic water level regimes of Poyang Lake.

5. Summary and Conclusions

This study presents detailed research that contributes to a better understanding of the flushing characteristics of the large floodplain of Poyang Lake with complex water level regimes. This study is the first to use hydrodynamic modeling combined with tracer release experiments to study transport time scales for this naturally hydraulic-linked lake system. The aim of this research was limited to revealing the spatial variability of the residence time, the seasonal behaviors of the travel time, and the water level dynamics controlling the residence and travel times. This study provides important information regarding the sustainable management of Poyang Lake given its deteriorating water quality. The main conclusions are summarized as follows:

(1) The residence times with highly spatial variability that differed in hydrographically sub-regions within Poyang Lake, which varies roughly from less than 10 days along the lake channels to 365 days in some bays adjacent to the lake shorelines. Relatively large spatial residence time variations between the periods of high (falling) and low (rising) water levels were observed in the lake depending on the seasonal water level regimes as a result of the combined effects of catchment rivers and the Yangtze River. Generally, the results indicate that Poyang Lake exhibits strong but spatially heterogeneous residence times that vary with its highly seasonal water level dynamics.

(2) The frequency distribution curve of the lake's residence time provides useful generic information regarding the degree of variation in flushing across the lake domain. The curves of the shallow gradients for Poyang Lake indicated higher spatial variability in residence times. Additionally, the characteristic 'step' of the frequency distribution curve demonstrates that the Poyang Lake domain is divided into fast (the main flow channels of the lake) and slow (some bays adjacent to the lake shorelines) flushing systems.

(3) The results indicated that the travel times from the potential pollution sources to the lake outlet varied seasonally. Nearly identical travel times occurred during the periods of high and falling water levels (up to 32 days), and these were four times longer than the travel times observed during the periods of low and rising water levels (less than 8 days), which is attributed to the advection-dispersion processes of the lake dependent on changes in the seasonal water level regime.

Our modeling effort represents a first step in integrating lake hydrodynamics to provide a better understanding of residence and travel times, information that is extremely important for spatially and temporally managing the chemical and ecological processes in Poyang Lake and for managing the responses of the lake to potential pollution sources. The findings of this study make it possible to conduct further ecological modeling experiments, as the transport of pollutants and their transport time scales depend largely on hydrodynamic fields. In addition, the future work should divide the lake domain into sub-regions with distinct flushing characteristics because this investigation may be beneficial for better understanding the local water environment of the lake. It is expected that the current study will eventually aid decision makers in creating a strategic lake management plan for Poyang Lake.

Acknowledgments

This work is jointly supported by the National Basic Research Program of China (973 Program) (2012CB417003), the Collaborative Innovation Center for Major Ecological Security Issues of Jiangxi Province and Monitoring Implementation (JXS-EW-00), the National Natural Science Foundation of China (41371062, 41201026 and 41301023), and the Science Foundation of Nanjing Institute of Geography and Limnology, Chinese Academy of Sciences (NIGLAS2012135001). The authors are grateful to Xia Liu of the Nanjing Institute of Geography and Limnology, CAS, for providing the algae biomass data used in this study.

Author Contributions

Yunliang Li performed the modelling simulations, results analysis and prepared the manuscript; Qi Zhang and Jing Yao contributed to the development and editing of the manuscript.

Conflicts of Interest

The authors declare no conflict of interest.

References

1. Lehner, B.; Döll, P. Development and validation of a global database of lakes, reservoirs and wetlands. *J. Hydrol.* **2004**, *296*, 1–22.
2. Pilotti, M.; Simoncelli, S.; Valerio, G. A simple approach to the evaluation of the actual water renewal time of natural stratified lakes. *Water Resour. Res.* **2014**, *50*, 2830–2849.
3. Dabrowski, T.; Hartnett, M.; Olbert, A.I. Determination of flushing characteristics of the Irish Sea: A spatial approach. *Comput. Geosci.* **2012**, *45*, 250–260.
4. Nguyen, T.; Thupaki, P.; Anderson, E.J.; Phanikumar, M.S. Summer circulation and exchange in the Saginaw Bay-Lake Huron system. *J. Geophys. Res.* **2014**, *119*, doi:10.1002/2014JC009828.
5. Monsen, N.E.; Cloern, J.E.; Lucas, L.V.; Monismith, S.G. A comment on the use of flushing time, residence time, and age as transport time scales. *Limnol. Oceanogr.* **2002**, *47*, 1545–1553.
6. Takeoka, H. Fundamental concepts of exchange and transport time scales in a coastal sea. *Cont. Shelf Res.* **1984**, *3*, 311–326.
7. Dyer, K.R. *Estuaries: A Physical Introduction*; John Wiley and Sons Ltd.: London, UK, 1997; pp. 1–210.
8. Cucco, A.; Umgiesser, G. Modeling the Venice Lagoon residence time. *Ecol. Model.* **2006**, *193*, 34–51.
9. Nyberg, L.; Rodhe, A.; Bishop, K. Water transit times and flow paths from two line injections of 3H and 36Cl in a micro-catchment at Gardsjon, Sweden. *Hydrol. Process.* **1999**, *13*, 1557–1575.
10. Botter, G.; Bertuzzo, E.; Rinaldo, A. Catchment residence and travel time distributions: The master equation. *Geophys. Res. Lett.* **2011**, *38*, doi:10.1029/2011GL047666.
11. Razmi, A.M.; Barry, D.A.; Lemmin, U.; Bonvin, F.; Kohn, T.; Bakhtyar, R. Direct effects of dominant winds on residence and travel times in the wide and open lacustrine embayment: Vidy Bay (Lake Geneva, Switzerland). *Aquat. Sci.* **2014**, *76*, 59–71.

12. Morales, T.; Valderrama, I.F.; Uriarte, J.A.; Antiguedad, I.; Olazar, M. Predicting travel times and transport characterization in karst conduits by analyzing tracer-breakthrough curves. *J. Hydrol.* **2007**, *334*, 183–198.
13. Dabrowski, T.; Hartnett, M. Modelling travel and residence times in the eastern Irish Sea. *Mar. Pollut. Bull.* **2008**, *57*, 41–46.
14. Shankman, D.; Heim, B.D.; Song, J. Flood frequency in China's Poyang Lake region: Trends and teleconnections. *Int. J. Climatol.* **2006**, *26*, 1255–1266.
15. Gao, J.; Jia, J.; Kettner, A.J.; Xing, F.; Wang, Y.P.; Xu, X.N.; Yang, Y.; Zou, X.Q.; Gao, S.; Qi, S.; *et al.* Changes in water and sediment exchange between the Changjiang River and Poyang Lake under nature and anthropogenic conditions, China. *Sci. Total Environ.* **2014**, *481*, 542–553.
16. Hu, Q.; Feng, S.; Guo, H.; Chen, G.Y.; Jiang, T. Interactions of the Yangtze River flow and hydrologic processes of the Poyang Lake, China. *J. Hydrol.* **2007**, *347*, 90–100.
17. Cui, L.; Wu, G.; Liu, Y. Monitoring the impact of backflow and dredging on water clarity using MODIS images of Poyang Lake, China. *Hydrol. Process.* **2009**, *23*, 342–350.
18. Guo, H.; Hu, Q.; Zhang, Q.; Feng, S. Effects of the Three Gorges Dam on Yangtze River flow and river interaction with Poyang Lake, China: 2003–2008. *J. Hydrol.* **2012**, *416*, 19–27.
19. Zhang, Q.; Ye, X.C.; Werner, A.D.; Li, Y.L.; Yao, J.; Li, X.H.; Xu, C.Y. An investigation of enhanced recessions in Poyang Lake: Comparison of Yangtze River and local catchment impacts. *J. Hydrol.* **2014**, *517*, 425–434.
20. Zhen, L.; Li, F.; Huang, H.; Dilly, O.; Liu, J.; Wei, Y.; Yang, L.; Cao, X. Households' willingness to reduce pollution threats in the Poyang Lake region, southern China. *J. Geochem. Explor.* **2011**, *110*, 15–22.
21. Wang, M.L.; Zhou, W.B.; Hu, C.M. Status of nitrogen and phosphorus in waters of Lake Poyang Basin. *J. Lake Sci.* **2008**, *20*, 334–338.
22. Wu, L.; Li, M.; Guo, Y.; Yang, X. Influence of Three Gorges Project on water quality of Poyang Lake. *Procedia Environ. Sci.* **2011**, *10*, 1496–1501.
23. Zhang, D.W.; Wei, Y.H.; Zhang, L.; Lin, G.G.; Chen, Y.W.; Tu, T.T. Distribution of heavy metals in water, suspended particulate matter and sediment of Poyang lake, China. *Fresenius Environ. Bull.* **2012**, *21*, 1910–1919.
24. Wei, Y.; Zhang, J.; Zhang, D.; Tu, T.; Luo, L. Metal concentrations in various fish organs of different fish species from Poyang Lake, China. *Ecotoxicol. Environ. Saf.* **2014**, *104*, 182–188.
25. Deng, X.; Zhao, Y.; Wu, F.; Lin, Y.; Lu, Q.; Dai, J. Analysis of the trade-off between economic growth and the reduction of nitrogen and phosphorus emissions in the Poyang Lake Watershed, China. *Ecol. Model.* **2011**, *222*, 330–336.
26. Wu, Z.; Lai, X.; Zhang, L.; Cai, Y.; Chen, Y. Phytoplankton chlorophyll a in Lake Poyang and its tributaries during dry, mid-dry and wet seasons: A 4-year study. *Knowl. Manag. Aquat. Ecosyst.* **2014**, *412*, 1–13.
27. Zhu, H.; Zhang, B. *The Poyang Lake*; University of Science & Technology of China Press: Hefei, China, 1997; pp. 1–349. (In Chinese)
28. Wang, S.; Dou, H. *Chinese Lakes*; Science Press: Beijing, China, 1998; pp. 1–580. (In Chinese)

29. Cheng, X.; Li, S. The typical lake eutrophication evolving process and its characteristic analysis in the middle and lower reaches of the Changjiang River. *Chin. Sci. Bull.* **2006**, *51*, 848–855. (In Chinese)
30. Rueda, F.; Moreno-Ostos, E.; Armengol, J. The residence time of river water in reservoirs. *Ecol. Model.* **2006**, *191*, 260–274.
31. Shaha, D.C.; Cho, Y.K.; Seo, G.H.; Kim, C.S.; Jung, K.T. Using flushing rate to investigate spring-neap and spatial variations of gravitational circulation and tidal exchanges in an estuary. *Hydrol. Earth Syst. Sci.* **2010**, *14*, 1465–1476.
32. Wan, Y.; Qiu, C.; Doering, P.; Ashton, M.; Sun, D.; Coley, T. Modeling residence time with a three-dimensional hydrodynamic model: Linkage with chlorophyll a in a subtropical estuary. *Ecol. Model.* **2013**, *268*, 93–102.
33. Shen, J.; Haas, L. Calculating age and residence time in the tidal York River using three-dimensional model experiments. *Estuar. Coast. Shelf Sci.* **2004**, *61*, 449–461.
34. Dargahi, B.; Stegn, S.G. Combined 3D hydrodynamic and watershed modeling of Lake Tana, Ethiopia. *J. Hydrol.* **2011**, *398*, 44–64.
35. Umgiesser, G.; Ferrarin, C.; Cucco, A.; Pascalis, F.D.; Bellafiore, D.; Ghezzo, M.; Bajo, M. Comparative hydrodynamics of 10 Mediterranean lagoons by means of numerical modeling. *J. Geophys. Res.* **2014**, *119*, 2212–2226.
36. Kanai, Y.; Ueta, M.; Germogenov, N.; Nagendran, M.; Mita, N.; Higuchi, H. Migration routes and important resting areas of Siberian cranes (Grus leucogeranus) between northeastern Siberia and China as revealed by satellite tracking. *Biol. Conserv.* **2002**, *106*, 339–346.
37. Li, Y.L.; Zhang, Q.; Yao, J.; Werner, A.D.; Li, X.H. Hydrodynamic and hydrological modeling of Poyang Lake-catchment system in China. *J. Hydrol. Eng.* **2014**, *19*, 607–616.
38. Feng, L.; Hu, C.; Chen, X.; Cai, X.; Tian, L.; Gan, W. Assessment of inundation changes of Poyang Lake using MODIS observation between 2000 and 2010. *Remote Sens. Environ.* **2012**, *121*, 80–92.
39. Danish Hydraulic Institute (DHI). *MIKE 21 Flow Model: Hydrodynamic Module User Guide*; DHI Water and Environment: Hørsholm, Denmark, 2007; pp. 1–90.
40. Barua, D.K.; Allyn, N.F.; Quick, M.C. Modeling tsunami and resonance response of Alberni inlet, British Columbia. In Proceedings of the 30th International Conference on Coastal Engineering, San Diego, CA, USA, 2–8 September 2006; pp. 1590–1602.
41. Martinelli, L.; Zanuttigh, B.; Corbau, C. Assessment of coastal flooding hazard along the Emilia Romagna littoral, IT. *Coast. Eng.* **2010**, *57*, 1042–1058.
42. Patgaonkar, R.S.; Vethamony, P.; Lokesh, K.S.; Babu, M.T. Residence time of pollutants discharged in the Gulf of Kachchh, northwestern Arabian Sea. *Mar. Pollut. Bull.* **2012**, *64*, 1659–1666.
43. Schoen, J.; Stretch, D.; Tirok, K. Wind-driven circulation patterns in a shallow estuarine lake: St Lucia, South Africa. *Estuar. Coast. Shelf Sci.* **2014**, *146*, 49–59.
44. Jin, G.; Xie, D.; Deng, H.; Yan, Y.; Liu, M.; Wang, Y. On seasonal hydrographic variety and environmental capacity of Poyang Lake. *Acta Agric. Univ. Jiangxiensis* **2011**, *33*, 388–393. (In Chinese)
45. Tang, C.; Xiong, X.; Zhang, X.; Zou, W. Simulation of the impact of the reverse flow from Yangtze River on the hydrodynamic process of Poyang Lake. *J. Lake Sci.* **2014**, in press.

46. Wu, Z.; He, H.; Cai, Y.; Zhang, L.; Chen, Y. Spatial distribution of chlorophyll a and its relationship with the environment during summer in Lake Poyang: A Yangtze-connected lake. *Hydrobiologia* **2014**, *732*, 61–70.
47. Gao, G.; Ruan, R.; Ouyang, Q. Water quality status and changing trend in Poyang Lake. *J. Nanchang Inst. Tech.* **2010**, *29*, 50–53. (In Chinese)
48. White, A.B.; Kumar, P.; Saco, P.M.; Rhoads, B.L.; Yen, B.C. Hydrodynamic and geomorphologic dispersion: Scale effects in the Illinois River Basin. *J. Hydrol.* **2004**, *288*, 237–257.
49. Jia, X.; Zeng, F.; Gu, Y. Semi-analytical solutions to one-dimensional advection-diffusion equations with variable diffusion coefficient and variable flow velocity. *Appl. Math. Comput.* **2013**, *221*, 268–181.
50. Zeinoddini, M.; Tofighi, M.A.; Vafaee, F. Evaluation of dike-type causeway impacts on the flow and salinity regimes in Urmia Lake, Iran. *J. Great Lakes Res.* **2009**, *35*, 13–22.

3

Hierarchical Classification of Groundwater Pollution Risk of Contaminated Sites Using Fuzzy Logic: A Case Study in the Basilicata Region (Italy)

Donatella Caniani †,*, Donata Serafina Lioi †, Ignazio M. Mancini † and Salvatore Masi †

School of Engineering, University of Basilicata, viale dell'Ateneo Lucano n. 10, Potenza 85100, Italy; E-Mails: saralioi@libero.it (D.S.L.); ignazio.mancini@unibas.it (I.M.M.); salvatore.masi@unibas.it (S.M.)

† These authors contributed equally to this work.

* Author to whom correspondence should be addressed; E-Mail: donatella.caniani@unibas.it;

Academic Editor: Miklas Scholz

Abstract: Groundwater is an important source of water. Since the control and removal of pollution are expensive, it is essential to identify the possible sources of contamination and to correctly classify groundwater on the basis of its intrinsic and integrated vulnerability. In the last years, many fuzzy models have been developed. There is a lack, however, of studies concerning the application of fuzzy logic for the assessment of the susceptibility of groundwater quality to anthropogenic activities and comparing these methods with traditional methods. The fuzzy model described in this paper assesses the aquifer integrated environmental vulnerability connected to the presence of illegal dumpsites and allows us to hierarchically classify them within a hazard scale. The results allowed us to carry out a comparison of the proposed fuzzy method with a traditional method. The fuzzy method proved to be a useful and objective tool for environmental planning.

Keywords: fuzzy logic; groundwater intrinsic vulnerability; groundwater integrated vulnerability; illegal open dumpsites

---

1. Introduction

In arid and semi-arid regions, groundwater is an important source of water. Since the control and removal of pollution are expensive, it is essential to identify the possible sources of contamination and to correctly classify groundwater on the basis of its intrinsic and integrated vulnerability. Sources of groundwater pollution are basically of two types: diffuse and point sources. Examples of this type of pollution are found everywhere. Groundwater pollution, in almost all cases, is due to human activities. The sources of groundwater pollution are in fact associated with a wide range of industrial, agricultural, commercial and domestic activities (such as industrial installations, landfills of waste, mining and quarrying activities, urban centres, lines of communication and infrastructure links, agricultural activities, livestock activities, and other activities that involve water systems). In the literature, two types of vulnerability are described: intrinsic and integrated (or specific) vulnerability. The latter differs from the first one, because it takes into account the sources of pollutants [1]. The assessment of aquifer pollution hazard should combine the vulnerability of the element at risk, *i.e.*, the aquifer, with the hypothetical contaminating element, *i.e.*, illegal dumpsites as in this study.

Methods of hierarchical classification of aquifer vulnerability can be used for a rapid assessment and mitigation of aquifer pollution hazard. These methods allow us to identify which sites immediately require remediation activities. In fact, integrated vulnerability ranking aims at developing policies and guidelines for the sustainable management of water and groundwater resources [1].

The intrinsic or natural vulnerability of aquifers is the specific susceptibility to pollution of the aquifer systems, in their different components and in different geometric and hydrodynamic situations. Vulnerability is the capacity to interact, disseminate and also to mitigate the effects of a pollutant that creates an impact on the quality of groundwater, in space and time.

In the last decades, groundwater vulnerability assessment has been the subject of many studies. Developed methods can be divided into three categories: statistical methods, process-oriented simulation models and index methods [1,2]. Process-oriented numerical models are used for site-specific studies and not for mapping vulnerability on a large scale. The main disadvantage of numerical models is not related to the mathematical complexity, but to the fact that most of them were made for soils of medium to high porosity. Numerical models, suitable for small portions of land, are numerous, such as MACRO 4.3 (dual-permeability model; [3,4]), RZWQM98 (dual-permeability model, [5]), Hydrus-1D (dual-porosity and dual permeability model [6,7]), and LEACHM [8]. Statistical methods used for aquifer vulnerability assessment comprise: descriptive statistics, multivariate regression analysis, and complex statistical analysis [2].

A number of index methods have been developed for the evaluation and mapping of the intrinsic aquifer vulnerability on a large scale. The most widely applied method to assess environmental vulnerability of groundwater is the DRASTIC method, adopted by the United States Environmental Protection Agency. The DRASTIC method is a parametric method, based on the allocation of scores and weights to different factors that influence the vulnerability of an aquifer [9,10].

In addition to the previously described methods, one of the methods of vulnerability assessment mostly used throughout the world is the geological approach. This method assesses the vulnerability by using a set of protocols for situations of varying complexity. This assessment is not influenced by the number and quality of data available but by the experience of the users. In the case of very large areas to be analysed, Civita M. [11,12] proposes the GNDCI-CNR method (National Group for the Defence from Hydrogeological Disasters-National Research Council) to assess the aquifer vulnerability. This method is suitable for large areas with different geological, hydrogeological and morphological conditions, in which data and information do not have a homogeneous distribution in the area.

In the last decade, a number of researchers started to develop and apply fuzzy and neuro-fuzzy models aiming at assessing groundwater intrinsic and integrated vulnerability [1,13–18]. In particular, a number of recent studies applied fuzzy methods for environmental risk analysis and management, whenever the availability and reliability of data are not enough for quantitative methods [19]. However, there is a lack, in literature, of studies proving the real advantage of fuzzy methods if compared with traditional methods. Moreover, there is a lack of studies concerning the application of fuzzy logic for the assessment of the specific vulnerability of groundwater connected to the presence of different pollution sources, and for hierarchically classifying these pollution sources within a hazard scale and thus understanding which one requires attention more urgently.

The objective of this study is to evaluate the pollution risk of the illegal dumpsites that are located in the Basilicata region (Southern Italy). In this study, the hydrogeological intrinsic vulnerability was evaluated with a method of zoning for homogeneous areas (GNDC-CNR method). While, fuzzy logic analysis was used to overcome the major drawbacks of conventional overlay analyses (*i.e.*, the subjective weightings and discrete ratings) in developing the integrated vulnerability index [14]. Fuzzy logic, associated with a sensitivity analysis, was applied to: identify the difference between fuzzy and conventional methods for the assessment of integrated groundwater vulnerability, determine a ranking of the pollution risk of each illegal dumpsite, and carry out a validation of the developed fuzzy method.

2. Materials and Methods

2.1. Intrinsic and Integrated Groundwater Vulnerability

In general terms, risk can be defined as a combination of hazard and vulnerability:

$$\text{Risk} = \text{Vulnerability} \times \text{Hazard}$$

where "vulnerability" indicates the degree of intrinsic weakness of the investigated natural system (intrinsic vulnerability) and "hazard" represents the probability that a detrimental event will occur in a given area in a period of time [18].

In this study, the considered sources of pollution are illegal dumpsites (Figure 1).

Since the study in question concerns the entire territory of the Basilicata region (Southern Italy), which has an area of about 10,000 km^2, the hydrogeological intrinsic vulnerability was evaluated with a method of zoning for homogeneous areas, the GNDC-CNR method (see Section 2.2.1).

A statistical approach can be used to evaluate the degree of hazard, in order to analyse the probability that a dangerous event may occur. In our case, the conditional probability is the probability that the concentration of a given pollutant is greater than a given threshold [18]. The major limits of this approach are the difficulty in finding detailed data, collected by monitoring a process, to evaluate the probability that the phenomenon considered might happen. Among the different models applied to the evaluation of groundwater pollution risk in a quantitative way, complex mathematical models can be considered the most reliable. However, a lot of detailed information and/or data often not available, inaccurate or doubtful are required, to run these models. To overcome the shortage of reliable information, an algorithmic approach has been integrated with a qualitative one. The hazard assessment was carried out using two methods (a traditional score and weight method and a fuzzy logic model). The same methods were used to integrate the intrinsic vulnerability of aquifers and the hazard of landfills, in order to calculate the risk of pollution of the illegal dumpsites. The parameters describing the possible sources of pollution are the same for both methods: water table depth, site acclivity, leachate production, waste volume and type, soil coverage, presence of monitoring system and river proximity index.

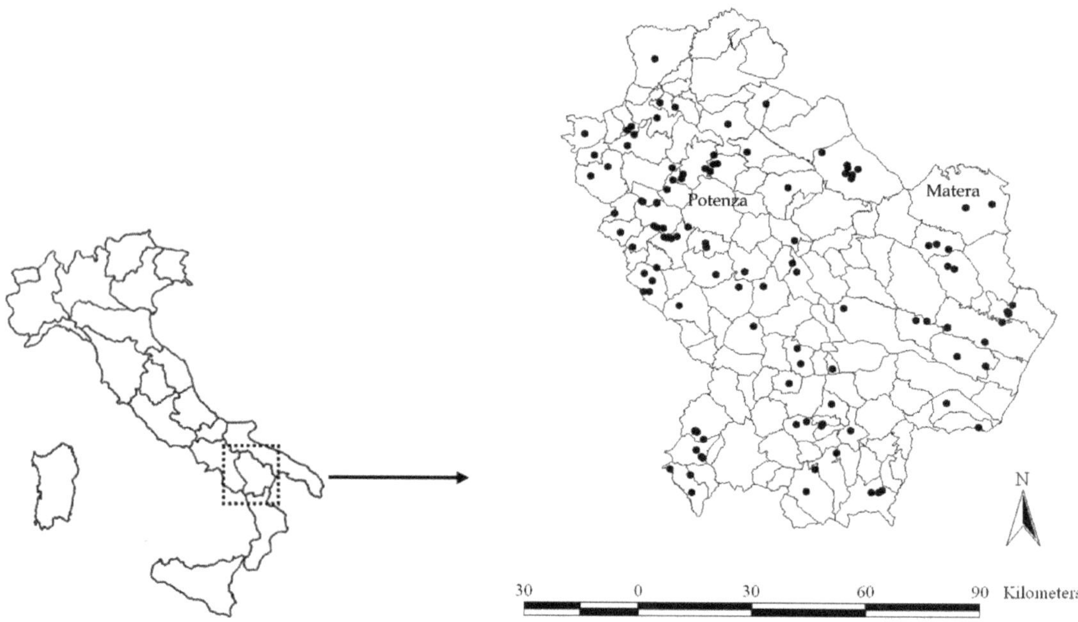

Figure 1. Location of the illegal dumpsites in the region of Basilicata (Italy).

Aquifer Intrinsic Vulnerability: The GNDCI-CNR Method

The GNDCI-CNR method [11,12] allows us a qualitative evaluation of the aquifer pollution vulnerability. It is a method of zoning for homogeneous areas and it requires no numerical parameters. The method uses a protocol in which each Italian hydrogeological situation has a degree of vulnerability.

The hydrogeological situations have been divided according to geometry, lithology and effective porosity of the aquifers.

Hence, the intrinsic vulnerability of aquifers can be defined by identifying the hydrogeological situation of interest among those listed. The method is very flexible and can be adapted to specific situations which are not listed in the protocol of methodology. This protocol must be used for assessing the vulnerability, which is characterized by all the typical situations. These situations are identified by the main factors that influence the aquifer vulnerability, such as depth, porosity, fractures, karst phenomena and piezometric level. The hydrological situations are placed in a legend composed of six columns, each of which is related to the degree of intrinsic vulnerability (Figure 2). This immediately makes the legend interpretative and usable for creation, interpretation and reading of the vulnerability map.

The vulnerability assessment is qualitative and based on the comparison between the situation of hydrogeological interest and the reference situation of hydrological conditions. The vulnerability map was obtained by taking into account different hydrological scenarios and by assigning the following degrees of intrinsic vulnerability: very high, high, medium, low, very low.

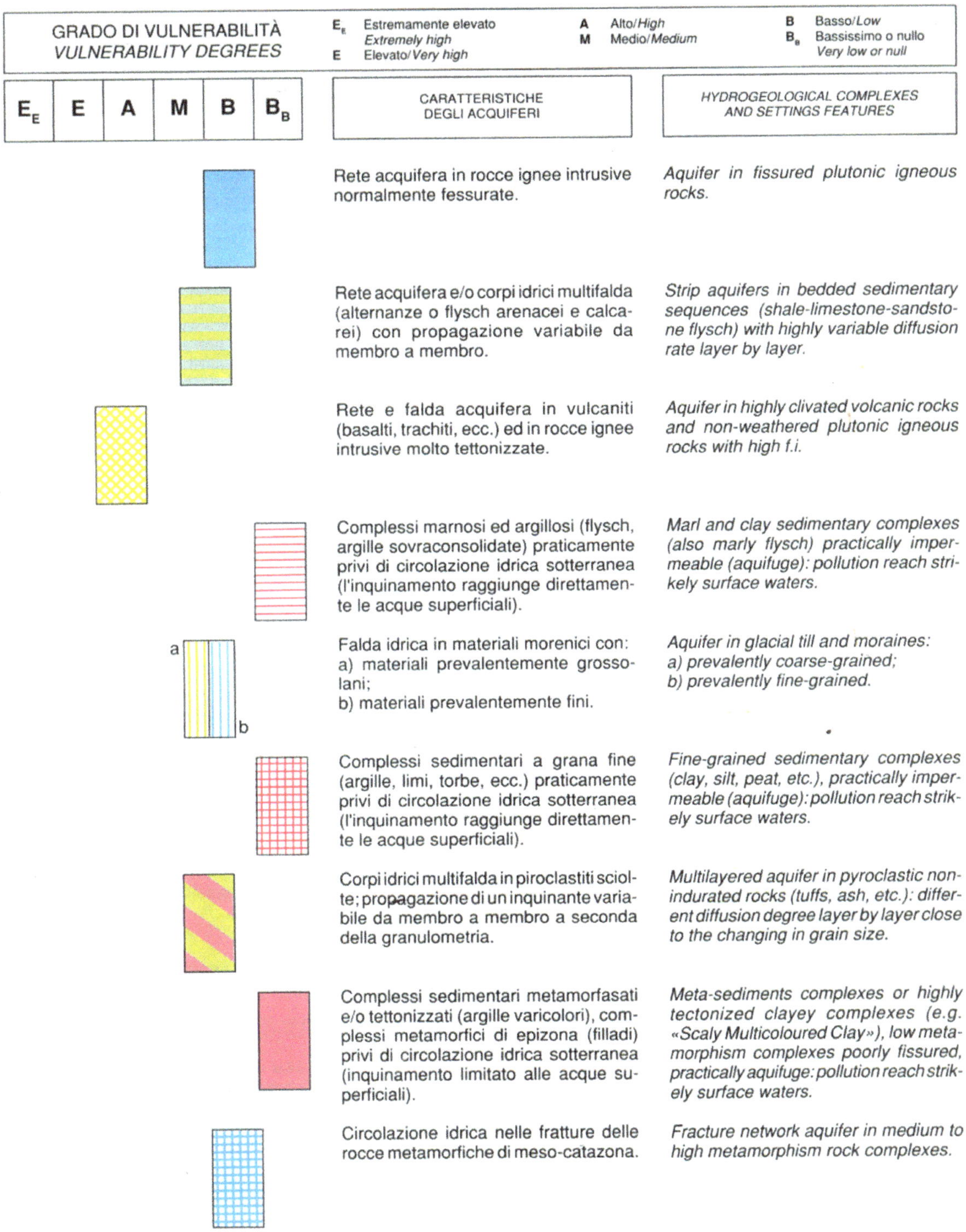

Figure 2. *Cont.*

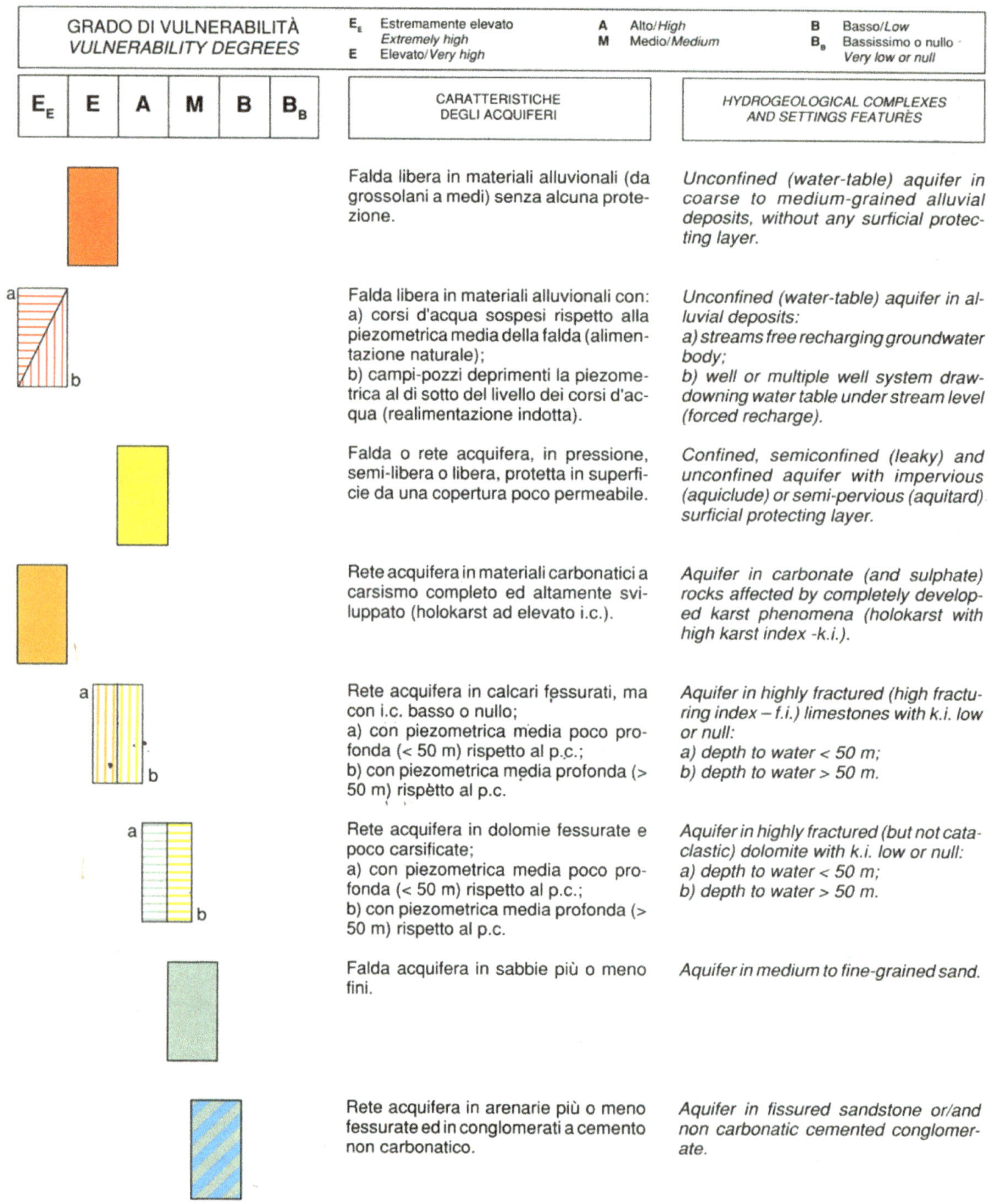

Figure 2. Protocol for assessing the intrinsic vulnerability of aquifers with the GNDCI-CNR method [11].

The application of the GNDCI-CNR method is based on map overlay of different levels of information and it provides some operational phases:

- identification and definition of hydrological and hydrodynamic characteristics of the area of interest;
- creation of the hydrogeological map and identification of scenario or hydrological and impact scenarios in the area of interest;

- identification, on the basis of the GNDCI-CNR protocol, of the reference scenario or scenarios adaptable to the hydrogeological situation of interest by the consequent assignment of the relative degrees of vulnerability.
- drafting of the aquifer intrinsic vulnerability map accompanied by an appropriate legend.

2.2. Case Study

The illegal dumpsites considered in this study are those found in the Basilicata region (Southern Italy) with the last census ("Corpo Forestale dello Stato" (Forest Rangers) and "Piano Regionale di Bonifica" (Regional Reclamation Plan, 2002), which identified 469 areas needing reclamation actions, environmental recovery and/or safety measures (315 in the province of Potenza and 204 in the province of Matera). Among these areas, 290 are illegal dumpsites: 122 in the province of Potenza and 168 in the province of Matera. In this study, we analysed 118 illegal dumpsites among all those identified (Figure 1), for which data related on the dumped waste were available to us. The assessment of risk factors of illegal dumpsites is a very complex issue. In fact, the information regarding the construction techniques of the plant, the managerial formalities and the types of waste are lacking. The characterization phase of the risk elements is therefore fundamental, since elements that affect the decision-making phase, and that can be easily reproducible and standardisable, must be identified. The elaboration of the method is set, in fact, as an objective to supply a decision making support for the prioritization during the phase of illegal dumpsite remediation.

2.2.1. Risk Factors

We measured the following parameters, for the definition of the input variables: waste volume and type, physical state of dumpsite and presence of monitoring system. Other important parameters, such as leachate production, site acclivity, proximity to river index and water table depth were calculated on the basis of other known information, by using GIS tools and historical hydrological data.

For the definition of the relative maps, the cell size was chosen to be 240 × 240 m, on the basis of the size of the study area.

2.2.2. Water Table Depth

The water table depth can be defined as the distance between the land surface and the groundwater table. This is a very important parameter, as it affects the travel time of the pollutant and the self-purification effectiveness of the unsaturated zone. The relative map was prepared interpolating punctual water table depth data, which was obtained from 506 wells located in the study area (about 10,000 km^2). The method of interpolation used is the Kriging method, with a spherical model of the experimental semivariogram.

2.2.3. Slope Gradient

Slope gradient plays an important role in the infiltration process and, consequently, in leachate production. Slope gradient Figure 3a was calculated by using the digital elevation model in a GIS environment.

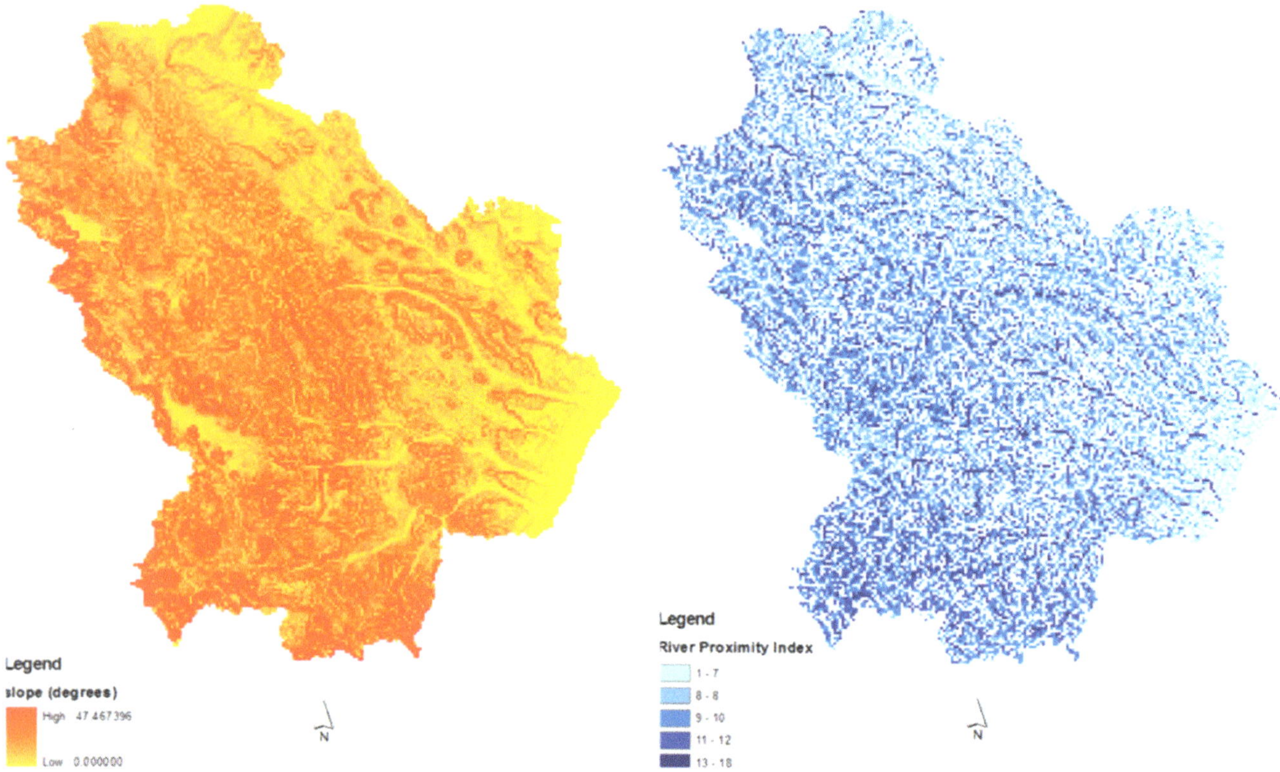

(**a**) (**b**)

Figure 3. Maps of slope gradient (**a**) and river proximity index (**b**).

2.2.4. River Proximity

The river proximity index (IC) was calculated by modifying the topographic index:

$$IC = FA \times \tan(\beta)$$

where FA is the flow accumulation (calculated as the number of all cells flowing into each downslope cell) and β is the site acclivity.

The river proximity index is shown in Figure 3 (on the right).

2.2.5. Leachate Production

Leachate production roughly calculated by considering the average values of rainfall in the involved area, since it directly depends on effective infiltration, estimated as proposed by Civita and De Maio [20]. Evapotranspiration was calculated with the Turc formula

The resulting map (Figure 4) was obtained by interpolating, with the Kriging method, the punctual average data collected between 1920 and 2009 in the 138 thermopluviometric stations of the Basilicata region.

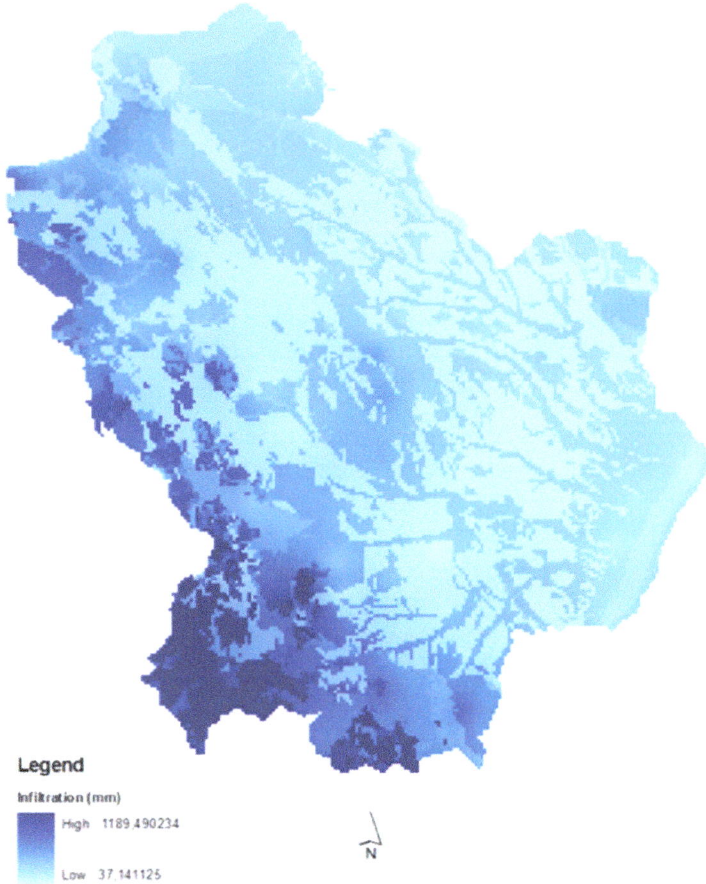

Figure 4. Map of leachate production.

2.3. Assessment of the Pollution Risk of the Illegal Dumpsites of the Basilicata Region: The "Scores and Weights" Method

In the first phase, the pollution risk of illegal dumpsites was assessed with a score and weight method [21]. This method can be used as a decision support tool for the identification and ranking of the sites that need to be reclaimed.

Every single factor was multiplied by a weight W that takes values on a scale of 1–5, depending on the potential risk. In addition, risk factor reduction in weight "R" was applied to each element. In this case, the factor is included in a scale of 0.1–1. The sum of products between the relative weights and reduction factors allows us to calculate the hazard index "IH" through Equation (1):

$$I_H = \sum W_i \times R_i \tag{1}$$

The hazard of each landfill has been evaluated on the basis of this hazard index, which is independent on the location of the site. The risk factors with the respective weights and reduction factors are shown in Table 1.

Table 1. Scores and coefficients of reduction (weights) of the considered risk factors.

Hazardous Factors of Landfills		Weights and Coefficient of Reduction of Risk Factors (*R*)
Waste input rate (waste volume, m^3)	<1,000	0.1
	1,000–5,000	0.3
	5,000–10,000	0.5
	10,000–20,000	0.7
	>20,000	1
Estimate of leachate production (infiltration value, mm)	0–100	0.1
	100–200	0.3
	200–300	0.5
	300–500	0.7
	>500	1
Waste types	MSW, MSW + IW	0.1
	non-MSW, non-MSW + MSW	0.5
	HW, MSW + HW, non-MSW + HW	1
Proximity to superficial water (Proximity river index value, adim)	0–5	0.1
	5–10	0.5
	>10	1
Water table depth (m)	>5	0.1
	2–5	0.5
	0–2	1
Site acclivity (degrees)	>20	0.1
	10–20	0.3
	5–10	0.7
	0–5	1
Monitoring system	Present	0.1
	Not present	1
Physical state of dumpsite	Without soil cover	0.3
	With soil cover	1

The hazard index I_H calculated with the score and weight method was normalized and then subdivided into five hazard classes (very high, high, medium, low, very low).

The obtained hazard of illegal open dumpsite was integrated with the intrinsic vulnerability of the aquifers, which was determined by using the GNDCI-CNR method described in Section 2.1.1, in order to give the pollution risk of illegal dumpsites.

2.4. Development of the Fuzzy Logic Model for Assessing Aquifer Integrated Vulnerability

The binary (or classical) logic is only concerned with what is completely true or what is completely false. Fuzzy logic, however, also extends its interest to that which is not completely true, which is probable or uncertain. Fuzzy logic is based on a linguistic approach, in which words or phrases of natural language are used instead of numbers.

Fuzzy logic operates on mathematical entities that are fuzzy sets, which obey rules, structures and axioms similar to those of classical logic. The only difference is that an object can simultaneously belong to several subsets in contrast to the classical theory. Belonging to a subset, in fuzzy logic, is associated with a degree of membership [22]. The set of deduction rules to be applied to a given system to achieve results through the use of fuzzy logic is the fuzzy inference process. Fuzzy inference is divided into three main phases: fuzzification, inference and defuzzification.

The fuzzification is the process by which the input variables are converted to fuzzy measures of their belonging to certain classes (such as, for example, Very Low, Low, Medium, High, Very High). This conversion from deterministic greatness to fuzzy greatness is done through membership functions (Figure 5). A membership function is a function that associates a value, usually numeric, to the degree of membership. By convention, the degree of membership assumes values ranging from zero, when the element does not belong to the subset, to one, when it belongs entirely to the subset. If you consider a membership function μA and a value $x°$, the conversion of $x°$ in a fuzzy set is a complement of set A corresponding to ordinate μA($x°$) (Equation (2)):

$$A' = \{x \mid \mu_{A'}(x) = \min(\mu_A(x°), \mu_A(x))\} \tag{2}$$

If the value $x°$ intercepts more than one membership function, you will have a degree of membership for each function equal to the intercept with the y-axis.

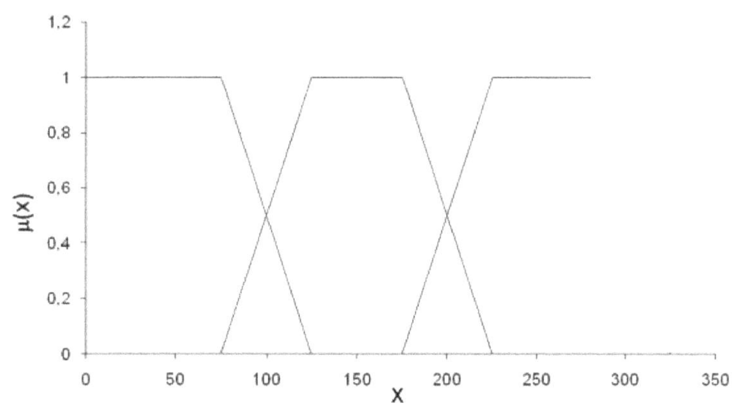

Figure 5. An example of membership function.

A membership function is a function which associates a value (usually numerical) with the level of membership to the set. By convention, the real number which represents the level of membership (μ(x)) takes a 0 value when the element does not belong to the set, and 1 when it belongs to it completely. Membership functions can have different shapes: the simplest are made up of straight lines, while the most used are the triangular (Figure 6a) and trapezoidal (Figure 6b) functions; the former are characterized by a triangular trend while the latter by a trapezoidal one. The advantage of these functions is in their simplicity. The triangular membership function depends on three scalar parameters a, b and c and is given by the following expression:

$$f(x, a, b, c) = max\left(min\left(\frac{x-a}{b-a}, \frac{c-x}{c-b}\right), 0\right) \tag{3}$$

while the trapezoidal one depends on four scalar parameters (a, b, c and d), as shown in the following formula:

$$f(x,a,b,c,d) = max\left(min\left(\frac{x-a}{b-a},1,\frac{d-x}{d-c}\right),0\right) \tag{4}$$

There are other more complex functions, *i.e.*, the Gauss function made up of a simple Gaussian curve (Figure 7a) which depends on the parameters σ and c (Equation (5)) and the Gauss2 function (Figure 7b) given by the fusion of two different Gaussian functions and depending on four parameters: $r1$ and $c1$, which define the shape of the function in the left part, and $r2$ e $c2$, which define the shape of the function in the right part. Moreover, between these types of functions, there is the bell membership function (Gbell) (Figure 7c) which is a hybrid of the Gaussian function; it is mainly used to manage non-fuzzy sets and depends on three parameters: a, b and c (Equation (6))

$$f(x;\sigma,c) = e^{\frac{-(x-c)^2}{2\sigma^2}} \tag{5}$$

$$f(x;a,b,c) = \frac{1}{1+\left|\frac{x-c}{a}\right|^{2b}} \tag{6}$$

Despite their simplicity, such functions cannot be used to represent asymmetry, which is important in some applications.

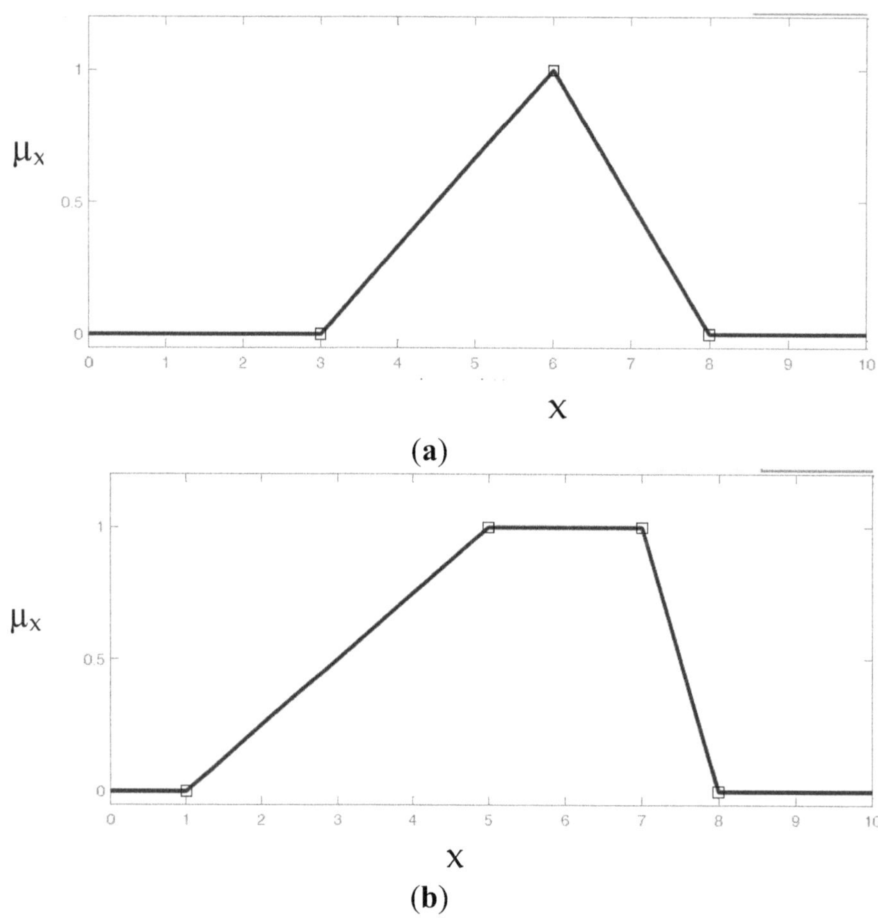

Figure 6. (**a**) Triangular membership function; (**b**) Trapezoidal membership function.

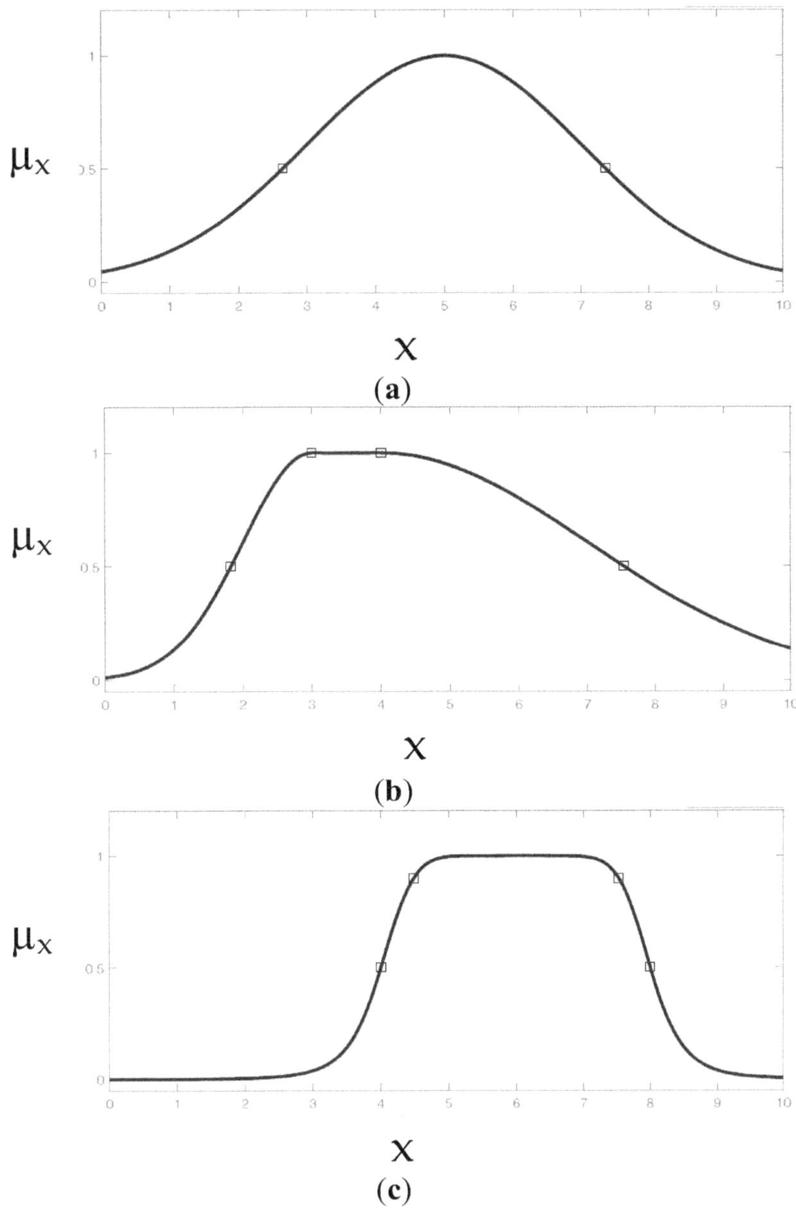

Figure 7. (**a**) Gauss membership function; (**b**) Gauss2 membership function; (**c**) Generalized bell (Gbell) membership function.

The sigmoid function (Figure 8a) may have left or right asymmetry and a horizontal asymptote. This function is ruled by parameters a and c (Equation (7)):

$$f(x;a,c)=\frac{1}{1+e^{-a(x-c)}} \tag{7}$$

Further asymmetric functions, Dsigm and Psigm membership functions, are represented in Figure 8b,c and described by Equation (8), depending on four parameters $a1$, $c1$, $a2$ and $c2$.

$$f(x;a,c)=\frac{1}{1+e^{-a(x-c)}} \tag{8}$$

The factors that discriminate the choice of the appropriate membership functions are of two types: the number of parameters of each curve and the shape of the function which represents the trend of the phenomenon.

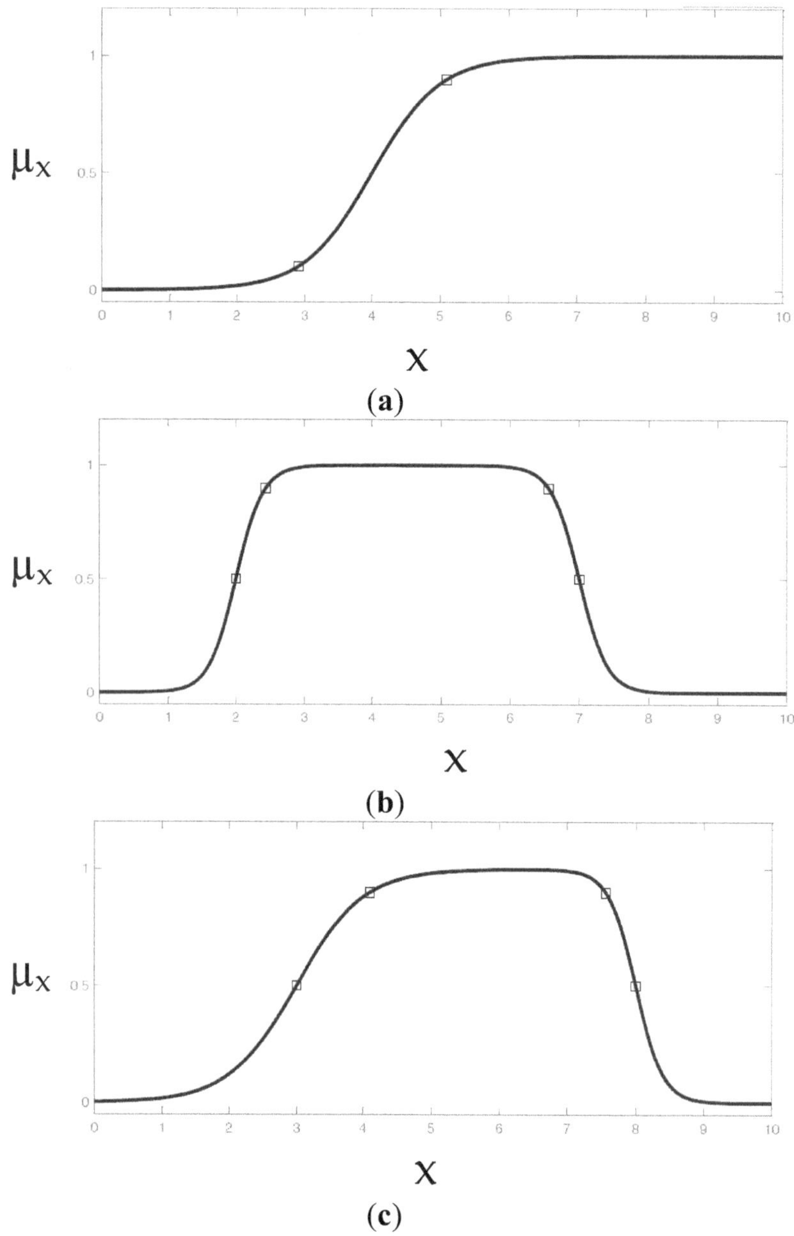

Figure 8. (**a**) Sigmoid (Sig) membership function; (**b**) Dsig membership function; (**c**) Psig membership function.

Fuzzy inference is the mapping process, starting from an input space and going to an output space. In order to obtain an output, you need to assign the fuzzy rules that integrate fuzzy sets through a series of antecedents and consequents. These rules are not represented by complex mathematical models but are constituted by simple linguistic expressions, which are converted in mathematical formalism in the language of fuzzy rules. The fuzzy rules, "if-then-else", are usually composed of two parts: the antecedent, which defines the condition, and the consequence, which defines the action. It is also possible to define a rule "Else", which enters in to question when none of the rules has been satisfied or when the "Then" is weakly satisfied. The output of the fuzzy inference is often unusable and, therefore, it needs to be converted into a deterministic value through the defuzzification operation.

The usually used defuzzification methods are the following:

- Centroid method: the chosen numerical value for the output is calculated as the centre of the mass of the fuzzy set;
- Bisector method: the output is the abscissa of the bisector of the area subtended to the fuzzy data set;
- Middle of maximum method: the output value is determined as the average of maximum values (Mom, middle of maximum) (Figure 4);
- Largest of maximum method: the output numerical value is calculated as the maximum of the maximum (Lom: Largest of maximum);
- Smallest of maximum method: the output value is represented by the output minimum value (Som: Smallest of maximum).

Among the methods found in literature, the most common are the centroid and Mom methods. In the Mom method, output is obtained as the arithmetic mean of the values of "*y*" where fuzzy set height is maximum (Equation (9)).

B' is the fuzzy set inferred by rules and

$$hgt(B') = \{y|\mu_{B'}(y) = Sup_{yeB'}\mu_{B'}(y)\} \tag{9}$$

is the values set of "*y*" for which height "$\mu_B'(y)$" is maximum.

Therefore, we have (Equation (10)):

$$y_{out} = \frac{\int_{hgt(B')} ydy}{\int_{hgt(B')} dy} \tag{10}$$

whose geometric significance is shown in Figure 9:

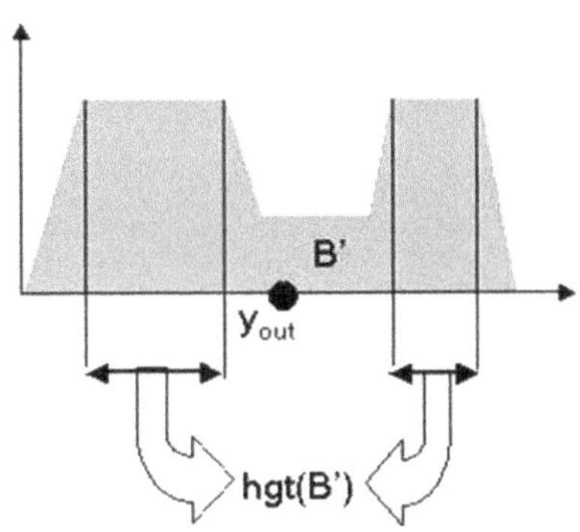

Figure 9. Geometric significance of the middle of maximum defuzzification method.

The output of the *Centroid method* is obtained as the abscissa of the centre of gravity inferred from the rules in the space of fuzzy sets of output of the algorithm. The formula in the case of continuous function is given by the following Equation (11):

$$y_{out} = \frac{\int y\mu(y)dy}{\int \mu(y)dy} \tag{11}$$

Whereas, the formula for discrete function is given by Equation (12):

$$y_{out} = \frac{\sum_i y_i\, \mu(y_i)}{\sum_i \mu(y_i)} \tag{12}$$

In *Largest of maximum*, the precise value of the output variable is one of which the fuzzy subset has the maximum value of truth. The main disadvantage of this method is that it does not consider the distribution of membership function.

The conceptual structure of the developed fuzzy model is represented in Figure 10. The intrinsic vulnerability of the aquifers and the hazard of the illegal dumpsites were integrated, in order to obtain the specific, or integrated, vulnerability, or pollution risk. The hazard of illegal dumpsites, however, was determined through the use of fuzzy logic, by considering the parameters describing the characteristics of the dumpsites.

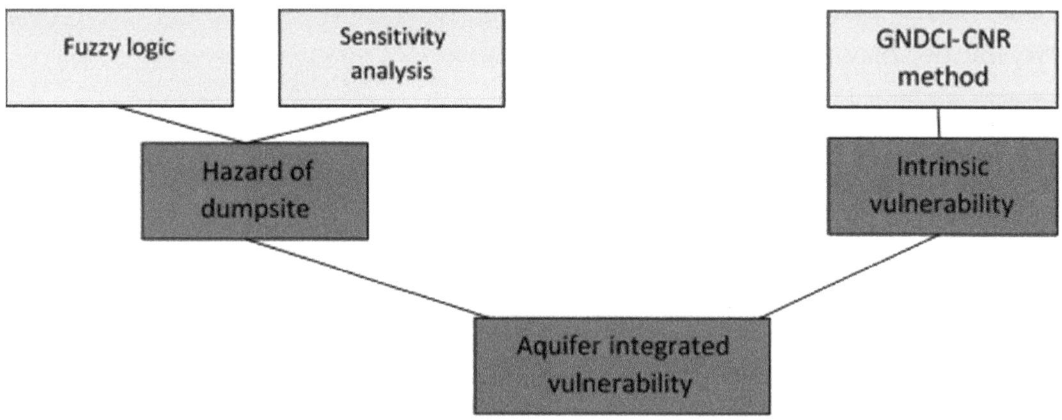

Figure 10. Conceptual model for the fuzzy aquifer specific vulnerability to pollution.

In order to reduce the number of fuzzy rules, the parameters listed in Section 2.2.1 were used to define three different fuzzy inferences, as shown in the conceptual diagram in Figure 11. The results obtained through the first two fuzzy inferences, defined as site vulnerability and dumpsite potentiality, respectively, were then aggregated to the crisp parameter called dumpsite conditions, by obtaining the hazard index of each dumpsite.

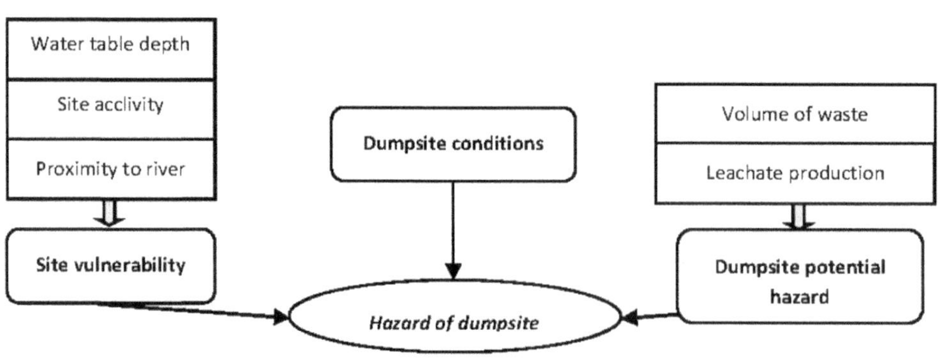

Figure 11. Conceptual diagram of the implemented fuzzy model for dumpsite risk assessment.

In the assignment of the rules, we assumed that the increase in vulnerability is favoured by low slopes, proximity to surface watercourses and low depth of the water table. On the contrary, potentiality of landfill evaluates the probability of a landfill to release contaminants depending on the waste volume and the leachate production. Therefore, with the increase of these two factors, this potentiality will increase as described by the fuzzy rules. The procedure to determine the integrated vulnerability index combines the results obtained by the two previous fuzzy diagrams with the addition of the landfill conditions. The landfill conditions parameter groups the non-fuzzy parameters, namely the type of waste, the presence of a coverage and the activity for each landfill. For the type of waste, a weight equal to 0.1 for municipal solid waste (MSW), 0.5 for special waste and 1 for hazardous ones were defined. A unitary weight was given to the absence of coverage, and a weight equal to 0.3 to the presence of coverage. A weight of 0.1 was given to the landfill inactivity, and a unitary one to its activity. Therefore, an increase of the landfill conditions parameter indicates an increase of the landfill vulnerability. For example, "if site vulnerability" is low, "if landfill potentiality" is high, and "if conditions of the landfill" is low then the hazard index of landfill is medium. In general, the basis of the fuzzy inference has been drawn up considering that with increasing values of the three parameters of the subset, increases the hazard of landfills.

The selection of the membership function and the defuzzification method was done with the application of the sensitivity analysis method described in [14]. The sensitivity analysis was performed by repeating the whole fuzzy inference procedure, modifying only one membership function or defuzzification method at a time, keeping all the other variables unvaried, thus performing 55 different simulations. We carried out an F test to compare more precisely the variances related to the distributions corresponding to each fuzzy model and to the starting data. What we observed from the analysis of deviation was confirmed by the F test which highlighted, as the best model, the Centroid defuzzification method associated with the trapezoidal or GBell membership functions gave comparable results (Table 2).

Table 2. F test results.

Defuzzification Method	Gauss2	Triangular	Trapezoidal	Gbell	Gauss	Disig	Psig	Pi
Centroid	0.692	0.594	0.969	0.966	0.280	0.001	0.001	0.479
Bisector	0.687	0.945	0.945	0.030	0.940	0.118	0.114	0.815
Mom	0.009	0.009	0.009	0.009	0.012	0.008	0.011	0.013
Lom	0.009	0.009	0.009	0.018	0.012	0.010	0.014	0.023
Som	0.026	0.025	0.025	0.012	0.021	0.087	0.094	0.093

The obtained fuzzy hazard index was successively revised for a classification in linguistic terms (Very low, Low, Medium, High, Elevated), by using a cumulative frequency distribution curve, and choosing, as the ends of the classes, the points of the curve where a clear slope variation is observed.

The intrinsic vulnerability and the landfill hazard, placed in a fuzzy model, provided the integrated vulnerability index. The fuzzy inference has been created using as input data the landfill hazard index and the classes of intrinsic vulnerability, which were converted into numerical values. Therefore, during the fuzzification, the trapezoidal-type membership function was defined for the intrinsic vulnerability of aquifers and the hazard of landfills.

In order, to determine the environmental risk index, the if-then fuzzy rules (Table 3) and the Centroid defuzzification method were defined. These rules have been established by considering that the aquifer pollution risk grows with an increase in both the intrinsic vulnerability of aquifers and the hazard of dumpsites.

Table 3. Fuzzy rules for the evaluation of aquifer integrated vulnerability.

Landfills	Aquifer Intrinsic Vulnerability				
Hazard	**Very Low**	**Low**	**Medium**	**High**	**Very High**
Very Low	Very Low	Low	Low	Medium	Medium
Low	Low	Low	Medium	Medium	High
Medium	Low	Medium	Medium	High	High
High	Medium	Medium	High	High	Very high
Very high	Medium	High	High	Very high	Very high

3. Results and Discussion

3.1. Aquifer Intrinsic Vulnerability

The map of the aquifer intrinsic vulnerability, determined with the GNDCI-CNR method, is shown in Figure 12. The map indicates that approximately all of the centre part of the Basilicata Region falls under the very low, low and medium vulnerability classes, except for the areas located at the edges of the region. The aquifer vulnerability was reclassified into six classes: extremely elevated, elevated, medium, high, medium, low, very low.

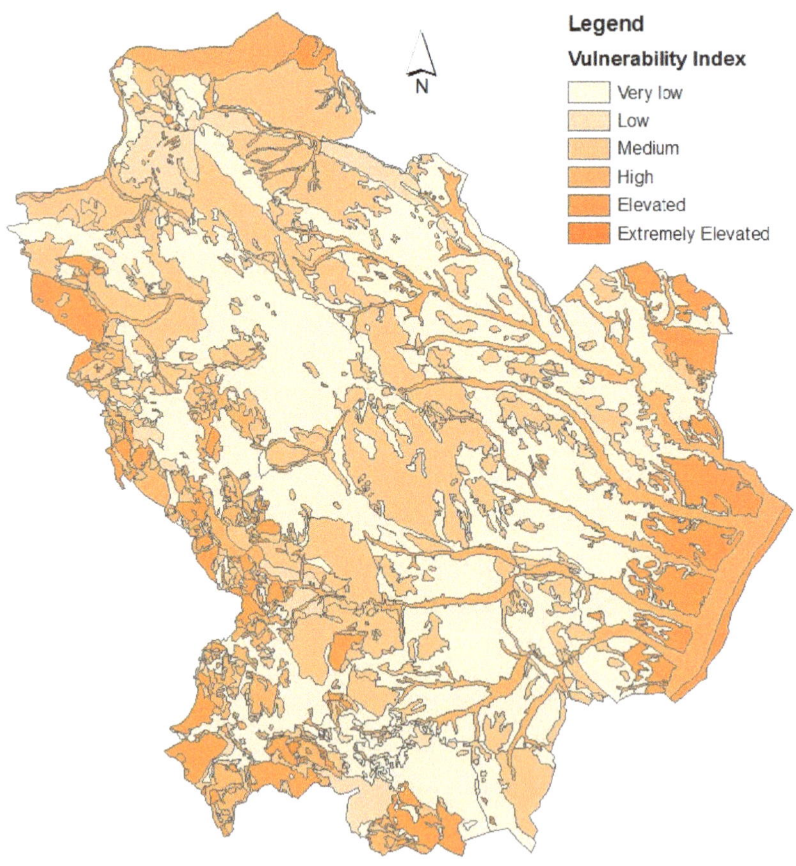

Figure 12. Aquifer intrinsic vulnerability map.

3.2. Results of the Pollution Risk of Dumpsites Obtained with the "Scores and Weights" and the Fuzzy Logic Method

The map related to the pollution risk of dumpsites, obtained with the "scores and weights" method is shown in Figure 13. The percentage distribution of the classes of risk is shown in Figure 14. Looking at the percentage results, we observed that most of the dumpsites show a medium-high vulnerability (about 70%).

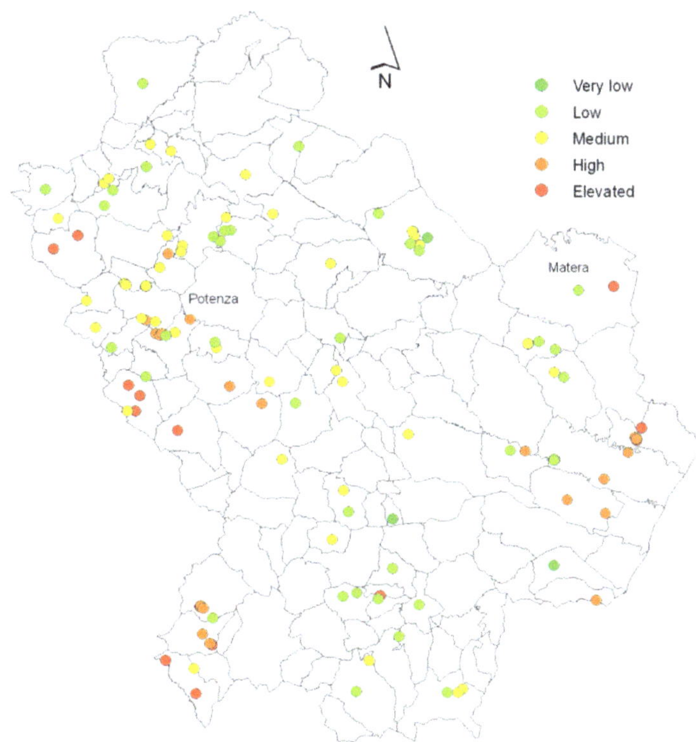

Figure 13. Map of the integrated vulnerability of illegal dumpsites ("scores and weights" method).

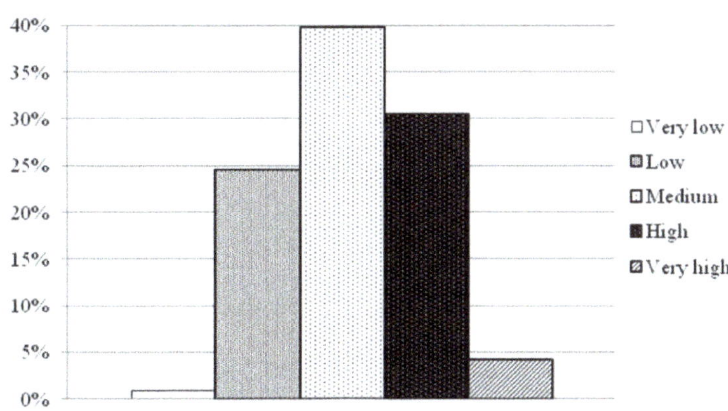

Figure 14. Distribution of the classes of integrated vulnerability of illegal dumpsite ("scores and weights" method).

The pollution risk obtained with the fuzzy logic method was reclassified in linguistic terms (Very low, Low, Medium, High, Very high), by using the cumulative frequency distribution curve (Figure 15) and by choosing, as the extremes of classes, the points of the curve where there is a distinct change of slope (Table 4).

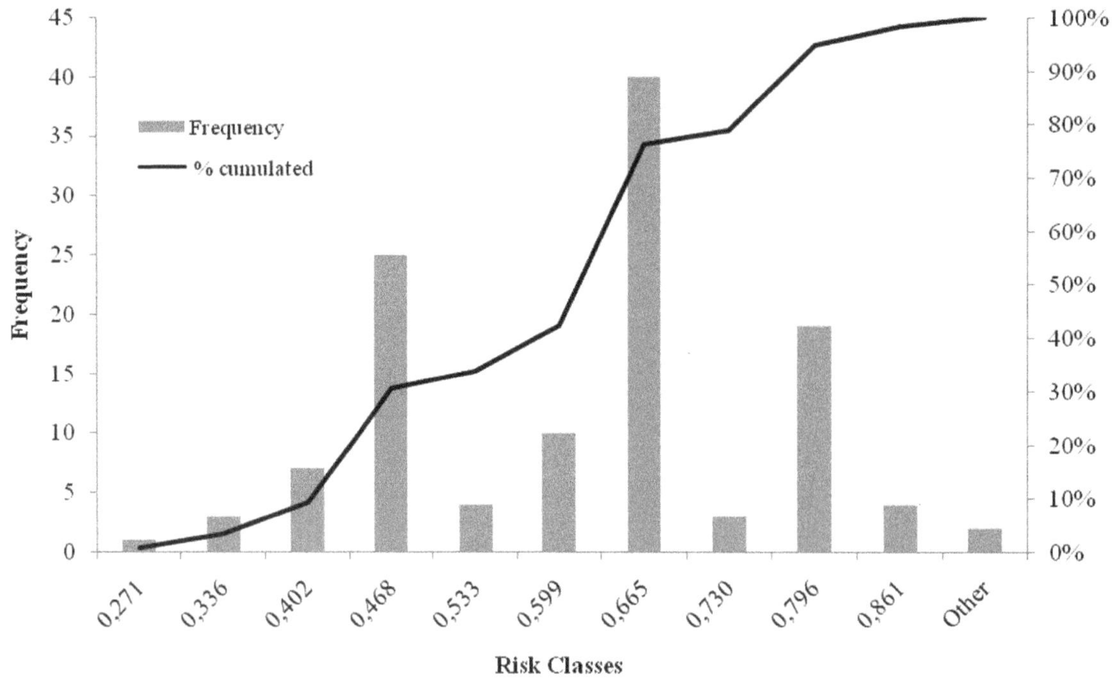

Figure 15. Cumulative frequency curve of risk classes (fuzzy logic method).

Table 4. Range classes of dumpsite pollution risk (fuzzy logic method).

Dumpsite Pollution Vulnerability Classes	Classes Range
Very low	0.402
Low	0.468
Medium	0.599
High	0.665
Very high	0.796

The classification identifies a large part of the dumpsites (about 60%) in the range of high and very high pollution risk, as shown in the graph of Figure 16. Figure 17 shows the spatial distribution of the risk.

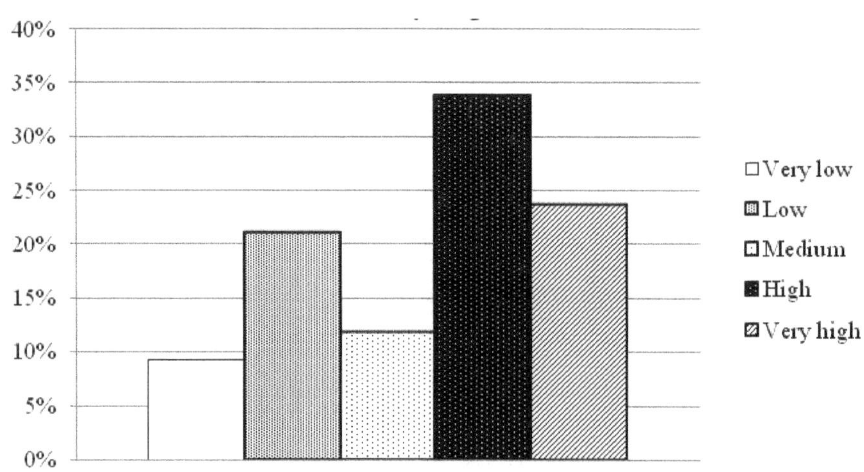

Figure 16. Distribution of the classes of integrated vulnerability of illegal dumpsite (fuzzy logic method).

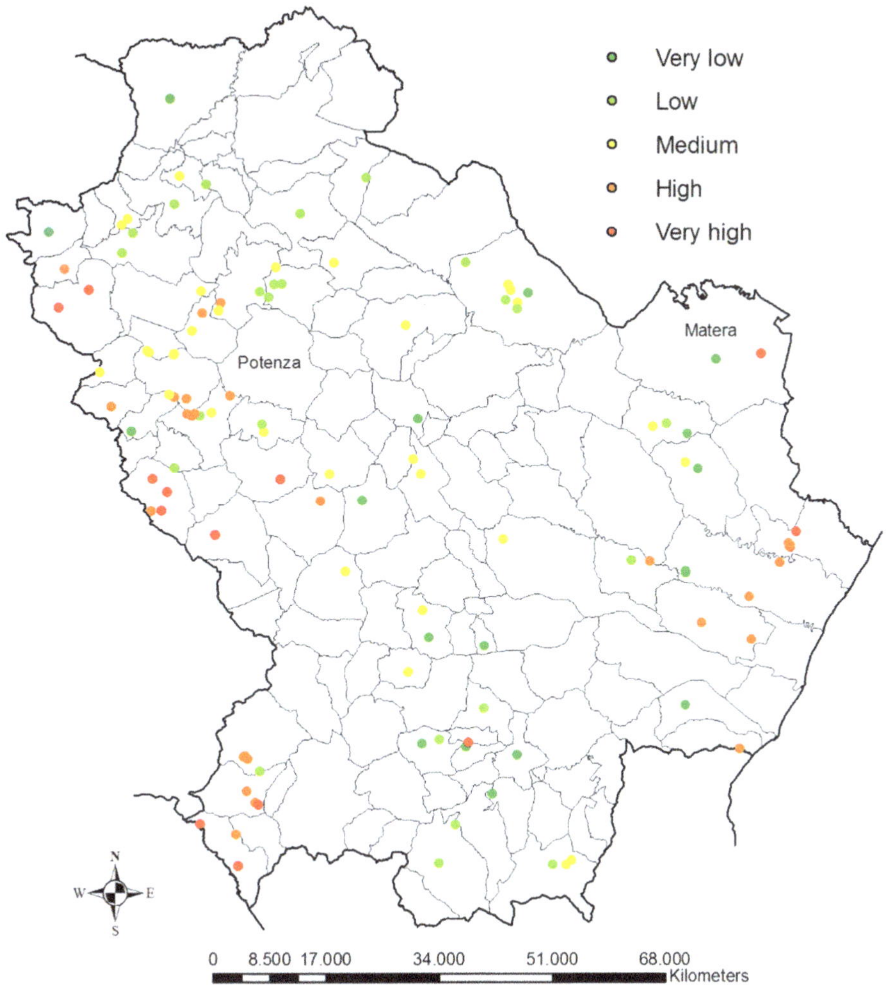

Figure 17. Map of dumpsite integrated vulnerability in the Basilicata region (fuzzy logic).

The analysis of the histograms (Figures 14 and 15) representing the percentage of dumpsites falling in different classes of risk by applying both "scores and weights" and fuzzy logic methods, shows that the both distributions are platykurtic because they present negative kurtosis (−1.16 for fuzzy logic results and −2.40 for "scores and weights" results). The skewness, which is a measure of the asymmetry of a distribution, is −0.45 for fuzzy logic results and 0.15 for "scores and weights" results. The comparison of the results obtained with both tested models indicates that 50% of the vulnerability of dumpsites is over-estimated by applying the fuzzy model while over 30% has a comparable vulnerability (Figure 18), because the fuzzy logic distribution moves to higher risk classes. This result should not be of concern because this means that the fuzzy logic method gives safer results and that the underestimated aquifers fall into the classes of very low and low risk.

For a more appropriate comparison, we calculated the statistical dispersion indices of the distributions of the risk indices obtained with the two methods. The comparison of the risk indices, without the division into classes, shows that the behaviour of the two distributions is very similar. The congruence of the two distributions is confirmed by the values of variance and standard deviation (Table 5).

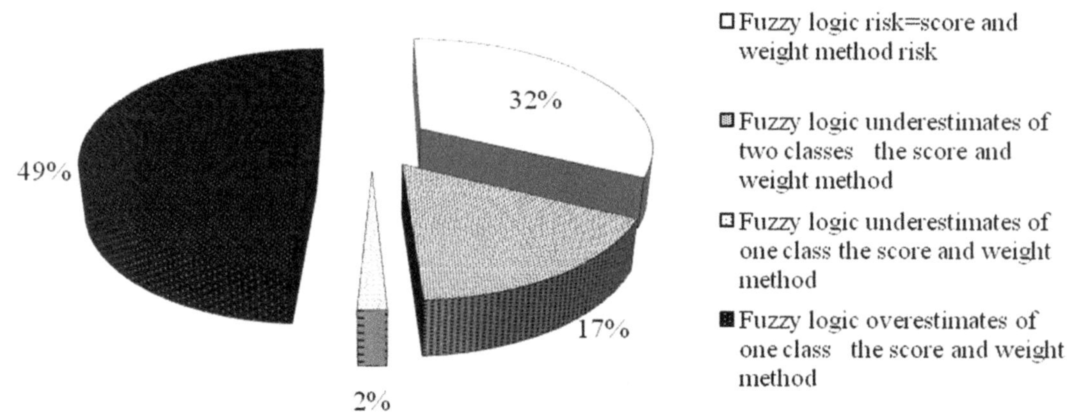

Figure 18. Comparison of the results obtained with both the tested models.

Table 5. Statistical dispersion indices for the comparison between the fuzzy logic method and the scores and weight method.

Statistical Dispersion Indices	Scores and Weight Method	Fuzzy Logic
Standard deviation	0.237	0.211
Variance	0.056	0.044
Mean	0.466	0.470

4. Conclusions

In this study, an index of environmental risk was introduced to evaluate both the intrinsic and integrated vulnerability of the aquifers of the Basilicata region (Southern Italy). The fuzzy method was developed and the results were used to make a comparison with a conventional "score and weight method". The aim of fuzzy logic was to reduce the subjectivity found in the score and weight method. The fuzzy methodology and the score and weight method allow us to identify contaminated sites that need prompt interventions. The evaluation is carried out on the basis of the integration of the characteristics of the dumpsite (hazard of dumpsite) and the intrinsic vulnerability of the studied aquifers.

It is possible to reduce the risk that a pollutant may leach along the soil profile and reach groundwater in areas with an intrinsic vulnerability by ranking different illegal dumpsites to their degree of hazard. For dumpsites characterized by a high degree of hazard, it would be necessary to propose actions aimed at mitigating their environmental impact. When investigating the risk of groundwater pollution due to the presence of illegal dumpsites, the system should be able to indicate the dumpsites in which we should concentrate efforts to mitigate or reduce the environmental impact. The evaluation of the pollution risk index, in fact, allowed us a classification of risk for a subsequent detailed planning of investigations for each site. The comparison of the two methods showed that the two obtained distributions of the results are congruent. This is confirmed by the similar values obtained for variances and standard deviations. The spatial distributions show that the two methods give very similar results. Besides, an analysis of the histograms representing the percentage of aquifers falling in different classes of risk, obtained by applying fuzzy logic and score and weight methods, show that the fuzzy logic has a distribution that moves at higher risk classes compared to the score and weight method.

The fuzzy method, unlike the traditional score and weight method, does not require a deep knowledge of the phenomenon and can be used by a non-expert in a screening context to better understand aquifer vulnerability for decision-making purposes. Moreover, this fuzzy method is transferable to other regions and other geology. This is very useful when it is necessary to make a broad classification of very numerous contaminated sites in order to decide which of them require immediate characterization and/or remediation procedures and which of them do not have characteristics of serious contamination hazard. The most important limitation of the model is that it is not possible to calculate an absolute value of pollution risk, but it is only possible to classify dumpsites in a hazard scale.

Acknowledgments

This work forms part of a research project supported by grant of the Basilicata Region through the Research project entitled "Innovative techniques for monitoring and management of municipal solid waste landfills and remediation of contaminated sites-POLLUTECH" (Contract Cod. No. 2/AP/05/2013REG-CUP G43G13000410009).

Author Contributions

Donatella Caniani and Ignazio Marcello Mancini developed the research idea, Salvatore Masi and Donatella Caniani planned the research activities, Donata Serafina Lioi and Donatella Caniani carried out the research activity, Donatella Caniani analysed the data; Donata Serafina Lioi and Donatella Caniani prepared the manuscript, Donatella Caniani, Salvatore Masi and Ignazio Marcello Mancini reviewed the final draft and supervised the research activity.

Conflicts of Interest

The authors declare no conflict of interest.

References

1. Rezaei, F.; Safavi, H.R.; Ahmadi, A. Groundwater vulnerability assessment using fuzzy logic: A case study in the zayandehrood aquifers, Iran. *Environ. Manag.* **2013**, *51*, 267–277.
2. Shahmaleki, N.; Behbahani, S.M.R.; Massah Bavani, A.R.; Khodaei, K. Comparison between logistic regression, modified DRASTIC and AHP-DRASTIC methods in groundwater vulnerability assessment. *J. Environ. Stud.* **2013**, *38*, 79–92.
3. Jarvis, N. *The MACRO Model Technical Description*; Swedish University of Agricultural Sciences (SLU); Department of Soil Sciences: Uppsala, Switzerland, 2002.
4. Merdun, H.; Quisenberry, V.L. Simulation of water and solute transport with MACRO model in Cecil loamysand soil. *Aust. J. Soil Res.* **2004**, *42*, 939–951.
5. Ma, L.; Ahuja, L.R.; Ascough, J.C.; Shaffer, M.J.; Rojas, K.W.; Malone, R.W.; Cameira, M.R. *Integrating System Modeling with Field Research in Agriculture: Application of the Root Zone Water Quality Model*; United States Department of Agriculture, Agricultural Research Service (USDA-ARS): Washington, DC, USA, 2000.
6. Castiglione, P.; Mohanty, B.P.; Shouse, P.J.; Simunek, J.; van Genuchten, M.T.; Santini, A. Lateral water diffusion in an artificial macroporous system. *Vadose Zone J.* **2003**, *2*, 212–221.

7. Simunek, J.; Sejna, M.; van Genuchten, M.T. *The HYDRUS-1D Software Package for Simulating the One-Dimensional Movement of Water, Heat, and Multiple Solutes in Variably-Saturated Media, Version 3.0.*; Department of Environmental Sciences, University of California Riverside: Riverside, CA, USA, 2005.
8. Hutson, J.L. *Leachm Model Description and User's Guide*; School of Chemistry, Physics and Earth Sciences, Flinders University of South Australia: Adelaide, Australia, 2003.
9. Rahman, A. A GIS based DRASTIC model for assessing groundwater vulnerability in shallow aquifer in Aligarh, India. *Appl. Geogr.* **2008**, *28*, 32–53.
10. Shouyu, C.; Guangtao, F. A DRASTIC-based fuzzy pattern recognition methodology for groundwater vulnerability evaluation. *Hydrol. Sci. J.* **2003**, *48*, 211–220.
11. Civita, M. *Legenda Unificata per le Carte Della Vulnerabilità dei Corpi Idrici Sotterranei/Unified Legend for the Aquifer Pollution Vulnerability Maps*; Studi sulla Vulnerabilità degli Acquiferi 1 Pitagora: Bologna, Italy, 1990.
12. Civita, M. Una metodologia per la definizione e il dimensionamento delle are di salvaguardia delle opere delle opere di presa delle sorgenti normali. *Boll. Ass. Min. SUBALPINA* **1988**, *25*, 423–440.
13. Afshar, A.; Mariño, M.A.; Ebtehaj, M.; Moosavi, J. Rule-based fuzzy system for assessing groundwater vulnerability. *J. Environ. Eng.* **2007**, *133*, 532–540.
14. Caniani, D.; Lioi, D.S.; Mancini, M.I.; Masi, S. Application of fuzzy logic and sensitivity analysis for soil contamination hazard classification. *Waste Manag.* **2011**, *31*, 583–594.
15. Dixon, B. Groundwater vulnerability mapping: A GIS and fuzzy rule based integrated tool. *Appl. Geogr.* **2005**, *25*, 327–347.
16. Gemitzi, A.; Petalas, C.; Tsihrintzis, V.A.; Pisinaras, V. Assessment of groundwater vulnerability to pollution: A combination of GIS, fuzzy logic and decision making techniques. *Environ. Geol.* **2006**, *49*, 653–673.
17. Mohammadi, K.; Niknam, R.; Majd, V.J. Aquifer vulnerability assessment using GIS and fuzzy system: A case study in Tehran-Karaj aquifer, Iran. *Environ. Geol.* **2009**, *58*, 437–446.
18. Uricchio, V.F.; Giordano, R.; Lopez, N. A fuzzy knowledge-based decision support system for groundwater pollution risk evaluation. *J. Environ. Manag.* **2004**, *73*, 189–197.
19. Sdao, F.; Lioi, D.S.; Pascale, S.; Caniani, D.; Mancini, I.M. Landslide susceptibility assessment by using a neuro-fuzzy model: A case study in the Rupestrian heritage rich area of Matera. *Nat. Hazards Earth Syst. Sci.* **2013**, *13*, 395–407.
20. Civita, M.; de Maio, M. *SINTACS. Un Sistema Parametrico per la Valutazione e la Cartografia Della Vulnerabilità Degli Acquiferi All'inquinamento: Metodologia e Automazione*; Pitagora Editrice: Bologna, Italy, 1997. (In Italian)
21. Rapti-Caputo, D.; Sdao, F.; Masi, S. Pollution risk assessment based on hydrogeological data and management of solid waste landfills. *Eng. Geol.* **2006**, *85*, 122–131.
22. Zadeh, L.A. Fuzzy sets. *Inf. Control* **1965**, *8*, 338–353.

The Stability of Revegetated Ecosystems in Sandy Areas: An Assessment and Prediction Index

Lei Huang [1,2,*] and Zhishan Zhang [1,2]

[1] Shapotou Desert Research and Experimental Station, Cold and Arid Regions Environmental and Engineering Research Institute, Chinese Academy of Sciences, Lanzhou 730000, China; E-Mail: zszhang@lzb.ac.cn

[2] Key Laboratory of Stress Physiology and Ecology in Cold and Arid Regions Gansu Province, Lanzhou 730000, China

* Author to whom correspondence should be addressed; E-Mail: mathecology@163.com;

Academic Editor: Lutz Breuer

Abstract: The stability and sustainability of revegetated ecosystems is a central topic in ecological research. In this study, long-term monitoring and focused research on vegetation, soil and soil moisture from 2006 to 2012 were used to develop a model for evaluating indices of ecosystem stability using the analytical hierarchy process method. The results demonstrated that rainfall (R), vegetation coverage (C), and surface soil moisture (S) were the three most influential factors among the 14 indicators considered in a revegetated desert area in the Tengger Desert, China. A stability index (*SI*) was defined as $SI = VAR(R) \times VAR(C)/VAR(S)$, and a comparative study was conducted to examine the stability index of the natural vegetation community. The *SI* was divided into three regimes: $SI < 0.006$ was stable, $0.006 \leq SI < 0.015$ was semi-stable, and $0.015 \leq SI$ was unstable. The stable, semi-stable and unstable periods of revegetated ecosystems in our simulations were 191, 17 and 11 years, respectively, within the total modeling period of 219 years. These results indicated that the revegetated desert ecosystem would be stable in most years during the vegetation succession, and this study presents new ideas for future artificial vegetation management in arid desert regions.

Keywords: sand-fixing vegetation; ecosystem stability; analytical hierarchy process; dynamical model

---

1. Introduction

Stability has frequently been considered an important characteristic of ecological systems because of its theoretical and practical significance [1–4]. However, the concept of ecosystem stability is complex and has often been discussed in vague terms because of the complex physical and biological structures or integrated functions [5]. This complexity also occurs in artificial ecosystems [6]. Grimm and Wissel (1997) [7] presented a review and analyzed ecological stability, addressing 163 definitions of 70 different ecological stability concepts. Constancy, resilience, persistence, resistance, elasticity, mathematical stability and other concepts have been discussed in the literature in relation to specific problems in various ecosystems [8,9]. In these studies, debates on complexity-stability and diversity-stability have provided points of contention [1,10]. Early studies suggested that simple ecosystems were less stable than complex ecosystems, but later studies came to the opposite conclusion [11]. Until recently, questions on the relationship between stability and complexity have not been answered [12–14]. However, for the diversity-stability debate, evidence from multiple ecosystems at a variety of temporal and spatial scales suggests that biological diversity acts to stabilize ecosystem functioning when presented with environmental fluctuations [1]. Moreover, variation among species in their response to such fluctuations is an essential requirement for ecosystem stability [15,16]. Thus, we can conclude that stability is a multi-dimensional concept that has scale-dependent features. Generally, ecosystem stability refers to the capacity of a natural system to apply self-regulating mechanisms and return to a steady state after an outside disturbance.

Ecologists have developed a variety of approaches to measure ecosystem stability [2,17–19]. In most cases, mathematical models or empirical methods are involved, with the former developed and expanded by Robert May in 1973 [20] using linear stability analysis on models constructed from a statistical universe (randomly constructed communities with randomly assigned interaction strengths). However, mathematically derived models are only suitable for characterizing the dynamic behavior of simple dynamic systems, whereas ecological systems are not usually uniform [21]. Most ecosystems operate in a variable environment that includes events at a wide range of frequencies and intensities, and it is often difficult to determine the degree of changes or disturbances. Empirical methods involve the development of a stability index that incorporates the main characteristics of ecosystem structure or environmental factors; the stability of an ecosystem can then be determined from these indicators [22,23]. However, this solution must still manage the problem of parameter selection [24]. For instance, in a forest ecosystem, stability may be described with biomass, diversity, dominant species density, nutrient cycling and soil characteristics, *etc*. Thus, developing a method that can combine quantitative measurements and qualitative descriptions is particularly pertinent when evaluating analyses of ecosystem stability. The best method for evaluating problems that involve a number of uncertain indices is the analytic hierarchy process (AHP) method [25–27]. The AHP is a multiple-criteria decision-making tool that has been widely applied in diverse fields, such as resource allocation, project design, maintenance management and policy evaluation. This method is particularly useful because it enables the decomposition of a given problem into a hierarchy of more easily comprehended sub-problems that can each be analyzed independently. The elements of the hierarchy can be related to any aspect of the qualitative and quantitative problems, including aspects that are tangible or intangible and carefully measured or roughly estimated. Once the hierarchy is built, the AHP systematically evaluates its various elements and derives numerical priorities for each of the decision alternatives [28,29]. Thus, one may easily apply the AHP to select promising technologies, and it provides a simple method for making decisions or performing environmental impact assessments [30–32]. However, relatively few studies have assessed the application of AHP to ecosystem stability [27,33,34], especially in revegetated desert areas.

Sand-binding vegetation has been widely used in arid desert regions and is considered one of the most effective methods of mitigating desertification [35]. However, in recent years, climate change and harsh natural environments have produced a series of problems, such as declining vegetation cover, poor plant growth and widespread water stress, which have led us to investigate the stability of artificial vegetation [6]. In this research, we hypothesized that the natural vegetation community is stable because natural vegetation in the study area is the result of long-term (*i.e.*, over thousands of years) evolution [36]. Therefore, this hypothesis is reasonable, and several studies have also illustrated that ecosystems with artificial vegetation would become stable over plant succession, with these stabilizing changes mainly reflected in the increased vegetation cover and soil texture improvements compared with that of the adjacent natural vegetation communities [37–39]. However, these improvements are only concomitant with the individual process of vegetation succession, and a comprehensive system for evaluating stability indices, including the integration of vegetation, soil and water factors, is still lacking. Furthermore, previous studies primarily focused on qualitative concept models, and a quantitative model is not available. In the present study, over 50 years of vegetation succession and long-term monitoring data (2006–2012) from the Shapotou Desert Research and Experimental Station on the southeastern edge of the Tengger Desert were used to investigate soil and vegetation characteristics at different stages of plant succession. The key influencing factors of the ecosystem stability were then determined with the AHP method, and a stability index was defined. Finally, the ecosystem's stability was predicted with a dynamical model of vegetation cover and soil moisture. Our results provide basic suggestions for sustainable ecosystem management and new hypotheses regarding vegetation succession models.

2. Materials and Methods

2.1. Study Site

The study was conducted at Shapotou Desert Research and Experimental Station of the Chinese Academy of Sciences, located in the Shapotou region at the southeastern margin of the Tengger Desert (37°32' N, 105°02' E). The climate at the site is characterized by abundant sunshine and low relative humidity. The average minimum monthly relative humidity is 33% in April, and the average maximum monthly relative humidity is 54.9% in August. The elevation of the area is 1330 m, and the mean annual precipitation is 188.2 mm according to meteorological records from 1956 to 2009, with rainfall occurring primarily between June and September [37]. The mean annual temperature is 9.6 °C, and the mean monthly temperatures are −6.9 °C in January and 24.3 °C in July. The evapotranspiration potential during the growing season (May to September) is 2300 to 2500 mm.

To ensure the smooth operation of the desert section of the Baotou-Lanzhou railway, a system involving sand-binding vegetation was established by the Chinese Academy of Sciences and other related departments in 1956. First, mechanical sand fences were installed at right angles to the prevailing winds, and then 1 m × 1 m straw sand barriers were erected in a checkerboard pattern behind the mechanical sand fences. Under non-irrigated conditions, xerophytic shrubs dominated by *Caragana korshinskii, Artemisia ordosica, Hedysarum scoparium, Caragana intermedia, Calligonum arborescens* and *Atraphaxis bracteata* were planted at a spacing of 1 m × 2 m or 2 m × 3 m using the checkerboard of straw barriers as a protective screen. This ecological shelter was extended in 1964, 1981 and 1987. As shown in Figure 1, a 16 km long protective system of vegetation was eventually established, and our research site was part of this protective system, which was 500 m wide on the north side and 200 m wide on the south side of the railway. Over the 50 years since the establishment of the vegetation, the environment in the area has improved, and the stabilized sand surface has created conditions that support the colonization of a number of species. The mass propagation of psammophytes has transformed the original moving sand into a complex man-made and natural desert vegetation landscape [37]. This ecological engineering project was viewed as a successful model for desertification control and ecological restoration along the transport line in the arid desert region of China. Because sites with different ages were stabilized using similar approaches, including the planting of shrub seedlings of the same species with the same density in similar straw checkerboards (see Table 1), they can represent the different successional stages of sand-binding vegetation.

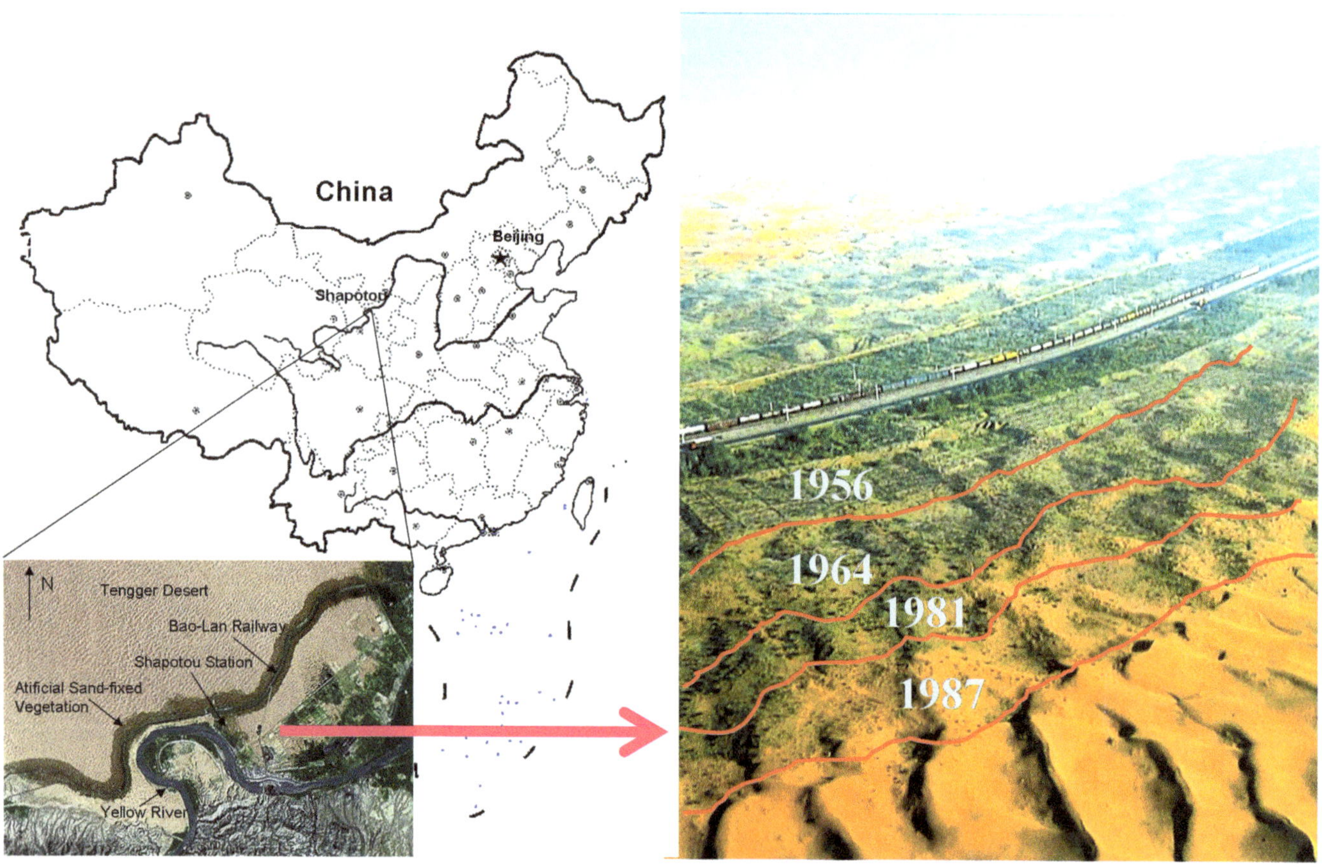

Figure 1. Schematic description of the study site (revegetated in 1956a, 1964a, 1981a and 1987a) in the Shapotou region of the Tengger Desert, northwestern China.

Table 1. Description of four revegetation sites with different ages and a natural community located in the southeastern fringe of the Tengger Desert, Northern China.

Year of Revegetation	Approaches to Sand Stabilization and Revegetation	Remaining Shrub Species of Revegetation	Native/Invasion Dominant Plant Species
1956	Straw-checkerboard of 1 m^2 planted with 10 xerophytic shrubs at a density of 16 individuals per 100 m^2	*Artemisia ordosica, Caragana korshinskii, Hedysarum scoparium*	*Artemisia ordosica, Scorzonera mongolica, Sonchus arvensis, Chloris virgata, Aristida adscensionis, Setaria viridis, Bassia dasyphylla, Chenopodium aristatum*
1964	Straw-checkerboard of 1 m^2 planted with 10 xerophytic shrubs at a density of 16 individuals per 100 m^2	*Artemisia ordosica, Caragana korshinskii, Hedysarum scoparium*	*Artemisia ordosica, Bassia dasyphylla, Eragrostis poaeoides, Sonchus arvensis, Scorzonera mongolica, Euphorbia humifusa*
1981	Straw-checkerboard of 1 m^2 planted with 10 xerophytic shrubs at a density of 16 individuals per 100 m^2	*Artemisia ordosica, Caragana korshinskii, C. microphylla, Hedysarum scoparium*	*Artemisia ordosica, Hedysarum scoparium, Bassia dasyphylla, Eragrostis poaeoides, Corispermum patelliforme*
1987	Straw-checkerboard of 1 m^2 planted with 10 xerophytic shrubs at a density of 16 individuals per 100 m^2	*Amorpha fruticosa, Artemisia ordosica, A. sphaerocephala, Caragana korshinskii, C. microphylla, Calligonum arborescens, Hedysarum scoparium*	*Hedysarum scoparium, Agriophyllum squarrosum, Bassia dasyphylla, Echinos gmelinii, Eragrostis poaeoides*
Natural	No	No	*Artemisia ordosica, Caragana korshinskii, Lespedeza davurica, Ceratoides latens, Oxytropis aciphylla, Stipa breviflora, Carex stenophylloides, Cleistogenes sogorica, Allium mongolicum, Oxytropis myriophylla, Enneapogon brachystachyus, Asparagus gobicus*

2.2. Methods

2.2.1. Sampling Method and Data Collection

Three 10 m × 10 m quadrats were established in each of the fixed observation plots in the sand-binding vegetation districts established in previous years (1956, 1964, 1981 and 1987) as well as in the adjacent natural vegetation zones for a total of 15 quadrats. The plant species number, height, and coverage for each species in the sand stabilization areas of different years were recorded or measured monthly from 2006 to 2012. Plant diversity (H') was estimated with the Shannon–Wiener index for each region and month according to the formula: $H' = -\sum p_i ln p_i$, where p is the proportion of each species i. Precipitation was recorded every 30 min using tipping bucket-type rain gauges (Casella) and Campbell CR30X data-loggers (Campbell Scientific, Logan, UT, USA). Samples were collected monthly with a soil auger, and the soil moisture of the samples was determined using the oven-drying method (0–40 cm) and neutron moisture probe method (40–300 cm). To avoid confusion between the surface and deep soil moisture, the gravimetric moisture content of the surface soil layer and bulk density measurements were used to calculate the volumetric moisture content.

Soil parameters were measured at depths of 0–20 cm at each site during the growing season in 2006 and 2010. In each plot, 100 soil sampling points were mechanically arranged in 10 m × 10 m vegetation plots of different ages. The transverse and longitudinal spacing were both 1 m. The surface of the sampling plots was flat, and composite samples were sieved through a 2 mm mesh screen and used for further analysis. Particle size was analyzed using the pipette method [40], and soil bulk density was measured using the ring-cutting method [41]. Soil organic carbon (SOC) was determined according to the dichromate oxidation method of Walkley–Black [42]. Total nitrogen was measured with a Kjeltec System 1026 Distilling Unit (Tecator AB, Hoganas, Sweden), and electrical conductivity (EC) was determined by preparing a suspension that consisted of a soil–water mixture in a ratio of 1:5 and was measured using a portable conductivity meter (Cole-Parmer Instrument Company, Vernon Hills, IL, USA). Topographic parameters (elevation, slope angle and aspect) were determined with a Real-Time Kinematic (RTK) global positioning system (GPS) (S86T, Southern Technology, Guangzhou, China) in 2006.

In this study, we chose the commonly applied method of space-for-time substitution, which assumes that simultaneous sampling of different sites of different ages is equivalent to resampling the same site through time [36]. The clay percentage was selected to represent the soil texture, and soil organic matter and total N content were used to reflect the soil nutrient regime. According to the depth distribution of the mass root systems of herbaceous plants and shrubs, the soil moisture content at 0–40 cm and 40–300 cm, respectively, were determined. The clay percentage was measured because soil texture is considered an important factor that determines the vegetation structure and composition under uniform climatic conditions, whereas soil moisture was measured because it is considered a driving force for ecological processes in arid zones. Thus, these indices can reflect the overall stability of the ecosystem [38].

2.2.2. Analytic Hierarchy Process Methodology

The AHP method was applied to select the major influencing factors from the above datasets. The AHP procedure involves three basic steps: (i) Design of the decision hierarchy; (ii) Pair-wise comparison of elements of the hierarchical structure; and (iii) Construction of an overall priority rating.

For more details, refer to Appendix A.1. In this study, fourteen important criteria were selected to evaluate the stability of the revegetated desert ecosystem (Figure 2). The top level of the diagram shows the overall goal of the hierarchy, "stability of revegetated desert ecosystems"; the second level lists the most influential factors obtained from other literature [36,37], such as soil moisture, soil characteristics, plant cover and topography; and the third level describes the attributes of each factor. After defining the criteria for selecting the evaluating index, five comparison matrices were developed: A–B, B1–C, B2–C, B3–C and B4–C. At each level, the criteria were compared pairwise according to their levels of influence and according to the specified criteria at the higher level. In AHP, multiple pairwise comparisons are based on a standardized comparison scale of nine levels, and ten experts were asked to perform pair-wise comparisons using a 1–9 preference scale that indicates the importance or dominance of one element over another. On this preference scale, 1 indicates equal preference and 9 indicates absolute preference. Intermediate values are used to express increasing preference/performance for one weight/alternative [43]. For example, if the criteria for soil moisture (B1) were judged as essential or of moderate importance, then the soil criteria (B2) with respect to the preservation of revegetated desert ecosystem stability would be given a score of 3. In addition, for other pairwise comparisons matrices, such as B2–C, the degree of importance was determined by the number of years of recovery required to reach the level of native ecosystem, such as desert steppe [36]. All of the important factors were then assigned appropriate weights, and a standardization index was calculated with the Z-score method. Finally, the integrated index was calculated, and the most influential factors in the revegetated ecosystem were determined.

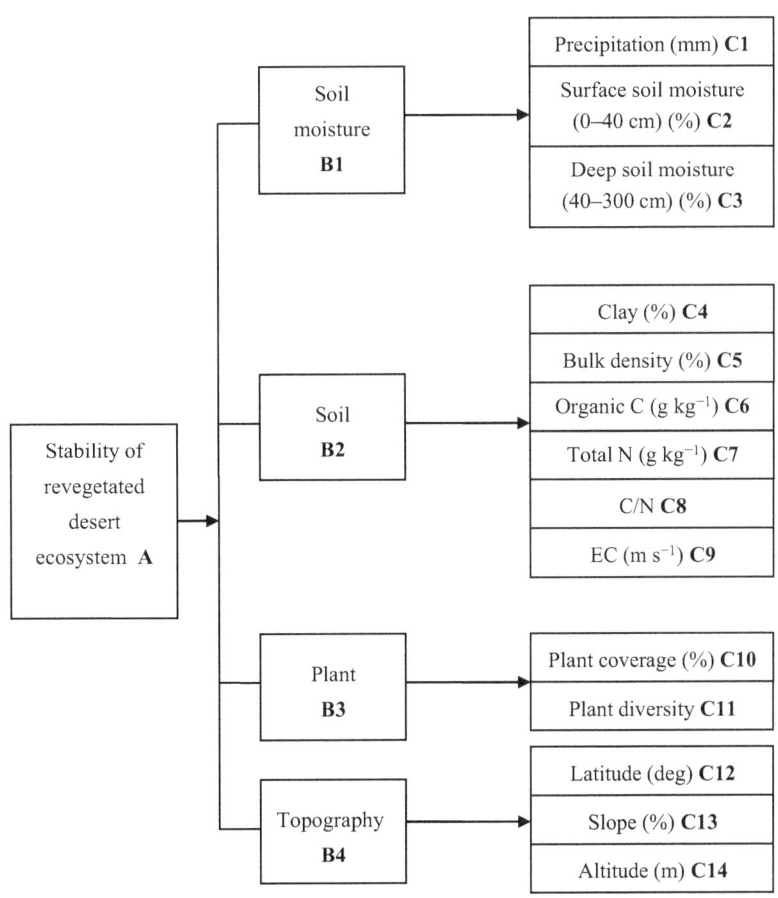

Figure 2. A hierarchy for the stability evaluation of revegetated desert ecosystems.

2.2.3. Coupled Dynamics of Soil Moisture and Vegetation

Soil water is the major driving force shaping vegetation patterns and processes in desert areas; however, plant growth, vegetation succession or landscape variability may also generate temporal and spatial heterogeneity of soil moisture [6]. Thus, to predict dynamic changes in soil water and vegetation succession in arid desert regions, a simplified ecohydrological box model that contains the coupled dynamics of vegetation and soil moisture was selected [44]. The model is as follows:

$$\frac{ds}{dt} = I(s,t) - [bx_b(s) + (1-b)x_0(s)] \tag{1}$$

$$\frac{db}{dt} = g(s)b(1-b) - u(s) \tag{2}$$

In Equation (1), s is the relative soil moisture averaged over the root zone ($0 \leq s \leq 1$) and b is the fraction of vegetated sites, where $0 < b < 1$. Variations in site occupancy depend on the rate at which empty sites are colonized and the rate at which vegetated sites become vacant as a result of mortality. $I(s, t)$ is the infiltration rate (mm day^{-1}); X_b and X_0 denote evapotranspiration (mm day^{-1}), which is distinct between vegetated and bare sites; $g(s)$ denotes the colonization rate; and u(s) denotes the local extinction rate (year^{-1}). All of the aforementioned rates were dependent on soil moisture. Each item in Equation (1) and the above parameter values are detailed in Appendix A.2. Matlab 7.0 (The MathWorks, Natick, MA, USA) and Origin 7.0 (OriginLab, Northampton, MA, USA) software were used for data simulation and analysis.

3. Results

3.1. Results of the Analytic Hierarchy Process (AHP) Application

Following the input of each factor and its importance into the expert choice and quantitative approach functions, the comparison results and weights of the four main criteria were calculated, and they are shown in Table 2. The results (principal vectors) show that the attributes have the following approximate priority weights: Soil moisture (0.91), soil (0.22), plant cover (0.23), and topography (0.28). The consistency ratio (*CR*) for this comparison was 0.086 < 0.10, which indicated that the weighted results were valid and consistent. In contrast, if the *CR* value were larger than the acceptable value of 0.10, the matrix results would be inconsistent and exempt from further analysis. Similarly, pair-wise comparisons of the sub-criteria indices with respect to the four criteria in the middle level were conducted. The comparison and weight results for this level are shown in Tables 3–6. The final stage of the AHP was to compute the contribution of each index to the overall goal, and the global weights were calculated by multiplying the local weights with criteria and sub-criteria. The final weights and ranking of the indices were then obtained as shown in Table 7. The ranking of critical ecosystem stability factors showed that soil moisture and plant factors are the most influential. The highest-ranked factor was precipitation (0.86), followed by plant coverage (0.25) and surface soil moisture (0.23). The *CR* values of all matrices were less than 0.10 and were therefore accepted. The largest value in the priority weight was the most important criterion, which means that precipitation, surface soil moisture and plant coverage were the three most influential factors determining the stability of the revegetated desert ecosystem.

Table 2. Judgment matrix of the objective hierarchy and the criterion hierarchy.

A	B1	B2	B3	B4	Priorities	AHP Criteria
B1	1.00	3.00	5.00	4.00	0.91	$\lambda_{max} = 4.23$;
B2	0.33	1.00	0.50	1.00	0.22	$CI = 0.078$;
B3	0.20	2.00	1.00	0.50	0.23	$RI = 0.900$;
B4	0.25	1.00	2.00	1.00	0.28	$CR = 0.086$

Notes: The weight of four evaluation criteria: Stability of revegetated desert ecosystem (A); Soil moisture (B1); Soil (B2); Plant (B3); Topography (B4).

Table 3. Judgment matrix of the criterion B1 and its related factors.

B1	C1	C2	C3	Priorities	AHP Criteria
C1	1.00	5.00	4.00	0.95	$\lambda_{max} = 3.09$; $CI = 0.047$;
C2	0.20	1.00	2.00	0.26	$RI = 0.58$; $CR = 0.081$.
C3	0.25	0.50	1.00	0.18	

Notes: The weight of three evaluation criteria: Soil moisture (B1); Precipitation (C1); Surface soil moisture (0–40cm) (C2); Deep soil moisture (40-300cm) (C3).

Table 4. Judgment matrix of the criterion B2 and its related factors.

B2	C4	C5	C6	C7	C8	C9	Priorities	AHP Criteria
C4	1.00	0.33	2.00	1.00	5.00	0.50	0.27	
C5	3.00	1.00	5.00	3.00	9.00	2.00	0.78	$\lambda_{max} = 6.04$;
C6	0.50	0.20	1.00	0.50	3.00	0.33	0.15	$CI = 0.008$;
C7	1.00	0.33	2.00	1.00	5.00	0.50	0.27	$RI = 1.24$;
C8	0.20	0.11	0.33	0.20	1.00	0.14	0.06	$CR = 0.006$.
C9	2.00	0.50	3.00	2.00	7.00	1.00	0.47	

Notes: The weight of nine evaluation criteria: Soil (B2); Clay (C4); Bulk density (C5); Organic C (C6); Total N (C7); C/N (C8); EC (C9).

Table 5. Judgment matrix of the criterion B3 and its related factors.

B3	C10	C11	Priorities	AHP Criteria
C10	1.00	5.00	0.98	$\lambda_{max} = 2$; $CI = 0$;
C11	0.20	1.00	0.20	$RI = 0.00$; $CR = 0$.

Notes: The weight of two evaluation criteria: Plant (B3); Plant coverage (C10); Plant diversity (C11).

Table 6. Judgment matrix of the criterion B4 and its related factors.

B4	C12	C13	C14	Priorities	AHP Criteria
C12	1.00	1.00	2	0.63254	$\lambda_{max} = 3.02$; $CI = 0.008$;
C13	1.00	1.00	3	0.72389	$RI = 0.58$; $CR = 0.013$.
C14	0.5	0.33	1	0.27546	

Notes: The weight of three evaluation criteria: Topography (B4); Latitude (C12); Slope (C13); Altitude (C14).

Table 7. Overall weight of the ecosystem stability evaluation index.

Indices	B1 0.91	B2 0.22	B3 0.23	B4 0.28	Overall Priorities
C1	0.95	–	–	–	0.86
C2	0.26	–	–	–	0.23
C3	0.18	–	–	–	0.16
C4	–	0.27	–	–	0.06
C5	–	0.78	–	–	0.18
C6	–	0.15	–	–	0.03
C7	–	0.27	–	–	0.06
C8	–	0.06	–	–	0.01
C9	–	0.47	–	–	0.11
C10	–	–	0.98	–	0.25
C11	–	–	0.20	–	0.05
C12	–	–	–	0.63	0.18
C13	–	–	–	0.72	0.20
C14	–	–	–	0.28	0.08
$CI = 0.052$; $RI = 0.955$; $CR = 0.055 < 0.1$					

3.2. Stability Index Definition, Measurement and Prediction

The above AHP analysis verified that precipitation (*C1*), surface soil moisture (*C2*) and plant coverage (*C10*) were the three main limiting factors that affected the stability of the ecosystem; the stability index (*SI*) was therefore defined as $SI = VAR\ (C1) \times VAR\ (C10)\ /\ VAR\ (C2)$. *VAR* denotes the variance of a random variable. Generally, we assumed that the natural vegetation has higher stability; thus, the greater the similarity of different successional stages of sand-binding vegetation, the higher its stability. As shown in Figure 3A–D, plant coverage and surface soil moisture depended on annual precipitation. In particular, the maximum annual rainfall in 2007 was 271.2 mm, which induced extensive plant growth; in contrast, in 2006, 2009 and 2010, the annual rainfall was only approximately 100 mm, and plant coverage and surface soil moisture were maintained at a relatively low level. However, the amplitude of plant coverage was less than that of surface soil moisture in different vegetative sites. The *SI* in the natural vegetation site was 0.005, and it was 0.011, 0.012, 0.013, 0.017 in the 1956a, 1964a, 1981a and 1987a vegetation sites, respectively. The above results suggest that with increasing years of sand-binding vegetation, the revegetated desert ecosystem would become more stable. However, when compared with the natural vegetation, ecological restoration in arid desert regions still occurred over a very long time scale.

At long time scales, we have simulated 219 years of rainfall, vegetation coverage and soil moisture changes from Equation (1), as shown in Figure 4A–C. Numerical results have shown that soil moisture and vegetation cover would increase in the future and that the soil moisture would be maintained at 2.5%, but the rate of increase was not very high. The vegetation coverage remained at 40%, with fluctuations, but it was still increasing, and the rate of increase of vegetation coverage was larger than that of soil moisture. The *SI* also fluctuated with ecosystem succession, most of which was closer to the natural vegetation, as shown in Figure 3D. Based on the above-measured data and its distribution patterns, the *SI* can be divided into several regimes, which we defined as follows: $SI < 0.006$ was stable,

$0.006 \le SI < 0.015$ was semi-stable and $0.015 \le SI$ was unstable. Thus, the stable, semi-stable and unstable periods in our simulations were 191, 17 and 11 years within the total 219 years, respectively, which indicated that the revegetated desert ecosystem would be stable the majority of the time.

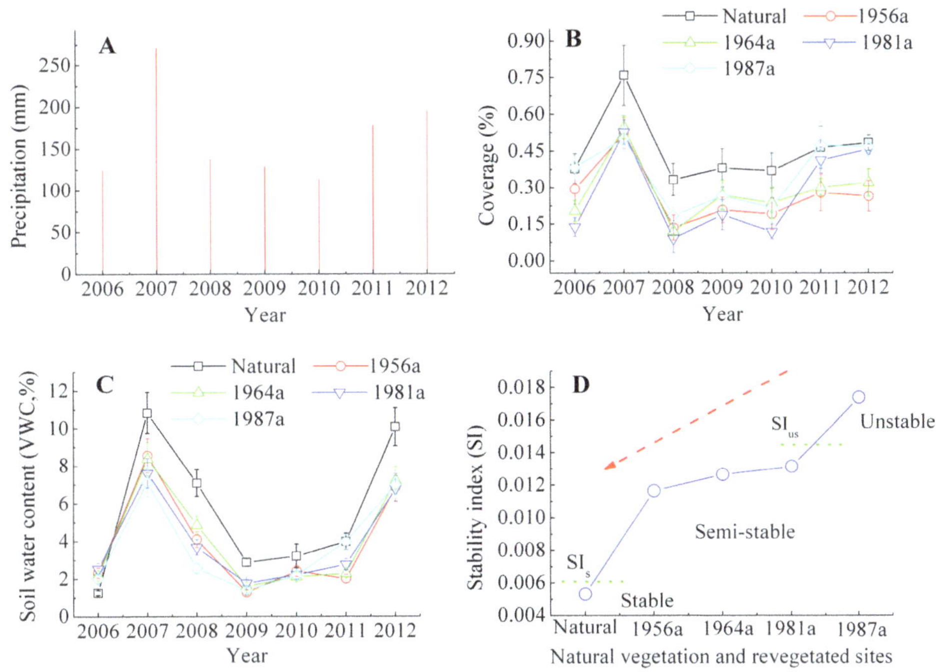

Figure 3. (**A**) Annual precipitation; (**B**) plant coverage; (**C**) soil moisture and (**D**) stability index of different vegetative sites and the control; The horizontal dashed lines represents the threshold of ecosystem stability, $SI_s = 0.004$, $SI_{us} = 0.015$.

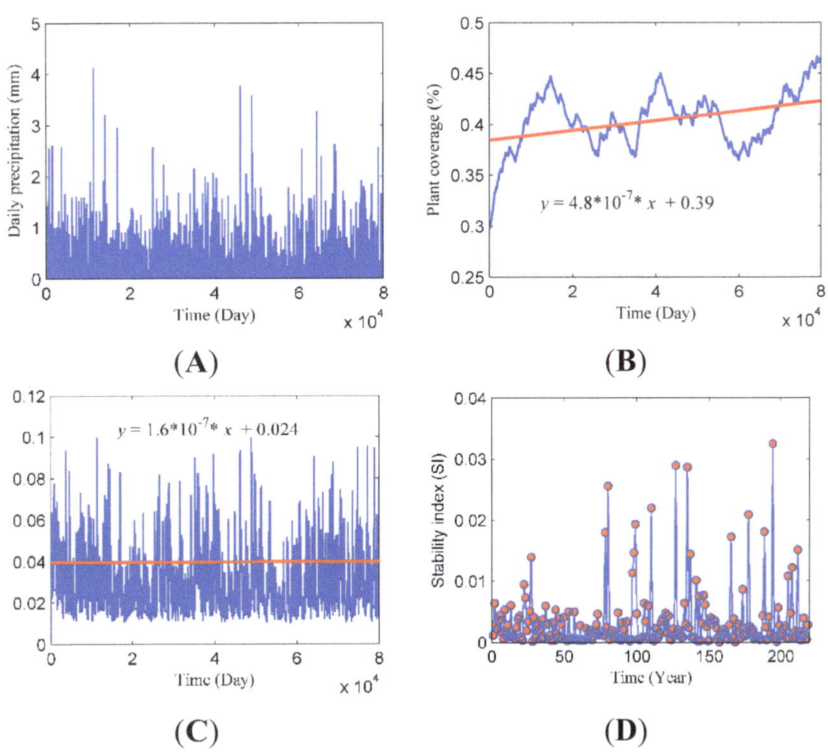

Figure 4. (**A**) Simulated results of daily precipitation; (**B**) plant coverage; (**C**) soil moisture and (**D**) stability index using Equation (1) over 219 years.

4. Discussion

Stability is an important indicator of ecosystem health and a necessary condition for ecosystem sustainability [4]. The assessment of ecosystem stability is helpful in revealing community dynamics and vegetation succession [2]. Debate is ongoing regarding the definition of stability [1], and each researcher has individual evaluation methods, which is because of regional differences or specific research objectives [14]. Therefore, in the present study, ecosystem stability was measured with the comparative method, and natural vegetation ecosystems were considered stable. In these ecosystems, natural vegetation indicators, such as plant coverage or soil moisture, were assumed to represent the standard, and ecosystems that are closer to this standard are more stable. This approach avoids theoretical controversy and provides a simple method for practical applications. In revegetated desert ecosystems, soil moisture, soil characteristics, plant cover and topography were selected as the integral influential factors for ecosystem stability. For the selection of useful variables, the AHP method was used because it effectively incorporates interdependent criteria and local problems involving both quantitative and qualitative issues. A key drawback in using the AHP method is the requirement of pair-wise comparison that must be completed by experts because such expert judgments may be affected by factors, such as fatigue and impatience during this process, especially when a large number of criteria or alternatives are involved [45]. To avoid this drawback, a reasonable and manageable number of criteria were contained in the model. Another alternative is the Bayesian inference [46], which works very well if opinions among the experts are strongly divergent, and different prior parameters can be used to test for robustness [47,48]. In this study, based on results from the literature and information gleaned from discussions with ten experts, who are authorities on eco-hydrology in arid desert areas and were unified in their feedback, the AHP method was chosen in this study, and a total of fourteen criteria were determined. Using the AHP method, we verified that precipitation, surface soil moisture and plant coverage were the key limiting factors for artificial vegetation stability in arid desert regions. Precipitation was considered the sole source of water replenishment in this area, and soil water was the main driving force of the ecosystem's patterns and processes. Additionally, the changes in the vegetation patterns also affected the redistribution of precipitation and infiltration depth in different soil layers. Therefore, these ecohydrological processes and their feedback mechanisms were identified as the main problems affecting the restoration and reconstruction of certain ecological engineering projects in arid regions [6]. However, for other ecosystems (such as agro-ecosystems, grasslands or forests), ecosystem structure and function are more complicated than in revegetated desert ecosystems [36–38]; therefore, evaluating ecosystem stability is more difficult [3,4,6]. Thus, a Bayesian methodology that provides for semiautomatic searches of consensus building should be considered instead of AHP.

To quantify the stability indicators, we defined the *SI* as the integration of variance of three variables. Because precipitation and surface soil moisture were inextricably linked, they always varied at the same time. Therefore, we defined the ratio of the two as a coefficient, and then multiplied it by the variance of vegetation cover, which directly represents the ecosystem restoration or degradation within a certain period. Thus, the stability of the ecosystem is specific, quantitative and verifiable. Simulated results from Equation (1) have shown that the ecosystem may become unstable in years with high precipitation because years with high annual precipitation are often followed by several years of continuous drought [49,50], which dramatically changes soil moisture and vegetation cover. This greater fluctuation leads to instability of the revegetated ecosystem. Compared with other empirical stability indices, such as community numbers, biomass or plant diversity [51], our stability index is simple and practical. Furthermore, it reflects the intrinsic characteristics of ecosystem stability, such as stochastic dynamics and temporal dependence, and accurately reflects the entire ecosystem with real environmental fluctuations. The *SI* values of different vegetation sites were then compared with the natural vegetation community, and stable or unstable regimes were determined. Based on field observations, the results showed that revegetated ecosystems increase in stability over time after the establishment of sand-binding vegetation. This conclusion is also supported by other studies in which the establishment of planted sand-binding vegetation in the Shapotou region is suggested to promote the improvement and restoration of regional habitats and provide suitable conditions for an increase in biodiversity in the desert ecosystem [37].

In terms of the mechanism by which sand-fixing vegetation promotes stability in the Tengger Desert, studies present inconsistent results on the formation mechanisms of stable plant communities. Shen (1986) [52] considered that *A. ordosica* may form a relatively stable climax community or plagioclimax, with the originally planted shrubs, such as *C. korshinskii*, *H. scoparium and A. ordosica*, degrading nearly 20 years later and withdrawing from the community. Zhao *et al*. (1988) [53] predicted that the next stage of vegetation succession would be herb-dominated plant communities. Li (2005) [38] advanced a conceptual model and stated that the revegetated plants would lead toward herb-dominated vegetation, which is similar to the primary vegetation types of the adjacent steppified desert and desert steppe. These results suggest that vegetation adapts to the habitat and revegetated ecosystems form a new equilibrium with vegetation succession. However, these previous studies were conceptual models, and vegetation succession assessments cannot be quantified. In the present study, a simplified dynamical model originally developed by Baudena *et al*. (2007) [44] that has been widely used in vegetation pattern analyses in arid and semi-arid areas [54,55] was applied, and it was capable of inferring the vegetation pattern features and useful information on underlying processes, including the susceptibility of the system to abrupt shifts to a desert state (*i.e*., unvegetated) as a result of climate change or anthropogenic disturbances [56]. Through our model simulations and the division of stability intervals, we predict that sand-fixing vegetation ecosystems will remain stable for a long period of time, although this stability will be interspersed with a number of semi-stable and unstable years. For certain theories or proposed mechanisms for the maintenance of ecosystem stability, such as the diversity theory or redundancy theory [57], self-organization of plant behaviors during a particular period of time is an essential component of vegetation succession over long time scales. Similar to banded or spot vegetation in North America, Africa and Australia [58–60], we could hypothesize that such self-organization reflects the normal vegetation successional pattern at different stages and suggest that vegetation may follow a “banded-spot-banded-spot (…)” pattern. Therefore, variations in vegetation composition, structure and responses to hydrological processes in the sand-binding areas are necessary stages of natural succession [61]. Thus, in artificial vegetation ecosystem management, the human should not be overly interfered with.

5. Conclusions

In this study, long-term monitoring and focused research was used to develop, measure and evaluate an index of ecosystem stability. Using the AHP method, we verified that rainfall (R), vegetation coverage

(C) and surface soil moisture (S) are the three most influential factors in a revegetated desert area in the Tengger Desert, China. Over short time scales, the stability of the revegetated sandy ecosystem increased with years of succession. However, the stability of the artificial vegetation ecosystem may fluctuate with vegetation succession on a timescale of hundreds of years. The revegetated desert ecosystem was mostly stable based on our established theoretical system of ecosystem assessment and prediction, thus verifying the success of this method for desertification control and ecological restoration along a transport line in arid desert regions. Furthermore, our results provide new ideas for future artificial vegetation management and sustainable development in arid revegetated desert areas.

Acknowledgments

This work was supported by the National Key Basic Research program (2013CB429905) and Chinese National Natural Scientific Foundation (41201084; 41201086; 31170385).

Author Contributions

Conceived and designed the experiments: Lei Huang; Performed the experiments: Zhishan Zhang; Analyzed the data: Lei Huang; Contributed reagents/materials/analysis tools: Zhishan Zhang; Wrote the paper: Lei Huang. All authors read and approved the final manuscript.

Appendix

A.1. Analytic process hierarchy (AHP) methodology

AHP is a multi-criteria analysis method that is based on pair-wise comparisons of the components of a particular problem. This method enables a complex problem to be broken into a goal, criteria, solutions and other levels to provide a simple method for decision-making. Applying the AHP procedure involves three basic steps [62]: (1) Decomposition, or the hierarchy construction; (2) Comparative judgments, or defining and executing data collection to obtain pair-wise comparison data on elements of the hierarchical structure and (3) Synthesis of priorities, or constructing an overall priority rating.

In the first stage, a complex decision problem is structured as a hierarchy. This structure comprises a goal or focus at the topmost level, multiple criteria that define alternatives in the middle, and decision alternatives at the bottom. The second step is the comparison of the alternatives and the criteria. Once the problem has been decomposed and the hierarchy is constructed, prioritization procedure starts in order to determine the relative importance of the criteria within each level. The pairwise judgment starts from the second level and finishes in the lowest level. In each level, a nominal scale is used for the evaluation. The scale used in AHP for preparing the pairwise comparison matrix is a discrete scale from 1 (the two choice options are equally preferred) to 9 (one choice option is extremely preferred over the other), as presented in Table A1.

Table A1. Evaluation Scale in AHP.

Intensity of Importance	Definition	Explanation
1	Equal Importance	Two elements have equal importance regarding the element in higher level
3	Moderate Importance	Experience or judgement slightly favours one element
5	Strong Importance	Experience or judgement strongly favours one element
7	Very Strong Importance	Dominance of one element proved in practise
9	Extreme Importance	The highest order dominance of one element over another
2,4,6,8	Compromises between the Above	When compromise is needed
Adverse	Adverse Comparisions	The adverse evaluation of the same criteria, adverse of the same point under multiplication

Elements in each level are compared in pairs with respect to their importance to an element in the next higher level. Starting at the top of the hierarchy and working down, the pair wise comparisons at a given level can be reduced to a number of square matrices $A=(a_{ii})_{n\times n}$ as in the following:

$$A=\begin{bmatrix} a_{11} & a_{12} & \cdots & a_{1n} \\ a_{21} & a_{22} & \cdots & a_{2n} \\ \cdots & \cdots & \cdots & \cdots \\ a_{n1} & a_{n2} & \cdots & a_{nn} \end{bmatrix} \quad a_{ii}=1,\ a_{ji}=1/a_{ij},\ a_{ij}\neq 0$$

After all pair wise comparison matrices are formed, the vector of weights, $w=(w_1,w_2,\ldots,w_n)$ is computed on the basis of Sattys eigenvector procedure. The computation of the weights involves two steps. First, the pair wise comparison matrix $A=(a_{ii})_{n\times n}$ is normalized by Equation (A1) and then the weights are computed by Equation (A2).

$$a_{ii}^* = \frac{a_{ij}}{\sum_{i=1}^{n} a_{ij}},\ \text{for all } j=1,2,\ldots,n \tag{A1}$$

$$w_i = \frac{\sum_{j=1}^{n} a_{ij}^*}{n},\ \text{for all } i=1,2,\ldots,n \tag{A2}$$

Satty (1980) showed that there is a relationship between the vector weights, w and the pair wise comparison matrix, *A*, as shown in Equation (A3).

$$Aw=\lambda_{\max} w \tag{A3}$$

The $\lambda_{\max}$ value is an important validating parameter in AHP and is used as a reference index to screen information by calculating the Consistency Ratio (*CR*) of the estimated vector. To calculate the *CR*, the Consistency Index (*CI*) for each matrix of order *n* can be obtained from Equation (A4).

$$CI=\frac{\lambda_{\max}-n}{n-1} \tag{A4}$$

Then, CR can be calculated using Equation (A5):

$$CR = \frac{CI}{RI} \tag{A5}$$

Where RI is the random consistency index obtained from a randomly generated pair wise comparison matrix. Table A2 shows the value of the RI from matrices of order 1 to 10 as suggested by Satty [28]. If $CR < 0.1$, then the comparisons are acceptable. If, however, $CR \geq 0.1$, then the values of the ratio are indicative of inconsistent judgments. In such cases, one should reconsider and revise the original values in the pair wise comparison matrix A.

Table A2. Random Indicators.

n	1	2	3	4	5	6	7	8	9	10
RI	0	0	0.58	0.90	1.12	1.24	1.32	1.41	1.45	1.49

A.2. The Coupled Soil-Vegetation Model

The coupled soil-vegetation system (B1) was original presented in Baudena *et al.* (2007) [44], and we have improved some items in the equation as following, which would give a better description in our research area.

$$\begin{aligned} \frac{ds}{dt} &= I(s,t) - [bx_b(s) + (1-b)x_0(s)] \\ \frac{db}{dt} &= g(s)b(1-b) - u(s)b \end{aligned} \tag{A6}$$

where s is relative soil moisture averaged over the root zone ($0 \leq s \leq 1$), The fraction of vegetated sites is denoted by b ($0 \leq b \leq 1$). The infiltration rate $I(s,t)$ is assumed to be equal to the rainfall rate, as long as the soil layer is not saturated; when rainfall exceeds the available water storage in the soil, the excess is converted into surface runoff. Thus,

$$I = \begin{cases} \dfrac{r}{nZr} & \dfrac{r\Delta t}{nZr} < 1-s \\ \dfrac{1-s}{\Delta t} & \dfrac{r\Delta t}{nZr} \geq 1-s \end{cases} \tag{A7}$$

where Δt = 1 day. n is soil porosity and Zr is the active soil depth in millimeters. r is daily rainfall, which is modeled as instantaneous events occurring according to a marked Poisson process of rate (mean frequency of rainfall events) λ, and with exponentially distributed depths with mean h. $x_b(s)$ was the water losses from vegetated soil, which include direct soil evaporation and plant transpiration, s^* is the critical soil moisture value below which plants start reducing transpiration by closing their stomata, and s_1 is the soil field capacity above which leakage occurs. The losses from evapotranspiration are assumed to increase linearly as a function of s until the moisture reaches a threshold s^*, above which the evapotranspiration takes place at a maximum value E_{max}, when the soil moisture exceeds the soil field capacity s_1, the leakage losses was start by an exponential growth and reaching the saturated hydraulic conductivity k_s at s =1.

$$x_b(s)=\begin{cases}\frac{E_{\max}}{s^*}s & 0\le s<s^*\\ E_{\max} & s^*\le s<s_1\\ E_{\max}+k_s\frac{e^{\beta(s-s1)}-1}{e^{\beta(1-s1)}-1} & s_1\le s<1\end{cases} \tag{A8}$$

$x_0(s)$ was the water losses in bare soil; E_{soil} is pure soil evaporation before the leakage occurs. It increases linearly up to field capacity s_1, above which the leakage losses was start with the same expression as for the vegetated soil.

$$x_b(s)=\begin{cases}\frac{E_{soil}}{s_1}s & 0\le s<s_1\\ E_{soil}+k_s\frac{e^{\beta(s-s1)}-1}{e^{\beta(1-s1)}-1} & s_1\le s<1\end{cases} \tag{A9}$$

$$x_b(s)=\begin{cases}\frac{E_{soil}}{s_1}s & 0\le s<s_1\\ E_{soil}+k_s\frac{e^{\beta(s-s1)}-1}{e^{\beta(1-s1)}-1} & s_1\le s<1\end{cases} \tag{A10}$$

the colonization and extinction rates $g(s)$ and $u(s)$ depend on s, as seen in the Figure A1, it can be modeled as:

$$g(s)=\frac{0.05s^2}{1+12.36s^2},\quad u(s)=0.0006e^{-4.69s} \tag{A11}$$

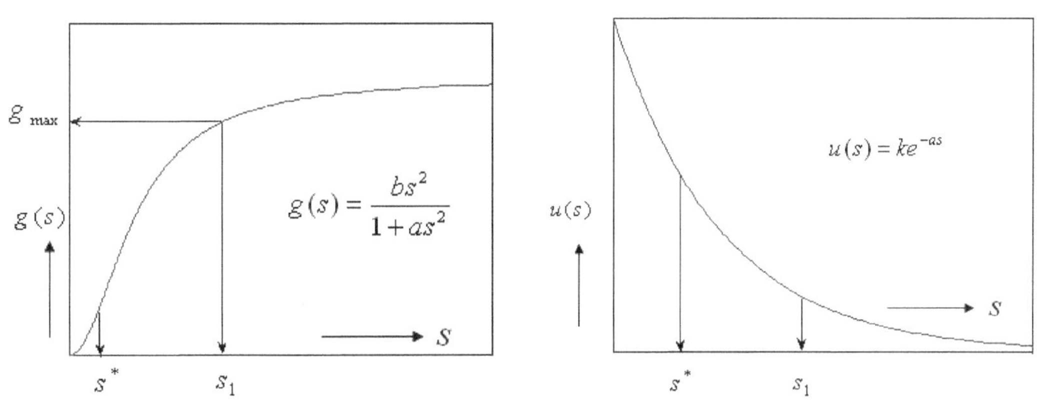

Figure A1. Sketch for vegetation colonization rate and extinction rate.

When the soil moisture was below the field capacity s_1, plant growth was slowly at first and then reached to its maximum $g_{\max}$ sharply with increased soil moisture, which have showed a typical "*S*" curve. The extinction rate is assumed to exponential decreased with soil moisture. The parameters for Matlab simulation were shown in Table A3.

Table A3. Parameters of the simplified ecohydrological box model.

Parameters	Symbol (Unit)	Value
Soil porosity	n	0.43
Active soil depth	Zr (cm)	40
Critical soil moisture below which plant undergoes water stress	s*	0.11
Field capacity	$s1$	0.56
Pore size distribution parameter	β	12.7
Saturated hydraulic conductivity	Ks (cm/d)	800
Pure soil evaporation	E_{soil} (mm/d)	0.1
Maximum evapotranspiration rate	E_{max} (mm/d)	3.67
Average rainfall frequency	λ (/d)	0.15
Average precipitation depth	h (mm/d)	0.61

The parameters in the model were obtained from the directly measured data. The cutting ring method was used to determine soil porosity n and field capacity s_1 of root zone (3 repeats at the 0, 20 and 40 cm soil depths, respectively, and take the average). The saturated soil hydraulic conductivity K_s was measure by using of the tension infiltrometer model SW 080B (SDEC, Paris, France) in undisturbed field conditions; $s*$ and another soil parameters β were determined according to related reference [63]. the historical precipitation information such as average rainfall frequency and depth were extracted from the receive data at a near weather station, and minor calibration was done by referring to the experiential relationship between precipitation and elevation. The depth of active soil or root zone depth, defined as the soil depth range in which 95% below-ground biomass were distributed, was determined by filed investigation. Pure soil evaporation E_{soil} were measured with micro-lysimeters, which were made using PVC pipes that were 30 cm high and 10 cm in internal diameter [64]. And maximum evapotranspiration rate E_{max} were determined with the combination of the stem heat balance technique (Dynamax Inc., Houston, TX, USA), the observations was measured continuous during the 2008-2012 growing season [65].

Conflicts of Interest

The authors declare no conflicts of interest.

References

1. McCann, K.S. The diversity-stability debate. *Nature* **2000**, *405*, 228–233.
2. Tilman, D.; Reich, P.B.; Knops, J.M.H. Biodiversity and ecosystem stability in a decade long grassland experiment. *Nature* **2006**, *441*, 629–632.
3. Allesina, S.; Tang, S. Stability criteria for complex ecosystems. *Nature* **2012**, *483*, 205–208.
4. Donohue, I.; Petchey, O.L.; Montoya, J.M.; Jackson, A.L.; McNally, L.; Viana, M.; Healy, K.; Lurgi, M.; O'Connor, N.E.; Emmerson, M.C. On the dimensionality of ecological stability. *Ecol. Lett.* **2013**, *16*, 421–429.
5. Hill, A.R. Ecosystem stability, some recent perspectives. *Prog. Phys. Geogr.* **1987**, *11*, 315–333.

6. Li, X.R.; Zhang, Z.S.; Huang, L.; Wang, X.P. Review of the ecohydrological processes and feedback mechanisms controlling sand-binding vegetation systems in sandy desert regions of China. *Chin. Sci. Bull.* **2013**, *58*, 1483–1496.
7. Grimm, V.; Wissel, C. Babel, or the ecological stability discussions, an inventory and analysis of terminology and guide for avoiding confusion. *Oecologia* **1997**, *109*, 323–334.
8. Cottingham, K.L.; Brown, B.L.; Lennon, J.T. Biodiversity may regulate the temporal variability of ecological systems. *Ecol. Lett.* **2001**, *4*, 72–85.
9. Deimeke, E.; Cohena, M.J.; Reissb, K.C. Temporal stability of vegetation indicators of wetland condition. *Ecol. Indic.* **2013**, *34*, 69–75.
10. Ives, A.R.; Carpenter, S.R. Stability and diversity of ecosystems. *Science* **2007**, *317,* 58–62.
11. Pimm, S.L. The complexity and stability of ecosystems. *Nature* **1984**, *307*, 321–326.
12. Haydon, D.T. Pivotal assumptions determining the relationship between stability and complexity: An analytical synthesis of the stability complexity debate. *Am. Nat.* **1994**, *144*, 14–29.
13. Townsend, S.E.; Haydon, D.T.; Matthews, L. On the generality of stability complexity relationships in Lotka-Volterra ecosystems. *J. Theor. Biol.* **2010**, *267*, 243–251.
14. Mougi, A.; Kondoh, M. Diversity of interaction types and ecological community stability. *Science* **2012**, *337*, 349–351.
15. Cleland, E.E. Biodiversity and ecosystem stability. *Nat. Education Knowl.* **2012**, *3*, 14.
16. Sterk, M.; Gort, G.; Klimkowska, A.; van Ruijven, J.; van Teeffelen, A.J.A.; Wamelink, G.W.W. Assess ecosystem resilience: Linking response and effect traits to environmental variability. *Ecol. Indic.* **2013**, *30*, 21–27.
17. Mitchell, R.J.; Auld, M.H.D.; Le Duc, M.G.; Marrs, R.H. Ecosystem stability and resilience, a review of their relevance for the conservation management of lowland heaths. *Perspect. Plant Ecol.* **2000**, *3*, 142–160.
18. Ives, A.R.; Dennis, B.; Cottingham, K.L.; Carpenter, S.R. Estimating community stability and ecological interactions from time series data. *Ecol. Monogr.* **2003**, *73*, 301–330.
19. Loreau, M.; de Mazancourt, C. Biodiversity and ecosystem stability, a synthesis of underlying mechanisms. *Ecol. Lett.* **2013**, *16,* 106–115.
20. May, R.M. Qualitative stability in model ecosystems. *Ecology* **1973**, *54*, 638–641.
21. Grimm, V.; Schmidt, E.; Wissel, C. On the application of stability concepts in ecology. *Ecol. Model.* **1992**, *63,* 143–161.
22. Wardle, D.A.; Bonner, K.I.; Barker, G.M. Stability of ecosystem properties in response to above-ground functional group richness and composition. *Oikos* **2000**, *89*, 11–23.
23. Hooper, D.U.; Chapin, F.S.I.; Ewel, J.J.; Hector, A.; Inchausti, P.; Lavorel, S.; Lawton, J.H.; Lodge, D.M.; Loreau, M.; Naeem, S.; *et al.* Effects of biodiversity on ecosystem functioning, a consensus of current knowledge. *Ecol. Monogr.* **2005**, *75*, 3–35.
24. Dale, V.H.; Beyeler, S.C. Challenges in the development and use of ecological indicators. *Ecol. Indic.* **2001**, *1*, 3–10.
25. Saaty, T.L. A scaling method for priorities in hierarchical structures. *J. Math. Psychol.* **1977**, *15*, 234–281.

26. Locantore, N.W.; Tran, L.T.; O'Neill, R.V.; McKinnis, P.W.; Smith, E.R.; O'Connell, M. An overview of data integration methods for regional assessment. *Environ. Monit. Assess.* **2004**, *94*, 249–261.
27. Zhang, R.Q.; Zhang, X.D.; Yang, J.Y.; Yuan, H. Wetland ecosystem stability evaluation by using analytical hierarchy process AHP, approach in Yinchuan Plain: China. *Math. Comput. Model.* **2013**, *57*, 366–374.
28. Saaty, T.L. *The Analytical Hierarchy Process: Planning, Priority Setting, Resource Allocation*; McGraw-Hill, New York, NY, USA, 1980.
29. Vaidya, O.S.; Kumar, S. Analytic hierarchy process, An overview of applications. *Eur. J. Oper. Res.* **2006**, *169*, 1–29.
30. Banai, R. Anthropocentric problem solving in planning and design, with analytic hierarchy process. *J. Archit. Plann. Res.* **2005**, *22*, 107–120.
31. Cheng, E.W.L.; Li, H.; Yu, L. The analytic network process (ANP), approach to location selection: A shopping mall illustration. *Onstr. Innov.* **2005**, *5*, 83–97.
32. Ho, W. Integrated analytic hierarchy process and its applications—A literature review. *Eur. J. Oper. Res.* **2008**, *186*, 211–228.
33. Barzekar, G.; Aziz, A.; Mariapan, M.; Ismail, M.H.; Hosseni, S.M. Using analytical hierarchy Pprocess (AHP), for prioritizing and ranking of ecological indicators for monitoring sustainability of ecotourism in northern forest: Iran. *Ecologia Balkanica* **2011**, *3*, 59–67.
34. Convertino, M.; Baker, K.M.; Vogel, J.T.; Lu, C.; Suedel, B.; Linkov, I. Multi-criteria decision analysis to select metrics for design and monitoring of sustainable ecosystem restorations. *Ecol. Indic.* **2013**, *26,* 76–86.
35. Le Houerou, H.N. Restoration and rehabilitation of arid and semiarid Mediterranean ecosystems in North Africa and West Asia, a review. *Arid Soil Res. Rehabil.* **2000**, *14*, 3–14.
36. Li, X.R.; He, M.Z.; Duan, Z.H.; Xiao, H.L.; Jia, X.H. Recovery of topsoil physiochemical properties in revegetated sites in the sand-burial ecosystems of the Tengger Desert: Northern China. *Geomorph* **2007**, *88*, 254–265.
37. Li, X.R.; Xiao, H.L.; Zhang, J.G.; Wang, X.P. Long-term ecosystem effects of sand-binding vegetation in the Tengger Desert: Northern China. *Restor. Ecol.* **2004**, *12*, 376–390.
38. Li, X.R. Influence of variation of soil spatial heterogeneity on vegetation restoration. *Sci. China Ser. D* **2005**, *35*, 361–370.
39. Eldridge, D.J.; Koen, T.B.; Harrison, L. Plant composition of three woodland communities of variable condition in the western Riverina, New South Wales, Australia. *Cunninghamia* **2007**, *10*, 189–198.
40. Loveland, P.J.; Walley, W.R. Particle size analysis. In *Soil and Environmental Analysis, Physical Methods*, 2nd ed.; Simth, K.A., Mullins, C.E., Eds.; Marcel Dekker, New York, NY, USA, 2001.
41. Nanjing Institute of Soil Research. *Analysis of Soil Physic-Chemical Features*; Shanghai Science and Technology Press, Shanghai, China, 1980.
42. Nelson, D.W.; Sommers, L.E. *Total Carbon, Organic Carbon and Organic Matter.* In *Methods of Soil Analysis Part 2: Chemical and Microbiological Properties*, 2nd ed.; Page, A.L., Miller, R.H., Keeney, D.R., Eds.; American Society of Agronomy, Soil Science Society of America, Madison, WI, USA, 1982; pp. 539–577.

43. Cinelli, M.; Coles, S.R.; Kirwan, K. Analysis of the potentials of multi criteria decision analysis methods to conduct sustainability assessment. *Ecol. Indic.* **2014**, *46*, 138–148.
44. Baudena, M.; Boni, G.; Ferraris, L.; von Hardenberg, J.; Provenzale, A. Vegetation response to rainfall intermittency in drylands, results from a simple ecohydrological box model. *Adv. Water Res.* **2007**, *30*, 1320–1328.
45. Lee, J.K.L.; Chan, E.H.W. The analytic hierarchy process AHP, approach for assessment of Urban renewal proposals. *Soc. Indicat. Res.* **2008**, *89*, 155–168.
46. Altuzarra, A.; Moreno-Jiménez, J.M.; Salvador, M. Consensus building in AHP-group decision making: A Bayesian approach. *Oper. Res.* **2010**, *58*, 1755–1773.
47. Cagno, E.; Caron, F.; Mancini, M.; Ruggeri, F. Using AHP in determining the prior distributions on gas pipeline failures in a robust Bayesian approach. *Reliab. Eng. Syst. Safe.* **2000**, *67*, 275–284.
48. Hahn, E.D. Decision making with uncertain judgments: A stochastic formulation of the analytic hierarchy process. *Decis. Sci.* **2003**, *34*, 443–466.
49. Mirza, M.Q.; Warrick, R.A.; Ericksen, N.J.; Kenny, G.J. Trends and persistence in precipitation in Ganges, Brahmaputra and Meghna river basins. *Hydrol. Sci. J.* **1998**, *43*, 845–858.
50. Wang, X.P.; Zhang, J.G.; Li, X.R.; Li, J.G. Distribution, trends and variability of precipitation in Shapotou region. *J. Desert Res.* **2001**, *4*, 260–264.
51. Lehman, C.L.; Tilman, D. Biodiversity, stability and productivity in competitive communities. *Am. Nat.* **2000**, *156*, 534–552.
52. Shen, W.S. The status of *Artemisia ordosica* in vegetation succession at Shapotou area. *J. Desert Res.* **1986**, *6*, 13–22.
53. Zhao, X.L. *Research on Problem of Controlling Sand of Vegetable*; Ningxia People's Publishing House: Yinchuan, China, 1988.
54. Baudena, M.; Provenzale, A. Rainfall intermittency and vegetation feedbacks in drylands. *Hydrol. Earth Syst. Sci.* **2008**, *12*, 679–689.
55. Baudena, M.; D'Andrea, F.; Provenzale, A. An idealized model for tree-grass coexistence in savannas. *J. Ecol.* **2010**, *98*, 74–80.
56. Borgogno, F.; D'Odorico, P.; Laio, F.; Ridolfi, L. Mathematical models of vegetation pattern formation in ecohydrology. *Rev. Geophys.* **2009**, *47*, RG1005.
57. Walker, B.H. Biodiversity and ecological redundancy. *Conserv. Biol.* **1992**, *6*, 18–23.
58. Klausmeier, C.A. Regular and irregular patterns in semiarid vegetation. *Science* **1999**, *284*, 1826–1828.
59. Rietkerk, M.; Dekker, S.C.; de Ruiter, P.C.; van de Koppel, J. Self-organized patchiness and catastrophic shifts in ecosystems. *Science* **2004**, *305*, 1926–1929.
60. Kefi, S.; RieterkI, M.; Alados, C.L.; Pueyo, Y.; Papanastasis, V.P.; ElAich, A.; de Ruiter, C. Spatial vegetation patterns and imminent desertification in Mediterranean arid ecosystems. *Nature* **2007**, *449*, 213–218.
61. Li, X.R. *Eco-hydrology of biological soil crusts in desert regions of China*; Higher Education Press: Beijing, China, 2012.
62. Lotfi, S.; Habibi, K.; Koohsari, M.J. An analysis of urban land development using multi-criteria decision model and geographical information system (A case study of Babolsar city). *Am. J. Environ. Sci.* **2009**, *5*, 87–93.

63. Laio, F.; Porporato, A.; Ridolfi, L.; Rodriguez-Iturbe, I. Plants in water-controlled ecosystems: active role in hydrologic processes and response to water stress—II Probabilistic soil moisture dynamics. *Adv. Water Resour.* **2001**, *24,* 707–723.
64. Zhang, Z.S.; Liu, L.C.; Li, X.R.; Zheng, J.G.; He, M.Z.; Tan, H.J. Evaporation properties of a revegetated area of the Tengger Desert North China. *J. Arid Environ.* **2008**, *72,* 964–973.
65. Huang, L.; Zhang, Z.S.; Li, X.R. The extrapolation of the leaf area-based transpiration of two xerophytic shrubs in a revegetated desert area in the Tengger Desert, China. *Hydrol. Res.* **2014**, doi:10.2166/nh.2014.171.

Model Calibration Criteria for Estimating Ecological Flow Characteristics

Marc Vis [1,†,*], Rodney Knight [2,†], Sandra Pool [1], William Wolfe [2] and Jan Seibert [1,3,†]

[1] Department of Geography, University of Zurich, Winterthurerstrasse 190, 8057 Zurich, Switzerland; E-Mails: sandra.pool@geo.uzh.ch (S.P.); jan.seibert@geo.uzh.ch (J.S.)
[2] U.S. Geological Survey Lower Mississippi—Gulf Water Science Center, 640 Grassmere Park, Suite 100, Nashville, TN 37211, USA; E-Mails: rrknight@usgs.gov (R.K.); wjwolfe@usgs.gov (W.W.)
[3] Department of Earth Sciences, Uppsala University, Villavägen 16, 752 36 Uppsala, Sweden

† These authors contributed equally to this work.

* Author to whom correspondence should be addressed; E-Mail: marc.vis@geo.uzh.ch;

Academic Editor: Lutz Breuer

Abstract: Quantification of streamflow characteristics in ungauged catchments remains a challenge. Hydrological modeling is often used to derive flow time series and to calculate streamflow characteristics for subsequent applications that may differ from those envisioned by the modelers. While the estimation of model parameters for ungauged catchments is a challenging research task in itself, it is important to evaluate whether simulated time series preserve critical aspects of the streamflow hydrograph. To address this question, seven calibration objective functions were evaluated for their ability to preserve ecologically relevant streamflow characteristics of the average annual hydrograph using a runoff model, HBV-light, at 27 catchments in the southeastern United States. Calibration trials were repeated 100 times to reduce parameter uncertainty effects on the results, and 12 ecological flow characteristics were computed for comparison. Our results showed that the most suitable calibration strategy varied according to streamflow characteristic. Combined objective functions generally gave the best results, though a clear underprediction bias was observed. The occurrence of low prediction errors for certain combinations of objective function and flow characteristic suggests that (1) incorporating multiple ecological flow characteristics into a single objective function would increase model accuracy, potentially benefitting decision-making processes; and (2) there may be a need to have different objective functions available to address specific applications of the predicted time series.

Keywords: hydrological modeling; ecological flow characteristics; objective functions; model calibration; parameter uncertainty; catchments

1. Introduction

The interactions between streamflow and aquatic ecosystems have occupied researchers across a range of disciplines for more than 50 years. Beginning with studies as early as Rantz [1] and continuing through Tennant [2] to the present day, numerous individual streamflow characteristics have been associated with various ecological responses [3]. More recently, studies have emphasized the importance of multiple streamflow characteristics operating simultaneously or interacting to influence ecological outcomes [4]. These streamflow characteristics are used to quantify relations between flow and ecological responses. At sites where streamflow records are available, the ecologically relevant streamflow characteristics (SFCs) can be derived directly from streamflow observations. However, many, probably most, sites of biological interest have few if any observed streamflow records.

Where streamflow records are unavailable, hydrological modeling is commonly used to derive flow time series, and these simulated time series are then used to derive streamflow characteristics. The basic assumption is that if a model is capable of reproducing observed streamflow with some accuracy, the simulated time series are also suitable to derive ecologically relevant flow characteristics. However, one has to note that flow simulations are never perfect and that they generally depend on the model and its parameterization. Therefore, the suitability of simulated flow series as a basis for the estimation of streamflow characteristics might vary considerably. Key issues that must be addressed include which aspects of the stream hydrograph (SFCs) should be estimated and which modeling approaches are best suited for estimating them.

At least two broad approaches to hydrologic modeling have been applied to ecological flow problems. Regional statistics have been used to predict ecologically relevant streamflow characteristics at ungauged sites to support the development of ecological response functions, with streamflow as the controlling variable [5–7]. Such statistical models depend on prior definition of the streamflow characteristics of interest and thus are of limited flexibility should other flow characteristics later emerge as important [8]. An alternative approach is the use of runoff models, which simulate an entire hydrograph for some period of interest from which any number of streamflow characteristics can subsequently be calculated [8]. Runoff models have been recommended by some authors as the tool of choice for ecological flow studies [4], while others have expressed reservations about their suitability for such applications [8,9].

There are two main criticisms related to using runoff models for application to ecological-flow studies. The first is the difficulty in transferring the calibrated model parameters from a gauged basin, where the model can be calibrated and verified, to an ungauged basin where model performance cannot be evaluated directly. This issue of predictions in ungauged catchments is an area of active research and can be addressed by different regionalization approaches [10]. However, even with perfectly estimated parameter values (*i.e.*, the estimated parameters for an ungauged catchment correspond to what had been achieved with local model calibration) a second issue remains. This is that the models are generally calibrated on some measure of overall model performance such as the model efficiency [8,9], while biological responses to streamflow are commonly associated with specific aspects of the hydrograph, such as the long-term mean or, often more important, high- or low-flow extremes [6,11–14]. This observation raises the question: Can alternative approaches to the design and calibration of runoff models improve their ability to estimate ecologically relevant flow characteristics with a level of accuracy and precision needed to provide useful insights to the interaction between streamflow and ecosystems?

In this study, we used the HBV-light model [15–19] for runoff simulations. This model is an example of a multi-tank catchment model, with 10–15 parameters which are typically estimated by calibration. Several objective functions, each focusing on a different aspect of the hydrograph, were used to calibrate HBV-light. The aim of this study was to evaluate different objective functions for their ability to produce simulated time series that adequately preserve ecologically important flow characteristics.

2. Materials and Methods

2.1. Study Catchments

The 27 catchments used in this analysis represent parts of four Level 3 Ecoregions [20], listed east to west: Blue Ridge, Ridge and Valley, Central Appalachians, and Appalachian (Cumberland) Plateau (Figure 1). The catchments have average basin area of 829 square kilometers (km^2) (range 104–4799 km^2) and average elevation of 491 m above the North American Vertical Datum of 1988 (NAVD 88) (range 174–937 m) (Table 1). Hardwood forest and pasture are the dominant land cover in the study area. Soils are deep in the Blue Ridge ecoregion which leads to increased baseflow in comparison to the relatively thinner soils of the Appalachian Plateau and Ridge and Valley ecoregions [20] Generally, topographic slope and regolith thickness decreases from east to west, while karst development is most prominent in the Ridge and Valley [21]. Combined, these catchment characteristics produce noticeable and documented regional variations in hydrologic response and streamflow regimes [21–24].

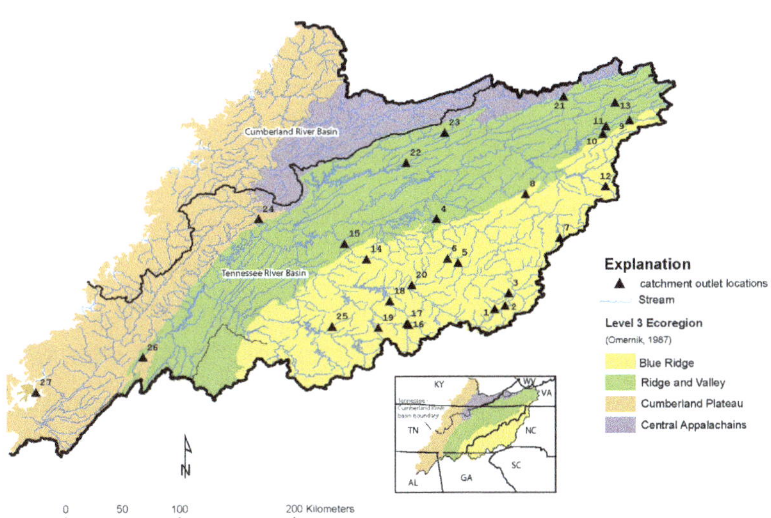

Figure 1. Catchment outlet locations for 27 basins modelled using 7 calibration schemes for HBV-light.

Table 1. U.S. Geological Survey (USGS) stream gaging sites used for model calibration and error evaluation. Latitude and longitude represent the basin outlet; ecoregion defined as the Level 3 ecoregion with the majority of the basin area; km^2, square kilometers; horizontal reference is North American Datum 1983; vertical reference is North American Vertical Datum 1988.

Map Number (Figure 1)	USGS Station Number	Latitude	Longitude	Average Elevation (m)	Primary Ecoregion (Omernik, 1987)	Basin Area (km^2)
1	03441000	35.2731	−82.7058	645	Blue Ridge	104
2	03443000	35.2992	−82.6239	628	Blue Ridge	766
3	03446000	35.3981	−82.5950	637	Blue Ridge	173
4	03455000	35.9816	−83.1611	308	Blue Ridge	4799
5	03459500	35.6350	−82.9900	712	Blue Ridge	906
6	03460000	35.6675	−83.0736	749	Blue Ridge	127
7	03463300	35.8314	−82.1842	810	Blue Ridge	112
8	03465500	36.1765	−82.4574	463	Blue Ridge	2082
9	03471500	36.7604	−81.6312	642	Blue Ridge	198
10	03473000	36.6518	−81.8440	546	Blue Ridge	785
11	03475000	36.7132	−81.8187	555	Ridge and Valley	534
12	03479000	36.2392	−81.8222	795	Blue Ridge	236
13	03488000	36.8968	−81.7462	519	Ridge and Valley	578
14	03497300	35.6645	−83.7113	337	Blue Ridge	271
15	03498500	35.7856	−83.8846	259	Blue Ridge	697
16	03500000	35.1500	−83.3797	612	Blue Ridge	361
17	03500240	35.1589	−83.3942	615	Blue Ridge	146
18	03503000	35.3364	−83.5269	537	Blue Ridge	1130
19	03504000	35.1275	−83.6186	937	Blue Ridge	135
20	03512000	35.4614	−83.3536	562	Blue Ridge	476
21	03524000	36.9448	−82.1549	457	Ridge and Valley	1382
22	03528000	36.4251	−83.3982	323	Ridge and Valley	3816
23	03531500	36.6620	−83.0949	384	Central Appalachians	828
24	03540500	35.9831	−84.5580	232	Cumberland Plateau	1815
25	03550000	35.1389	−83.9806	474	Blue Ridge	268
26	03568933	34.8975	−85.4631	202	Ridge and Valley	379
27	03574500	34.6243	−86.3064	174	Cumberland Plateau	814

Temperature and precipitation in the study area vary with longitude and elevation. Average annual temperature in the area is 13.9 degrees Celsius (°C). The warmest months of the year are July and August, and the coldest are typically January and February [25]. The Blue Ridge averages about 1350 millimeters per year (mm/y) of precipitation annually, compared to 1450 mm/y in the Cumberland Plateau and Ridge and Valley [26]. Locally, precipitation in the Blue Ridge can exceed 2000 mm/y at the highest elevations. Less than 2 percent of the precipitation comes as snow (based on 1:10 ratio of rain to snow). The streamflow regime in the study area is characterized by peak runoff typically between December and April as the result of frozen or saturated soils and low evapotranspiration rates. Summer months typically have lower streamflows because of increased temperatures and evapotranspiration rates, though occasional convective or tropical storm systems may produce locally severe flooding. Lowest flows occur in the late-summer through the fall coinciding with continuing high temperatures and evapotranspiration rates combined with decreased precipitation (October is the driest month generally). Annual runoff for the study area varies from approximately 450 to more than 760 mm [27].

The Tennessee and Cumberland River basins (considered as one aquatic ecoregion by Abell *et al.* [28]) have the highest level of freshwater diversity in North America and possibly the most diversity for any temperate freshwater ecoregion in the world [29,30]. Included in this measure are 231 fish species (with 67 (29 percent) being endemic) along with a globally outstanding unionid mussel and crayfish fauna. Many of these species are restricted to the Tennessee and Cumberland River basins [28] (pp. 212–213). A wide range of human activities threaten these populations, including urbanization, mining, logging, agriculture, and other forms of land disturbance that alter hydrologic response [28]. In addition, the entire main channels of the Tennessee and Cumberland Rivers, together with many of their tributaries, have been impounded. Flow alteration as a result of these activities has degraded or destroyed stream habitat according to Abell *et al.* [28], with more than 57 fish species and 47 mussel species at risk in the Tennessee–Cumberland aquatic ecoregion [31] (cited in Abell *et al.* [28], p. 213).

2.2. HBV Model

The HBV model [15,16] is a simple multi-tank-type model for simulating runoff. Rainfall and air temperature data [32] as well as estimated potential evaporation data based on the American Society of Civil Engineers Penman–Monteith method [33–36] are inputs to the model, which consists of four commonly used routines: (1) snow; (2) soil moisture; (3) response; and (4) routing. These routines, or slight modifications, are commonly used in other similar models (for example PRMS; Leavesley, Lichty, Troutman, and Saindon, 1983). In the snow routine, snow accumulation and snow melt are calculated by a degree-day method [37]. The soil moisture routine represents soil–water storage, which is used in conjunction with temperature and precipitation to drive evaporation and groundwater recharge. Evaporation from the soil tank equals the potential evaporation if the relative soil moisture storage is above a certain fraction, while below that fraction a linear reduction is applied. The response routine consists of connected shallow and deep groundwater storage terms and simulates runoff by summing up three linear outflow equations representing peak, intermediate and base flow. The routing routine delivers simulated runoff to the catchment outlet based on a triangular weighting function in the routing routine.

Catchments can be separated into different elevation and vegetation zones as well as into subbasins in HBV. In this study, however, catchments were disaggregated using only different elevation zones to reduce problems of over-parameterization. Calculations were performed separately for each elevation zone according to catchment for the snow and soil-moisture routines. Groundwater storage was treated as a lumped representation for each catchment. The version of HBV used in this study, HBV-light [18], corresponds to a slightly modified version of HBV-6. HBV-light uses a warming-up period of normally one year to set state variable values according to the preceding meteorological conditions and parameter sets. A more detailed description of HBV-light can be found in [18].

2.3. Calibration

The HBV-light model was applied to the 27 catchments using a daily time step. Each catchment was separated into elevation zones of 200 m, which cover at least 5 percent of the area of their respective catchment. Elevation zones covering less than 5 percent of the catchment area were merged with neighboring elevation zones. Rainfall and temperature data were compiled for the different elevation zones with a lapse rate of 10 percent/100 m and 0.6 °C/100 m, respectively. The long-term monthly potential evaporation data were linearly interpolated to daily values and corrected by using the deviations of the temperature to its long-term mean.

For all catchments, the first three years of input data measurements were used for the "warming-up" of the model to estimate the initial state variables. The rest of the data were divided into two equal time periods (14 years) covering the hydrological years (1 October through 30 September) from 1983 to 1996 and from 1996 to 2009. Each time period served both as calibration and validation period; when using the first time period for calibration the second time period was used for validation, and vice versa. This approach to calibration, validation, and parameterization allows us to consider distributions of parameter values derived from multiple independent realizations of the model, providing a generally robust evaluation. To address parameter uncertainty and equifinality [38], each calibration was repeated 100 times (here called calibration trials), which because of the random elements of the Genetic Algorithm and Powell optimization (GAP, [39]) used for calibration, resulted in 100 different parameterizations. The feasible parameter value ranges were defined based on previous studies (Table 2) [40].

Table 2. Parameter ranges used during the Genetic Algorithm and Powell optimization (GAP) calibrations within HBV-light. (°C, degrees Celsius; mm, millimeter; D, day).

Parameter	Explanation	Minimum	Maximum	Unit
Snow Routine				
TT	Threshold temperature	−2	2.5	°C
CFMAX	Degree-day factor	0.5	10	$mm \cdot °C^{-1} \cdot D^{-1}$
SFCF	Snowfall correction factor	0.5	1.2	-
CFR	Refreezing coefficient	0	0.1	-
CWH	Water holding capacity	0	0.2	-
Soil Routine				
FC	Maximum storage in soil box	100	550	mm
LP	Threshold for reduction of evaporation (relative storage in the soil box)	0.3	1	-
BETA	Shape coefficient	1	5	-
Response Routine				
PERC	Maximal flow from upper to lower box	0	4	$mm \cdot D^{-1}$
UZL	Maximal storage in the soil upper zone	0	70	mm
K0	Recession coefficient (upper box, upper outflow)	0.1	0.5	D^{-1}
K1	Recession coefficient (upper box, lower outflow)	0.01	0.2	D^{-1}
K2	Recession coefficient (lower box)	0.00005	0.1	D^{-1}
Routing Routine				
MAXBAS	Routing, length of weighting function	1	5	D

We considered seven different objective functions for calibration, which consisted of either single or combined statistical criteria evaluating the fit between observed and simulated values (Tables 3 and 4) to assess the influence of an objective function on the value of the simulated ecological indicators. The objective functions were chosen to represent different statistical aspects of streamflow. The combinations of criteria were defined to evaluate different aspects simultaneously; for example, combination 2 (C2) included Reff, MARE, Spearman, and Volume Error (see Table 3 for a description of the criteria). Reff and MARE are sensitive to peaks and low flows, respectively, and therefore help evaluate performance with respect to extreme discharge values. Volume Error expresses how well the model predicts overall runoff volume for the simulation period, whereas the Spearman rank coefficient reflects the model's success in replicating the overall timing and magnitude of discharge. Each objective function was used to calibrate the model for each time period, resulting in 14 simulated time series (seven objective functions for two different calibration periods) of streamflow for each catchment modeled.

Table 3. Definitions criteria used in objective functions for the automatic calibration trials using the Genetic Algorithm and Powell optimization (GAP) algorithm.

Criterion	Description	Definition
Reff	Model efficiency	$1-\frac{\sum(Q_{\text{obs}}-Q_{\text{sim}})^2}{\sum(Q_{\text{obs}}-\overline{Q_{\text{obs}}})^2}$
LogReff	Efficiency for log(Q)	$1-\frac{\sum(\ln Q_{\text{obs}}-\ln Q_{\text{sim}})^2}{\sum(\ln Q_{\text{obs}}-\ln \overline{Q_{\text{obs}}})^2}$
Lindström	Lindström measure	$Reff-0.1\frac{\lvert\sum(Q_{\text{obs}}-Q_{\text{sim}})\rvert}{\sum(Q_{\text{obs}})}$
MARE	Measure based on the Mean Absolute Relative Error [(1)]	$1-\frac{1}{n}\sum\frac{\lvert Q_{\text{obs}}-Q_{\text{sim}}\rvert}{Q_{\text{obs}}}$
Spearman	Spearman rank correlation [(2)]	$\frac{\sum(R_{\text{obs}}-\overline{R_{\text{obs}}})(S_{\text{sim}}-\overline{S_{\text{sim}}})}{\sqrt{\sum(R_{\text{obs}}-\overline{R_{\text{obs}}})^2}\sqrt{\sum(S_{\text{sim}}-\overline{S_{\text{sim}}})^2}}$
VolumeError	Volume error	$1-\frac{\lvert\sum(Q_{\text{obs}}-Q_{\text{sim}})\rvert}{\sum(Q_{\text{obs}})}$

[(1)] Where n is the number of days; [(2)] Where R_{obs} and S_{sim} are the ranks of Q_{obs} and Q_{sim}, respectively.

Table 4. The three combination objective functions used during the Genetic Algorithm and Powell optimization (GAP) calibrations within HBV-light. The criteria were weighted equally in each case. See Table 3 for a more detailed specification of each of the criteria.

Combined Objective Function	Criteria
C1	Reff, LogReff, VolumeError
C2	Reff, MARE, Spearman, VolumeError
C3	Spearman, VolumeError

2.4. Evaluation

The choice of the SFCs is based on studies of Knight *et al.* [6], which identified 12 specific streamflow characteristics, from a larger suite identified in Knight *et al.* [41], as most appropriate indicators for fish species richness in the study area (Table 5). All SFCs were computed using the simulated runoff of each catchment that was calibrated with one of the seven objective functions and for the two different calibration and validation time periods. The value of each streamflow characteristic was determined for both time periods based on the measurement data. All indices were computed using the free EflowStats R-Package [42].

Table 5. Definition of streamflow characteristics used in this study (adapted and modified from Knight *et al.*, 2014 and Thomson and Archfield, 2014) (mm/day, millimeters per day; -, no units; %, percent).

Streamflow Characteristic	Abbreviation	Description	Units
Magnitude			
Mean annual runoff	MA41	Annual mean daily streamflow	mm/day
Maximum October runoff	MH10	Mean maximum October streamflow across the period of record	mm/day
Lowest 15% of daily runoff	Flowperc	85% exceedance of daily mean streamflow for the period of record	mm/day
Rate of streamflow recession	RA7	Median change in log of streamflow for days in which the change is negative across the period of record	mm/day
Ratio			
Average 30-day maximum runoff	DH13	Mean annual maximum of a 30-day moving average streamflow divided by the median for the entire record	–
Stability of runoff	TA1	Measure of the constancy of a flow regime by dividing daily flows into predetermined flow classes	–
Frequency			
Frequency of moderate floods	FH6	Average number of high-flow events per year that are equal to or greater than three times the median annual flow for the period of record	number/year
Frequency of moderate floods	FH7	Average number of high-flow events per year that are equal to or greater than three times the median annual flow for the period of record	number/year
Variability			
Variability of March runoff	MA26	Standard deviation for March streamflow divided by the mean streamflow for March	–
Variability in high-flow pulse duration	DH16	100 times the standard deviation for the yearly average high-flow pulse durations (daily flow greater than the 75th percentile) divided by the mean of the yearly average high pulse durations	%
Variability of low-flow pulse count	FL2	100 times the standard deviation for the average number of yearly low-flow pulses (daily flow less than the 25th percentile) divided by the mean low-flow pulse counts	%
Date			
Timing of annual minimum runoff	TL1	Julian date of annual minimum flow occurrence	Julian day

For each objective function, 100 calibration trials were accomplished per catchment for both periods (1983–1996 and 1996–2009), producing 100 independently optimized parameter sets per catchment per simulation period. For each objective function and streamflow characteristic, the sources of uncertainty in the results were analyzed. The spread reflects both differences in behavior among the 27 catchments and uncertainty among the parameter sets, but the relative importance of these two sources of variability is not uniform. The variability because of differences between catchments was analyzed by computing the medians of the streamflow characteristics over the 100 runs per catchment. To be able to compare the median values, normalization was carried out by dividing the median values by the corresponding observed flow characteristic value. For analyzing the spread resulting from parameter uncertainty, the ranges over 100 runs per catchment were divided by the range over the median values of the different catchments. The spread because of parameter uncertainty was compared to the variation between the different catchments.

To quantify the performance of objective functions in representing the different flow characteristics, Spearman rank correlation coefficients and Nash-Sutcliffe efficiencies (NSEs) were computed between the (median) simulated and observed flow characteristic values of the 27 different catchments. Where NSE of 1.0 corresponds to identical flow characteristic values between simulated and observed runoff time series for each catchment, a Spearman rank correlation coefficient of 1.0 only requires the order of observed and simulated flow characteristic values to be the same.

3. Results

The model efficiencies that could be achieved for the different catchments varied from 0.64 to 0.91 (calibration) and 0.61 to 0.90 (validation), indicating reasonably good runoff simulation with the calibrated HBV-light model. As an example of the performance of the simulations with regard to the streamflow characteristics, the results for two indices (DH16 (variability in high-flow pulse duration) and MA41 (mean annual runoff)) for one catchment (03455000) are shown in Figure 2. Each plot contains 28 boxplots (one for each combination of an objective function, time period and calibration or validation). Each of the boxplots is based on 100 streamflow characteristic values obtained by using the 100 different parameter sets per catchment for the simulations. In both cases, there were clear deviations of the flow characteristics computed from the simulated time series compared to the observed runoff series as indicated by the red lines (red line represents observed SFC value). The streamflow characteristic DH16 was largely underestimated, especially for period 1 (1983–1996) (Figure 2a). The spread among the 100 different simulations was considerably larger for period 2 (1996–2009) than for period 1. For SFCs such as MA41 (Figure 2b), the performance differences in predicting the streamflow characteristic were prominent between the four combinations of calibration and validation periods.

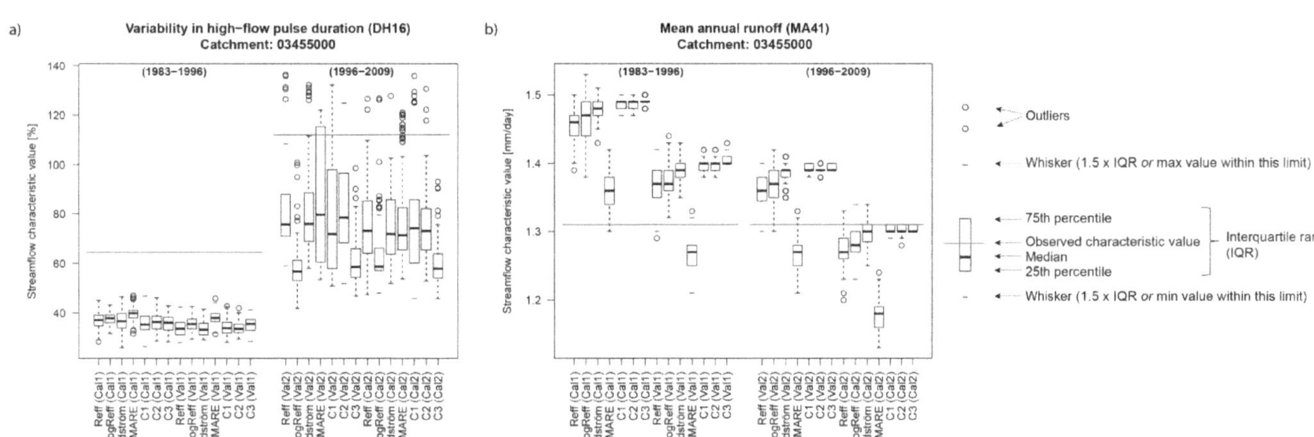

Figure 2. Boxplots for catchment 4 (03455000) and (**a**) streamflow characteristic DH16 (Variability in high-flow pulse duration); (**b**) streamflow characteristic MA41 (Mean annual runoff). Cal1 and Cal2 are calibration of period 1, respectively period 2, whereas Val1 and Val2 are validation of period 1, respectively period 2.

The agreement between observed and simulated flow characteristics varied considerably among the different catchments (Figure 3). Each plot contains 28 boxplots (one for each combination of an objective function, time period and calibration or validation). Each boxplot is based on 27 values (one value per catchment), which were normalized by dividing the median streamflow characteristic value based on simulated runoff by the corresponding streamflow characteristic value computed based on the observed runoff time series. The spread between the different catchments is much smaller for the streamflow characteristic MA41 (mean annual runoff) than for the other flow characteristics. Except for the criteria LogReff and MARE, MA41 was reproduced well for both calibration periods, whereas values were slightly underestimated when being validated on period 1 and slightly overestimated when validated on period 2. Both MA41 (mean annual runoff) and MH10 (maximum October runoff) were reproduced less well for parameter sets derived by calibration based on the criteria LogReff and MARE, both of which are more sensitive to low flow conditions than the other criteria.

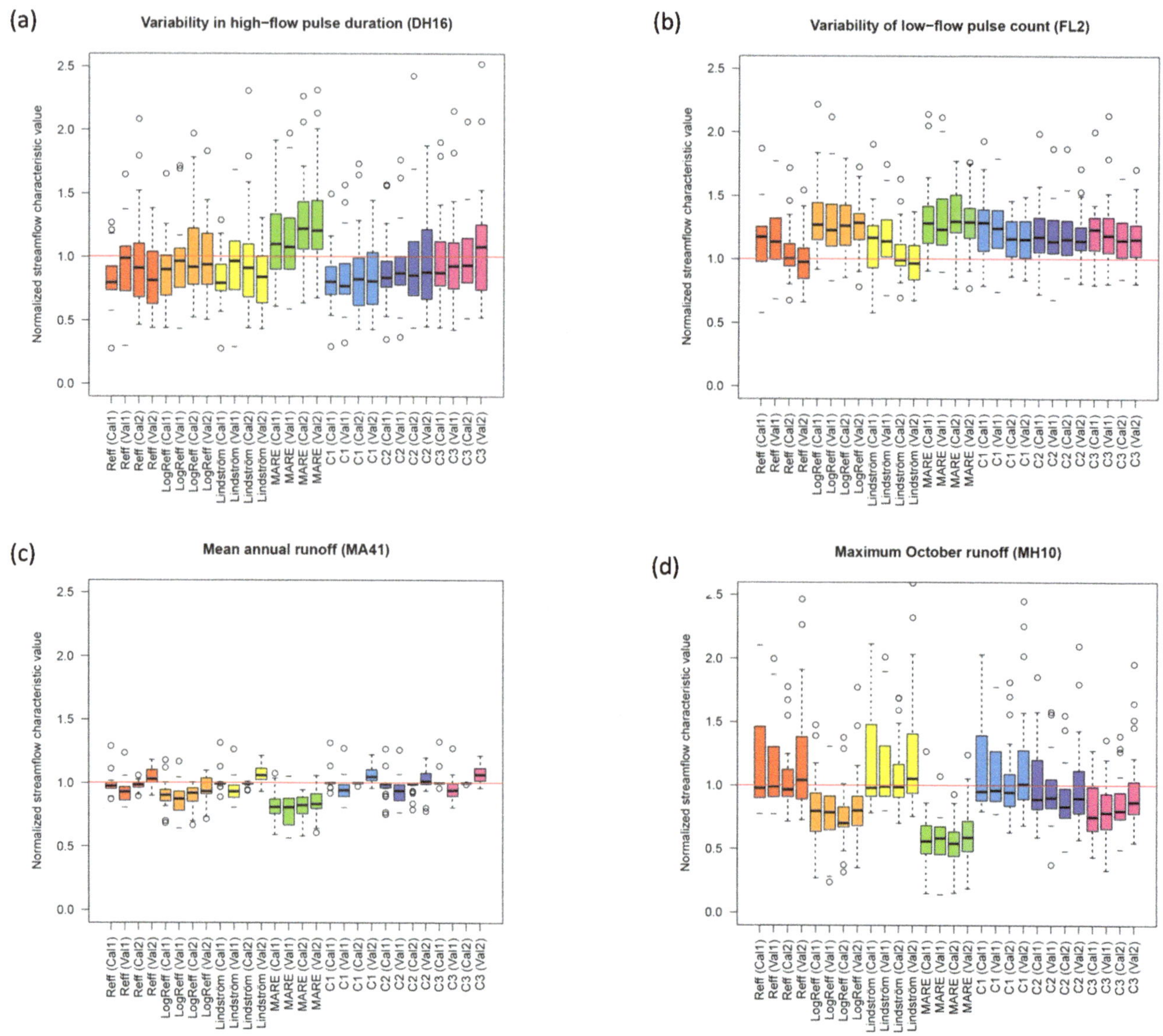

Figure 3. *Cont.*

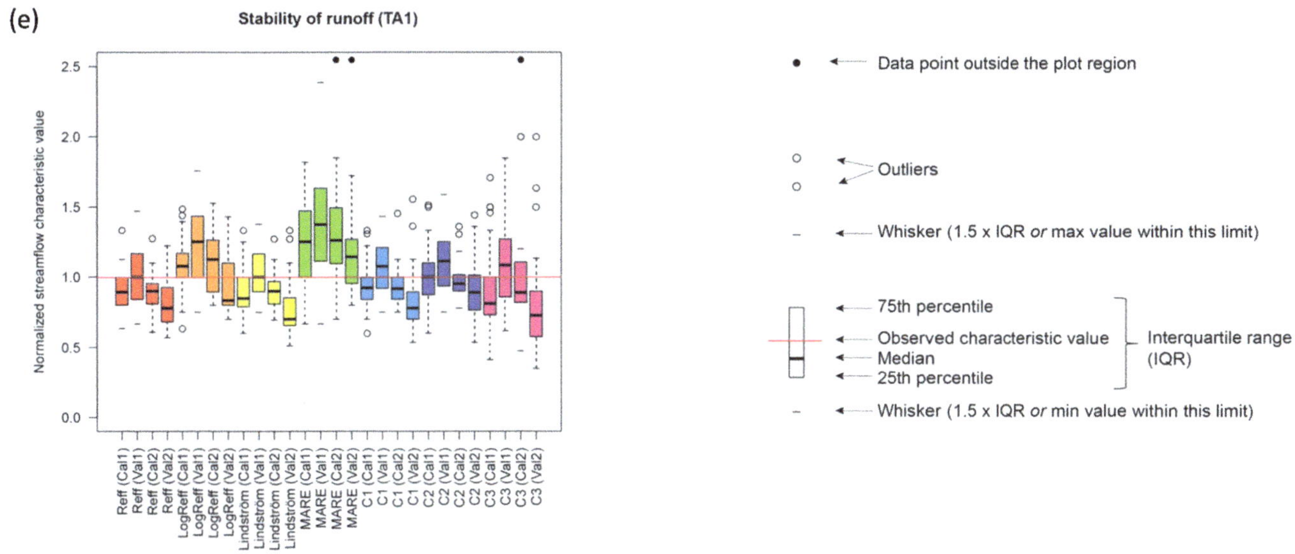

Figure 3. Normalized median flow characteristic values for five different flow characteristics: (**a**) DH16 (Variability in high-flow pulse duration); (**b**) FL2 (Variability of low-flow pulse count); (**c**) MA41 (Mean annual runoff); (**d**) MH10 (Maximum October runoff) and (**e**) TA1 (Stability of runoff). Each color corresponds to an objective function. Per objective function, the four boxplots represent (from left to right) calibration period 1 (Cal1), validation period 1 (Val1), calibration period 2 (Cal2) and validation period 2 (Val2). Each boxplot is based on 27 normalized median flow characteristic values, one value for each of the 27 catchments. Medians were computed over 100 runs per catchment. Normalization was carried out by dividing the median values by the corresponding observed flow characteristic value.

The distribution of the 27 relative ranges (per catchment—Dividing the range over the 100 runs per catchment by the range over the 27 median catchment values) is a measure for the consistency over the different catchments (Figure 4). While for some cases there was a low variation (indicated by narrow distributions of relative range), for many cases a considerable variation was observed. For calibrations based on the Nash-Sutcliffe efficiency, for instance, the median relative range varied from around 0.1 for MA41 (mean annual runoff) to above 1 for FL2 (variability of low-flow pulse count).

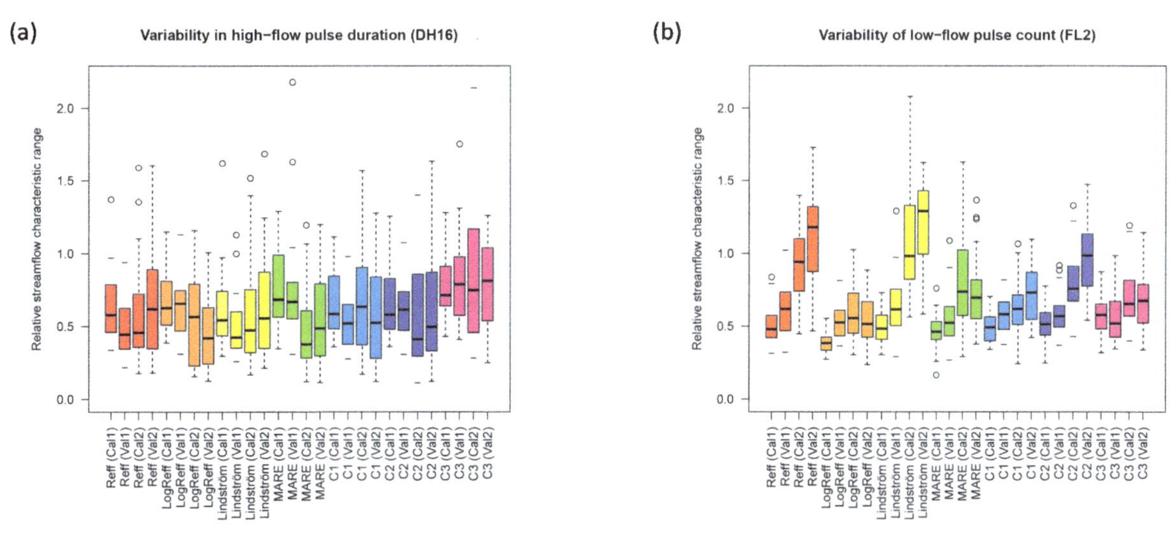

Figure 4. *Cont.*

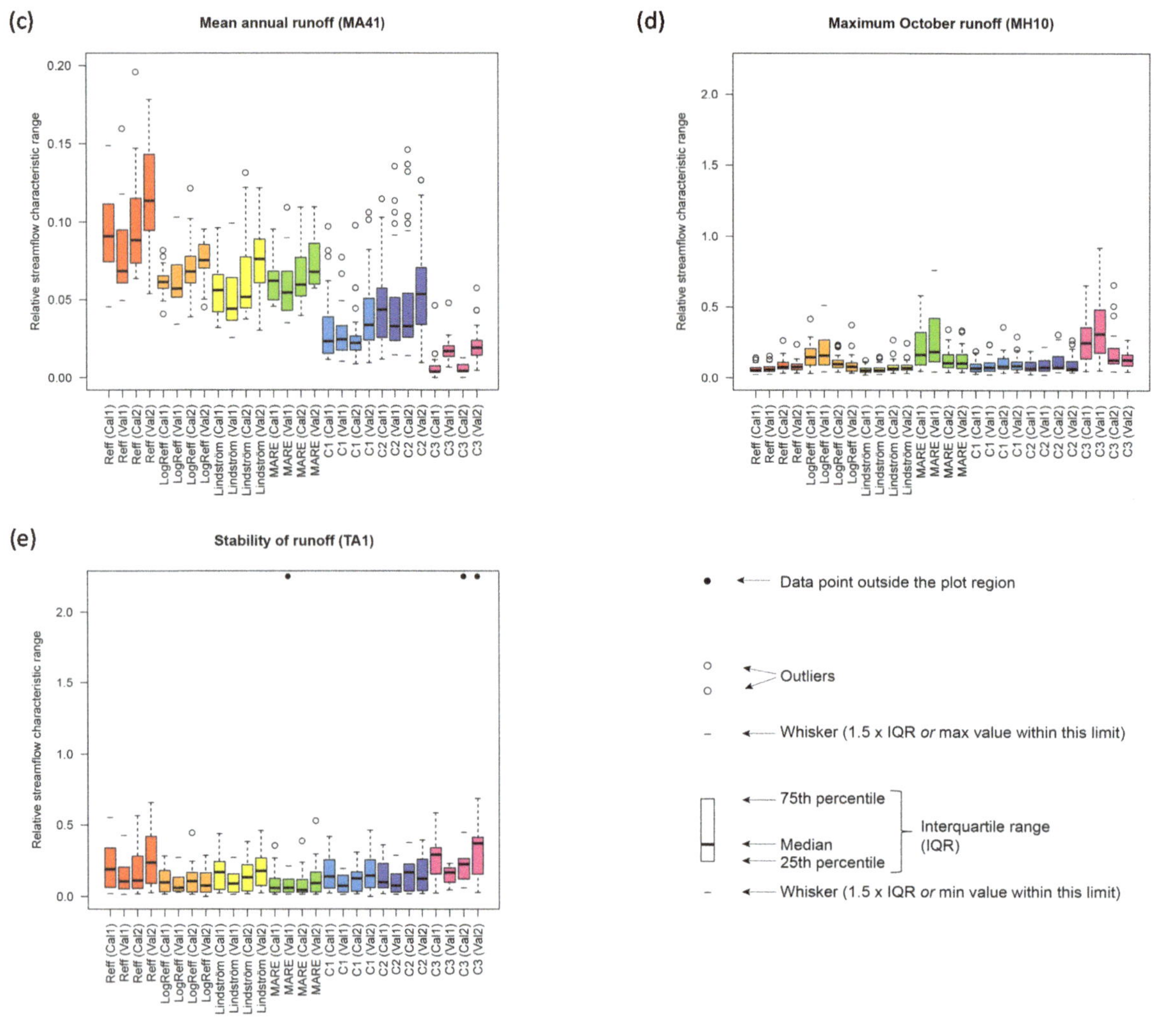

Figure 4. Relative ranges as a measure for parameter uncertainty for streamflow characteristics (**a**) DH16 (Variability in high-flow pulse duration); (**b**) FL2 (Variability of low-flow pulse count); (**c**) MA41 (Mean annual runoff); (**d**) MH10 (Maximum October runoff) and (**e**) TA1 (Stability of runoff). Each color corresponds to an objective function. Per objective function, the four boxplots represent (from left to right) calibration period 1 (Cal1), validation period 1 (Val1), calibration period 2 (Cal2) and validation period 2 (Val2). Each boxplot is based on 27 values, one value for each of the 27 catchments. Relative ranges were computed by dividing the range over the 100 runs per catchment by the range over the 27 median catchment values. Note that the Mean annual runoff (MA41) has been plotted on a different scale.

Agreement among the different streamflow characteristics and the different objective functions varied considerably (Figure 5). Comparison of streamflow characteristics based on observed runoff series against the medians of those obtained from simulated time series allows evaluating the agreement in relation to the variation between catchments. These scatter plots show that the agreement varied considerably among both the different streamflow characteristics and the different objective functions. While only plots with flow characteristics calculated for the first calibration period are shown, results were similar for the other calibration and validation periods. The performance for all streamflow characteristics and all combinations of calibration/validation periods were evaluated using the Spearman rank correlation coefficients (Table 6), which evaluates how well the relative ranking of the indices between the catchments is captured, and the model efficiencies (Table 7), which evaluate how well the exact values were predicted. Typically, the values were similar for periods 1 and 2, when the parameterizations obtained by calibration for the respective period were used, resulting in a median difference of 0.015 for the Spearman Rank correlation and 0.0855 for NSE. In general, results are expected to be poorer for the validation period in comparison to the calibration period; however, for the respective validation periods the values were only slightly lower (median difference of −0.0215 (Spearman) and −0.029 (NSE)). This indicates that results were similar for the two periods and were similar when looking at the validation periods. The average median percent error for estimated streamflow characteristics was almost always less than zero, indicating that the objective functions used for model calibration typically underestimated each of the 12 streamflow characteristics being evaluated (Table 8).

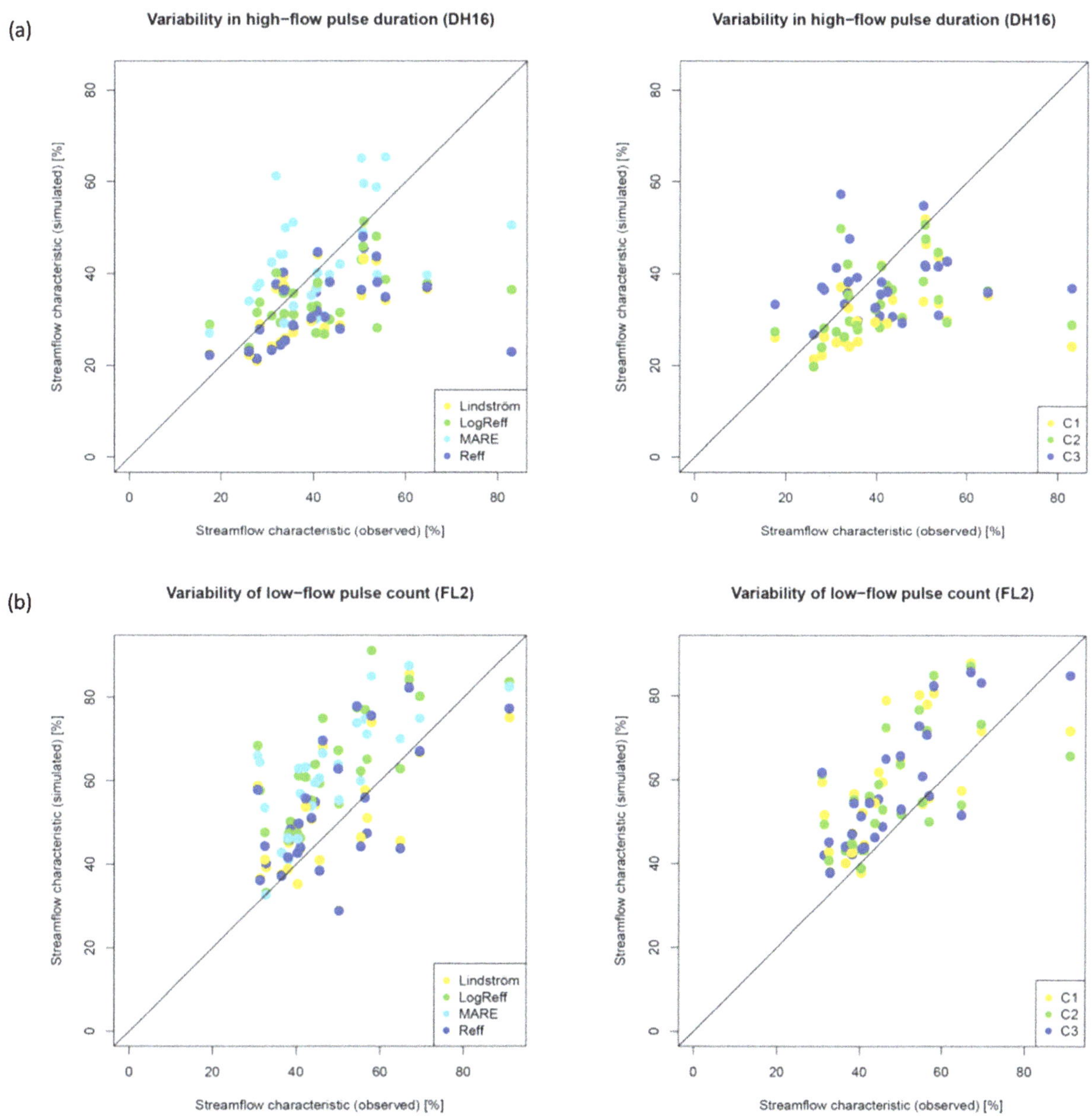

Figure 5. *Cont.*

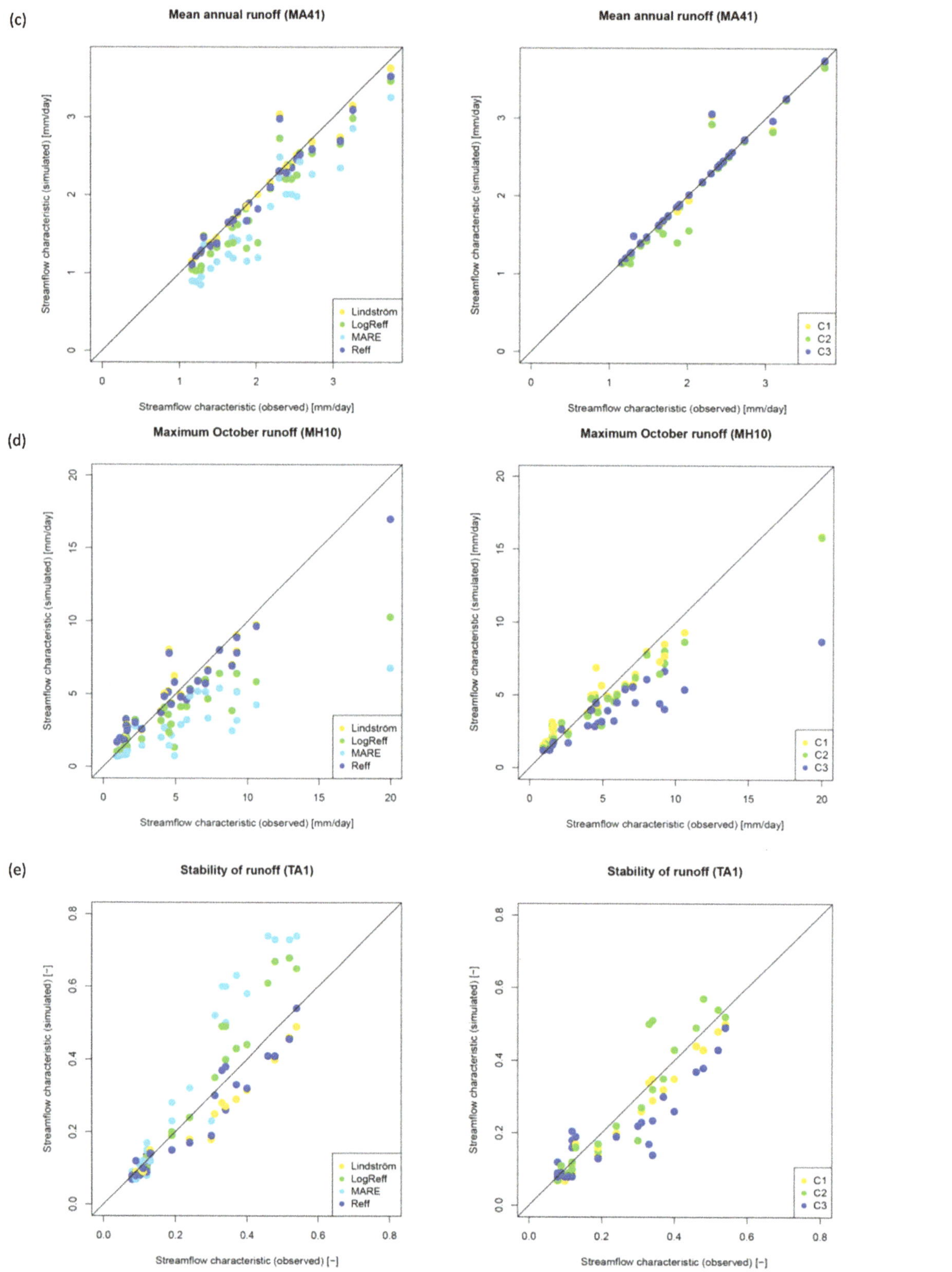

Figure 5. Scatterplots for the streamflow characteristics (**a**) DH16 (Variability in high-flow pulse duration); (**b**) FL2 (Variability of low-flow pulse count); (**c**) MA41 (Mean annual runoff); (**d**) MH10 (Maximum October runoff) and (**e**) TA1 (Stability of runoff) for calibration period 1. The points represent the median value of all 100 calibration trials in each catchment based on single criteria objective functions (**left column**) and multi-criteria objective functions (**right column**).

Table 6. Spearman rank correlation coefficients between objective functions (horizontal) and streamflow characteristics (vertical) based on observed respective simulated streamflow (for each group of four values: upper − left = calibration period 1 (Cal1), upper − right = validation period 2 (Val2), lower − left = validation period 1 (Val1), lower − right = calibration period 2 (Cal2)). Colors are ranging from white (for a Spearman rank correlation of 0) to dark green (for a Spearman rank correlation of 1).

	Reff		LogReff		Lindström		MARE		C1		C2		C3	
MA41	0.973	0.978	0.930	0.927	0.980	0.983	0.919	0.918	0.980	0.981	0.947	0.928	0.981	0.986
	0.957	0.991	0.929	0.947	0.961	0.998	0.926	0.950	0.961	1.000	0.952	0.979	0.962	1.000
MH10	0.930	0.831	0.874	0.853	0.916	0.837	0.834	0.829	0.941	0.837	0.958	0.874	0.918	0.898
	0.960	0.940	0.862	0.868	0.958	0.934	0.822	0.829	0.957	0.918	0.942	0.903	0.885	0.933
Flowperc	0.796	0.978	0.810	0.986	0.790	0.961	0.814	0.979	0.808	0.980	0.810	0.983	0.685	0.867
	0.778	0.985	0.808	0.996	0.781	0.980	0.804	0.996	0.803	0.995	0.806	0.996	0.683	0.897
RA7	0.736	0.724	0.877	0.885	0.726	0.735	0.888	0.896	0.870	0.873	0.851	0.892	0.696	0.797
	0.756	0.836	0.930	0.930	0.719	0.775	0.848	0.902	0.878	0.919	0.880	0.917	0.744	0.789
DH13	0.977	0.938	0.974	0.948	0.971	0.908	0.960	0.960	0.981	0.945	0.976	0.945	0.926	0.691
	0.955	0.866	0.976	0.937	0.955	0.877	0.964	0.957	0.971	0.910	0.978	0.885	0.871	0.573
TA1	0.972	0.929	0.968	0.943	0.977	0.906	0.947	0.974	0.968	0.884	0.960	0.899	0.875	0.766
	0.936	0.956	0.933	0.966	0.952	0.942	0.884	0.936	0.958	0.948	0.942	0.964	0.904	0.924
FH6	0.943	0.851	0.916	0.906	0.935	0.875	0.728	0.863	0.953	0.916	0.900	0.921	0.569	0.663
	0.926	0.888	0.853	0.931	0.931	0.898	0.634	0.855	0.942	0.930	0.901	0.919	0.498	0.613
FH7	0.948	0.933	0.881	0.889	0.949	0.935	0.810	0.887	0.967	0.945	0.965	0.952	0.688	0.563
	0.927	0.951	0.842	0.889	0.941	0.960	0.763	0.805	0.945	0.967	0.944	0.967	0.480	0.520
MA26	0.849	0.917	0.789	0.906	0.855	0.920	0.704	0.858	0.894	0.923	0.903	0.915	0.631	0.856
	0.752	0.932	0.699	0.894	0.782	0.935	0.672	0.829	0.821	0.933	0.831	0.928	0.381	0.769
DH16	0.534	0.645	0.443	0.662	0.503	0.673	0.402	0.471	0.510	0.745	0.525	0.683	0.145	0.482
	0.429	0.549	0.421	0.654	0.410	0.514	0.346	0.645	0.526	0.659	0.511	0.650	0.094	0.518
FL2	0.521	0.443	0.740	0.628	0.609	0.449	0.734	0.703	0.709	0.602	0.684	0.668	0.755	0.594
	0.548	0.617	0.659	0.604	0.579	0.659	0.641	0.626	0.672	0.711	0.620	0.695	0.616	0.628
TL1	0.477	0.394	0.643	0.520	0.471	0.347	0.612	0.753	0.603	0.330	0.531	0.428	0.574	0.418
	0.407	0.112	0.646	0.546	0.418	0.065	0.623	0.777	0.497	0.362	0.531	0.201	0.600	0.280

Table 7. Nash-Sutcliffe efficiencies between objective functions (horizontal) and streamflow characteristics (vertical) based on observed respective simulated streamflow (for each group of four values: upper − left = calibration period 1 (Cal1), upper − right = validation period 2 (Val2), lower − left = validation period 1 (Val1), lower − right = calibration period 2 (Cal2)). Colors are ranging from white (for Nash-Sutcliffe efficiencies of 0 or lower) to dark green (for a Nash-Sutcliffe efficiency of 1).

	Reff		LogReff		Lindström		MARE		C1		C2		C3	
MA41	0.917	0.936	0.840	0.881	0.936	0.933	0.584	0.626	0.946	0.939	0.922	0.927	0.949	0.930
	0.858	0.967	0.746	0.835	0.900	0.993	0.490	0.554	0.914	0.999	0.875	0.965	0.916	1.000
MH10	0.848	0.820	−0.627	0.570	0.841	0.796	−3.942	−1.220	0.820	0.871	0.796	0.879	−1.630	0.663
	0.859	0.934	−0.931	0.332	0.874	0.926	−5.692	−2.258	0.848	0.926	0.756	0.850	−1.367	0.667

Table 7. *Cont.*

	Reff		LogReff		Lindström		MARE		C1		C2		C3	
Flowperc	0.416	0.749	0.611	0.837	0.356	0.660	0.647	0.960	0.463	0.680	0.614	0.804	0.170	0.477
	0.484	0.868	0.569	0.967	0.491	0.820	0.465	0.966	0.538	0.848	0.591	0.939	0.373	0.669
RA7	0.209	0.281	0.071	0.193	0.279	0.370	−0.420	−0.284	−0.043	−0.063	−0.229	−0.197	−9.226	−7.224
	−0.628	−0.230	0.369	0.385	−0.608	−0.277	0.156	0.186	0.276	0.252	0.190	0.231	−5.173	−4.088
DH13	0.372	−0.164	0.884	0.472	−0.601	−1.895	0.910	0.858	0.797	0.522	0.770	0.874	−7.603	−20.044
	0.638	0.427	0.919	0.748	0.437	−0.030	0.814	0.914	0.902	0.813	0.672	0.817	−4.235	−14.891
TA1	0.898	0.432	0.856	0.882	0.829	0.108	0.672	0.803	0.918	0.477	0.886	0.749	0.502	-1.020
	0.863	0.912	0.718	0.845	0.892	0.926	0.548	0.685	0.881	0.974	0.839	0.953	0.806	0.705
FH6	0.709	0.628	−1.354	−0.967	0.660	0.559	−7.331	−4.461	0.513	0.502	0.210	0.282	−3.781	−5.629
	0.714	0.622	−0.788	−0.465	0.717	0.612	−4.768	−3.426	0.736	0.680	0.533	0.522	−2.536	−4.020
FH7	0.746	0.756	−0.440	−1.246	0.585	0.600	−0.752	−1.837	0.769	0.725	0.842	0.820	−13.413	−22.837
	0.813	0.826	0.290	−0.242	0.801	0.820	−0.260	−0.612	0.912	0.930	0.932	0.954	−9.425	−11.728
MA26	0.618	0.849	0.080	0.033	0.582	0.832	−0.418	−1.114	0.789	0.882	0.848	0.872	−4.116	−4.256
	0.331	0.862	0.184	0.320	0.324	0.886	0.178	−0.513	0.500	0.894	0.564	0.878	−1.898	−2.343
DH16	−3.044	−0.329	−3.375	0.050	−3.323	−0.307	−0.463	−0.371	−3.727	−0.006	−2.768	0.192	−3.474	−0.562
	−0.937	−0.182	−2.056	0.186	−1.012	−0.234	−1.025	0.006	−1.535	−0.092	−1.562	0.119	−2.785	−0.309
FL2	0.118	−1.176	−0.469	−1.557	0.201	−0.931	−0.556	−1.448	−0.266	−0.827	−0.167	−1.773	0.139	−0.948
	−0.040	−1.198	−0.530	−1.841	0.056	−1.123	−0.759	−1.703	−0.203	−0.409	−0.132	−1.246	-0.104	−1.018
TL1	−0.376	−4.676	−0.211	−3.016	−0.310	−5.502	−0.361	−2.672	−0.017	−4.483	−0.196	−4.053	−0.023	−2.708
	−0.505	−4.322	−0.250	−3.892	−0.518	−4.338	−0.557	−2.218	−0.400	−4.503	−0.489	−5.932	0.021	−3.529

Table 8. Median percent error for streamflow characteristics by model objective function for calibration period 1 (Cal1).

Objective Function	MA41	MH10	RA7	TA1	DH13	FH7	FH6	FL2	MA26	DH16	TL1	E85	Average Median Error (Percent)
Lindström	−0.6	−1.8	−25.0	−15.2	−18.1	−23.0	−12.0	16.8	9.1	−20.8	3.7	19.1	−5.6
LogReff	−9.5	−20.0	−50.0	7.7	−9.5	−37.5	−27.0	26.9	−7.3	−10.0	4.8	15.2	−9.7
MARE	−18.9	−44.0	−57.1	25.0	−7.4	−44.4	−41.4	28.2	−19.6	9.9	5.5	−7.3	−14.3
Reff	−2.5	−2.1	−18.2	−10.8	−14.7	−20.0	−12.0	17.5	9.8	−20.2	4.2	9.8	−4.9
C1	0.0	−4.8	−50.0	−7.7	−13.1	−19.0	−14.1	28.6	4.9	−19.7	3.4	29.9	−5.1
C2	−0.8	−10.6	−42.9	0.0	−7.5	−14.0	−18.2	17.7	2.2	−16.4	4.0	13.2	−6.1
C3	0.0	−24.5	−44.4	−18.9	−18.9	−69.3	−37.6	23.6	−28.1	−12.5	3.4	24.1	−16.9
Average Median Percent Error	−4.6	−15.4	−41.1	−2.8	−12.7	−32.5	−23.2	22.8	−4.2	−12.8	4.1	14.9	–

4. Discussion

In the absence of observed data, environmental flow studies necessarily rely on some form of streamflow estimation to model the response of aquatic ecology to alteration of the streamflow regime. Knight *et al.* [23] and Murphy *et al.* [8] raised the question of validity and began evaluation of model accuracies for predicting known ecologically-relevant streamflow characteristics. Murphy *et al.* [8] and Shrestha *et al.* [9] highlight that typical calibration approaches, often focused on daily, monthly, or annual mean values, are inadequate when predicting more subtle aspects of the flow regime. An increasing body of work is making use of statistical modeling approaches to address hydrologic and hydro-ecological questions [5,7,43–45]. However, as already stated by Murphy *et al.* [8] and Shrestha *et al.* [9], runoff models have advantages as well as limitations, particularly in regard to developing streamflow time series reflecting land cover, human population, or climatic projections. As such, runoff models should be closely evaluated to better understand if the calibration approaches and predictive accuracies yield results amenable to their end use.

While the HBV-light model was used in this study, there is little reason to assume that results would be discernibly different if another calibrated runoff model were used. Partly this reflects the fact that most mechanistic runoff models are fundamentally similar in concept and application, using more or less the same or similar routines. Fundamentally, if calibration is used, the simulated series are fitted to the observed series according to some objective function, and regardless of the specific model being used, this fit does not ensure agreement in all possible aspects of the hydrograph shape.

The accuracy of prediction and appropriateness of calibration is important in the context of environmental flow application as error of predicting flow-regime components will be translated and probably amplified as error in estimating ecological response. A given approach to model calibration will lead to accurate prediction of the runoff with regard to the used objective function measure, however accurate prediction of other aspects may be lacking. For example, Knight *et al.* [41] (Figure 2) published linear functions representing the 80th quantile upper-bound relationship of specialized insectivore scores to three streamflow characteristics (TA1, FH6, and RA7; see Table 5 for definitions). Following Murphy *et al.* [8], we use these relations to evaluate the accuracy of streamflow characteristic predictions as well as predicted ecological response based on the seven calibration approaches discussed herein for a single model (catchment 03488000). Using the equations from Knight *et al.* [41] and simulated streamflow presented in this paper, values of insectivore scores varied from 0.49 to 0.87 for RA7, 0.53 to 0.8 for TA1, and 0.58 to 0.84 for FH6 (Table 9; Figure 6). While median percent difference error for estimated specialized insectivore score for RA7 was a modest 8.2 percent under the estimate using observed data, individual departures from the observed values ranged from −19.7 to 42.6 percent for RA7, −13.1 to 31.1 percent for TA1, and −10.8 to 29.2 percent for FH6. Model results in this example are similar to those for a regional regression model reported by Murphy *et al.* [8] (9 percent difference for streamflow characteristic and 16 percent over estimation for insectivore score using HBV-light. Results presented here are considerably different than those for a rainfall-runoff model example from Murphy *et al.* [8], showing 90 percent overestimated for the same ecological score.

The objective functions used for model calibration resulted overall in an underprediction of the 12 streamflow characteristics being evaluated (Table 8). The general underprediction of the flow characteristics is a result similar to that seen in Murphy *et al.* [8] where a TOPMODEL application calibrated on mean annual flow was evaluated in the context of predicting the same streamflow characteristics. The median errors presented here are within plus-or-minus 30 percent of observed values, proposed by Kennard *et al.* [46] as an acceptable band of uncertainty, for 8 to 12 streamflow characteristics (out of 12) depending on the objective function (Figure 7, Table 8). This is in stark contrast to the rainfall runoff model evaluated in Murphy *et al.* [8]) where 13 of 19 streamflow characteristics were outside this band. While similar patterns are seen in overall model results, the calibration approaches evaluated in this paper appear to have provided more accurate estimates across the flow regime as defined by these characteristics. These results can be attributed both to the use of 100 parameter sets, which resulted in more robust flow characteristic estimations, and the use of different objective functions. Parameter uncertainty was substantial for many streamflow characteristics depending on which objective function was used. Despite this, high model efficiencies could still be achieved in many cases when using the median of 100 calibration trials as a more robust prediction for streamflow characteristics.

Table 9. Comparison of selected streamflow characteristics based on simulated and observed streamflow time series for a single model location (site 13 (03488000)) and calibration period 1 (Cal1). (TA1, RA7, and FH6, defined in Table 5; values in parentheses represent the specialized insectivore score using the associated streamflow characteristic value based on linear equations presented in Knight *et al.* [41], Figure 2; hydro, percent error for streamflow characteristic derived from simulated and observed streamflow time series; eco, percent error for specialized insectivore score based on streamflow characteristic derived from simulated and observed streamflow time series).

Objective Function (see Table 3 for Definitions)	RA7		Percent Error	TA1		Percent Error	FH6		Percent Error
	Simulated	Observed	Hydro/Eco	Simulated	Observed	Hydro/Eco	Simulated	Observed	Hydro/Eco
Lindström	0.14 (0.49)		27.3/−19.7	0.4 (0.55)		−16.7/−9.8	13 (0.59)		13.4/−9.2
LogReff	0.1 (0.66)		−9.1/8.2	0.67 (0.75)		39.6/23	10.08 (0.7)		−12/7.7
MARE	0.06 (0.83)		−45.5/36.1	0.73 (0.8)		52.1/31.1	6.62 (0.84)		−42.2/29.2
Reff	0.125 (0.55)		13.6/−9.8	0.41 (0.56)		−14.6/−8.2	13.38 (0.58)		16.8/−10.8
C1	0.12 (0.57)	0.11 (0.61)	9.1/−6.6	0.43 (0.57)	0.48 (0.61)	−10.4/−6.6	12.92 (0.59)	11.46 (0.65)	12.7/−9.2
C2	0.09 (0.7)		−18.2/14.8	0.57 (0.68)		18.8/11.5	12.38 (0.62)		8/−4.6
C3	0.05 (0.87)		−54.5/42.6	0.38 (0.53)		−20.8/−13.1	6.54 (0.84)		−42.9/29.2

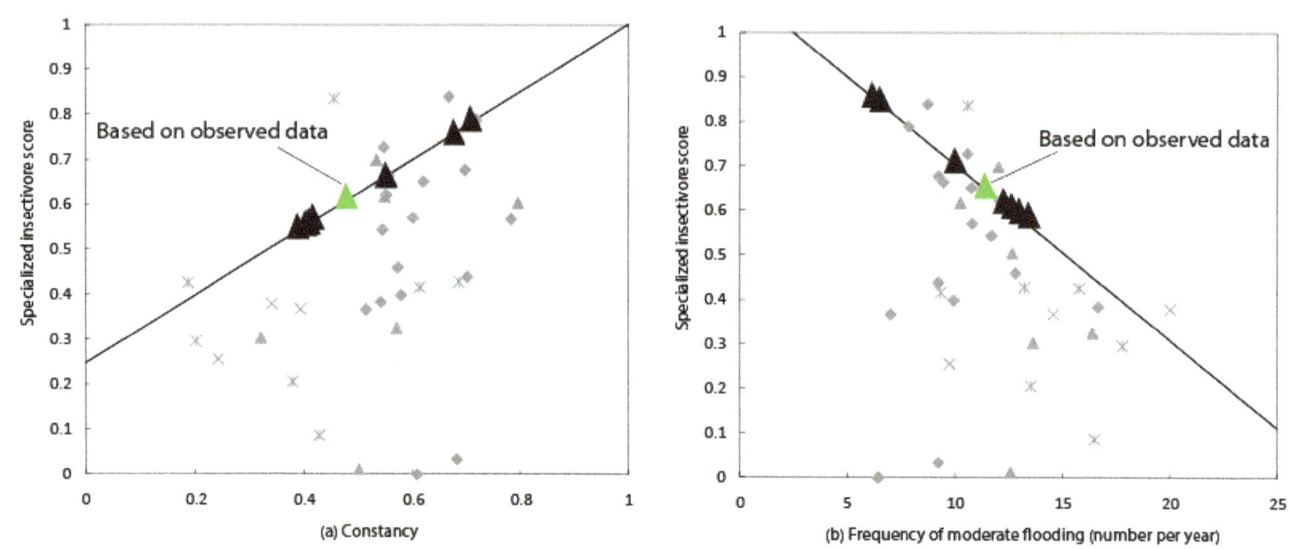

Figure 6. *Cont.*

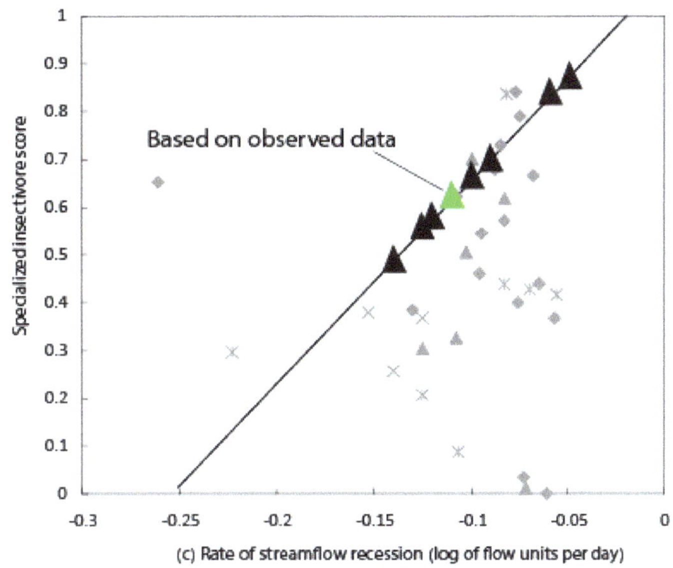

Figure 6. Example of an ecological flow application by comparison of estimated values for three streamflow characteristics for site 13 (03488000) (Table 1, Figure 1) and calibration period 1 (Cal1). (**a**) Constancy; (**b**) Frequency of moderate flooding (number per year) and (**c**) Rate of streamflow recession (log of flow units per day). Black triangles represent model estimated values based on the seven objective functions. Green triangle represents streamflow characteristics based on observed data. Values for RA7 (Rate of streamflow recession) were multiplied by negative 1 to convert values to those in the original analysis. Thin black lines represent 80th percentile quantile regression lines based on the 33 data point (grayed) in the background used by Knight *et al.* [41]. (Figure modified from Knight *et al.* [41]).

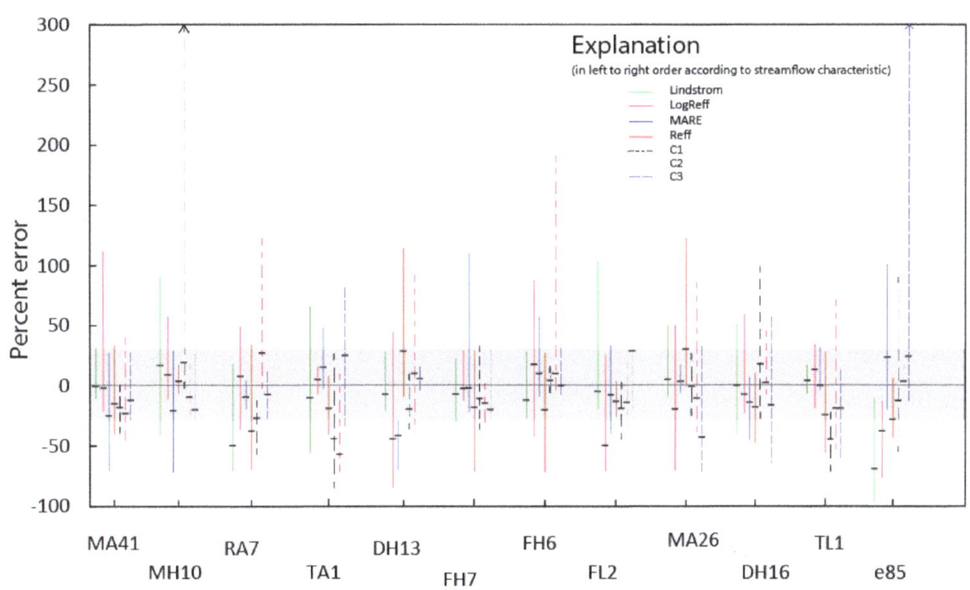

Figure 7. Minimum, maximum, and median percent errors according to objective function and streamflow characteristic for calibration period 1 (Cal1). Each vertical bar is based on the median error for the 27 catchments. The gray band in the center of the figure represents ±30 percent difference [46] Vertical bars with arrows indicate the maximum percent error exceeded the axis scale.

While the low average median percentage error would indicate a good performance with regard to the estimated flow characteristics, the scatter plots and computed Nash-Sutcliffe efficiencies and Spearman rank correlations reveal a slightly different picture. Spearman rank correlations were rather high for many of the objective functions and streamflow characteristics. For many of those objective function and flow characteristic combinations, however, Nash-Sutcliffe efficiencies were much lower. This shows that, although a clear bias might be observed in the predicted streamflow characteristic values, the order between the catchments was preserved quite well. In practice it might be more important to determine how well the flow characteristics are reproduced relative to the variation among catchments in the region than to determine the relative error value. When evaluating the scatter plots (Figure 5), low values of the Nash-Sutcliffe efficiencies indicated that the represented variability was relatively low, and the low Spearman rank correlations indicated that some flow characteristics that were not similar on a ranking scale were estimated correctly for the different catchments.

Considering individual streamflow characteristics, a pattern in predictive accuracy is evident. Most notably, streamflow characteristics that reflect average conditions (MA41, MA26, TA1, and TL1) were predicted quite well, with average median percent errors ranging from 2.8 to 4.6 percent absolute (Table 8). However, for some of these characteristics, especially TL1, the relative variation of the simulated values among the catchments were rather poor (Tables 6 and 7). Aspects of the hydrograph representative of high-flow conditions (MH10, FH7, FH6, DH13, DH16, and RA7) were underpredicted consistently (between 12.7 and 41.1 percent), with individual model calibrations underpredicting values up to 70 percent under observed. Low-flow characteristics were overpredicted (FL2 and E85) by 22.8 and 14.9 percent respectively. This appears to indicate that the model, regardless of calibration, may be retaining water during high-flow periods and allowing it to release during low-flow periods. The considerable underprediction of RA7 (rate of streamflow recession) indicates that higher flow events receded at a slower rate, which is suggestive of water stored in groundwater, and subsequently abundant groundwater discharge. The underprediction of RA7 and overprediction of low-flow characteristics are complementary.

MA41 (mean annual runoff) was predicted extremely well, particularly when using those calibrations where the objective function included the volume error as criterion, which is expected as this criterion is equivalent to the mean annual runoff. Predictions of MA41 also performed quite well when calibrated using the Nash-Sutcliffe efficiency. This performance might be attributed to the sensitivity of the Nash-Sutcliffe efficiency for high flows, which could reduce the error in the estimation of mean annual runoff. As noted by Murphy *et al.* [8], inclusion of ecological flow characteristics as criteria in calibrations may yield better simulations.

5. Conclusions

The accuracy of simulated runoff resulting from seven objective functions was evaluated in this paper by comparing streamflow characteristics based on observed and predicted streamflow time series. While the ultimate goal is to produce the most accurate simulated streamflow time series at ungauged catchments based on the transfer of calibrated parameter sets from gauged to ungauged catchments, the comparison in this study addresses an important part of the total uncertainty, namely the uncertainty related to the prediction accuracy specific streamflow characteristics that were not part of the calibration routine. The primary conclusion is that good model performance in terms of objective functions, such as the frequently used Nash-Sutcliffe model efficiency, does not ensure that all flow characteristics computed from these simulations will correspond to those derived from observed runoff. This is an important consideration that is often overlooked by users of model output who use simulated time series for various analyses, supporting resource allocation decisions, or establishing flow policy. While expecting simulated runoff series to agree with the observed in all possible aspects is unreasonable, this analysis serves as a further reminder of the substantial errors possible, using ecological flow characteristics as the example.

Two novel approaches were used in this study. First, we evaluated the effectiveness of seven objective functions for simulating streamflow time series and subsequent streamflow characteristic calculations. This allowed for critical examination of the importance of the objective function choice, as results differed substantially among objective functions. Results indicate there was no single best calibration strategy, but not surprisingly, different strategies provided better predictions for different streamflow characteristics. However, there was some indication that the combined objective functions, which evaluate the runoff simulations in different aspects, might be generally more suitable across a range of flow characteristics. Second, parameter uncertainty was explicitly considered by using the combination of 100 different equally possible parameter sets for each calibration trial instead of the typical single optimal calibrated parameter set. Our results confirmed the value of this approach by showing that different parameter sets can be similar with respect to the objective function used (similarity between the Nash-Sutcliffe for example) but differ greatly with respect to other characteristics. We demonstrated that using only one parameter set could result in substantial uncertainties, which can be reduced by using the values based on several parameter sets as more robust estimation.

More research is needed to determine which objective functions are most useful to ensure acceptable simulations of ecological flow characteristics, or other regime-defining characteristics. One suitable approach beyond the objective functions used in this paper might be to include streamflow characteristics of particular interest as objective functions in the calibration. This corresponds to the suggestion to include various hydrological signatures as diagnostic tools [47]. The fact that simulation-based flow characteristics varied largely depending upon which objective functions were used indicates that there is a considerable potential to improve model calibrations by considering specific flow characteristics when evaluating model performance during calibration. While it can be expected that performances improve when a certain streamflow characteristic is explicitly included in the objective function, it is less clear which criteria should be included to ensure acceptable simulations for calculation of streamflow characteristics in general. Further research is therefore motivated to explore which criteria to include in the objective function to obtain streamflow simulations that preserve as many streamflow characteristics as possible.

Acknowledgments

This paper is a product of discussions and activities that took place at the U.S. Geological Survey John Wesley Powell Center for Analysis and Synthesis as part of the workgroup focusing on Water Availability for Ungauged Rivers (https://powellcenter.usgs.gov/). Funding for this research was provided by the Tennessee Wildlife Resources Agency, the National Park Service, the U.S. Geological Survey Cooperative Water Program, and the University of Zurich. The use of trade, firm, or product names is for descriptive purposes only and does not imply endorsement by the U.S. Government.

Author Contributions

Rodney Knight and Jan Seibert conceived the initial ideas for this study; Marc Vis performed the simulations; Jan Seibert, Marc Vis, Sandra Pool and Rodney Knight analyzed the results; all authors contributed to writing of the paper.

Conflicts of Interest

The authors declare no conflict of interest.

References

1. Rantz, S.E. *Stream Hydrology Related to the Optimum Discharge for King Salmon Spawning in the Northern California Coast Ranges*; U.S. Geological Survey Water-Supply Paper 1779-AA: Washington, DC, USA, 1964; p. 15.
2. Tennant, D.L. Instream Flow Regimens for Fish, Wildlife, Recreation and Related Environmental Resources. *Fisheries* **1976**, *1*, 6–10.
3. Olden, J.D.; Poff, N.L. Redundancy and the choice of hydrologic indices for characterizing streamflow regimes. *River Res. Appl.* **2003**, *19*, 101–121.
4. Poff, N.L.; Richter, B.D.; Arthington, A.H.; Bunn, S.E.; Naiman, R.J.; Kendy, E.; Acreman, M.; Apse, C.; Bledsoe, B.P.; Freeman, M.C.; *et al.* The ecological limits of hydrologic alteration (ELOHA): A new framework for developing regional environmental flow standards. *Freshw. Biol.* **2010**, *55*, 147–170.
5. Carlisle, D.M.; Wolock, D.M.; Meador, M.R. Alteration of streamflow magnitudes and potential ecological consequences: A multiregional assessment. *Front. Ecol. Environ.* **2011**, *9*, 264–270.
6. Knight, R.R.; Murphy, J.C.; Wolfe, W.J.; Saylor, C.F.; Wales, A.K. Ecological limit functions relating fish community response to hydrologic departures of the ecological flow regime in the Tennessee River basin, United States. *Ecohydrology* **2014**, *7*, 1262–1280.
7. Sanborn, S.C.; Bledsoe, B.P. Predicting streamflow regime metrics for ungauged streamsin Colorado, Washington, and Oregon. *J. Hydrol.* **2006**, *325*, 241–261.
8. Murphy, J.C.; Knight, R.R.; Wolfe, W.J.; Gain, W.S. Predicting Ecological Flow Regime at Ungaged Sites: A Comparison of Methods. *River Res. Appl.* **2013**, *29*, 660–669.
9. Shrestha, R.R.; Peters, D.L.; Schnorbus, M.A. Evaluating the ability of a hydrologic model to replicate hydro-ecologically relevant indicators. *Hydrol. Process.* **2014**, *28*, 4294–4310.
10. Hrachowitz, M.; Savenije, H.H.G.; Blöschl, G.; McDonnell, J.J.; Sivapalan, M.; Pomeroy, J.W.; Arheimer, B.; Blume, T.; Clark, M.P.; Ehret, U.; *et al.* A decade of Predictions in Ungauged Basins (PUB)—A review. *Hydrol. Sci. J.* **2013**, *58*, 1198–1255.
11. Clausen, B.; Biggs, B. Relationships between benthic biota and hydrological indices in New Zealand streams. *Freshw. Biol.* **1997**, *38*, 327–342.
12. Clausen, B.; Biggs, B.J. Flow variables for ecological studies in temperate streams: Groupings based on covariance. *J. Hydrol.* **2000**, *237*, 184–197.
13. Poff, N.L.; Ward, J.V. Implications of Streamflow Variability and Predictability for Lotic Community Structure: A Regional Analysis of Streamflow Patterns. *Can. J. Fish. Aquat. Sci.* **1989**, *46*, 1805–1818.
14. Puckridge, J.T.; Walker, K.F.; Costelloe, J.F. Hydrological persistence and the ecology of dryland rivers. *Regul. Rivers Res. Manag.* **2000**, *16*, 385–402.

15. Bergström, S. *Development and Application of a Conceptual Runoff Model for Scandinavian Catchments*; SMHI: Norrköping, Sweden, 1976; No. RHO 7, p. 134.
16. Bergström, S. *The HBV Model: Its Structure and Applications*; SMHI Hydrology: Norrköping, Sweden, 1992; p. 35.
17. Lindström, G.; Johansson, B.; Persson, M.; Gardelin, M.; Bergström, S. Development and test of the distributed HBV-96 hydrological model. *J. Hydrol.* **1997**, *201*, 272–288.
18. Seibert, J.; Vis, M.J.P. Teaching hydrological modeling with a user-friendly catchment-runoff-model software package. *Hydrol. Earth Syst. Sci.* **2012**, *16*, 3315–3325.
19. Singh, V.P. *Computer Models of Watershed Hydrology*; Water Resources Publications: Highlands Ranch, CO, USA, 1995.
20. Omernik, J.M. Ecoregions of the Conterminous United States. *Ann. Assoc. Am. Geogr.* **1987**, *77*, 118–125.
21. Wolfe, W.; Haugh, C.; Webbers, A.; Diehl, T. Preliminary Conceptual Models of the Occurrence, Fate, and Transport of Chlorinated Solvents in Karst Regions of Tennessee. Department of Interior, US Geological Survey, Water Resources Investigations Report 97-4097. Available online: http://pubs.usgs.gov/wri/wri974097/new4097.pdf (accessed on 3 April 2015).
22. Hoos, A.B. Recharge Rates and Aquifer Hydraulic Characteristics for Selected Drainage Basins in Middle and East Tennessee. Department of the Interior, US Geological Survey, Water Resources Investigations Report 90-4015, 34. Available online: http://pubs.water.usgs.gov/wri904015/ (accessed on 24 June 2010).
23. Knight, R.R.; Gain, W.S.; Wolfe, W.J. Modelling ecological flow regime: An example from the Tennessee and Cumberland River basins. *Ecohydrology* **2012**, *5*, 613–627.
24. Law, G.S.; Tasker, G.D.; Ladd, D.E. Streamflow-Characteristic Estimation Methods for Unregulated Streams of Tennessee. US Geological Survey, Scientific Investigations Report 2009–5159, 212. p, 1 Plate. Available online: http://pubs.usgs.gov/sir/2009/5159/ (accessed on 16 June 2010).
25. U.S. Department of Commerce Climatography of the United States No. 85 Divisional Normals and Standard Deviations of Temperature, Precipitation, and Heating and Cooling Degree Days 1971–2000 (And Previous Normals Periods) Section 1: Temperature. United States Department of Commerce, Washington, DC, USA, 2007.
26. U.S. Department of Commerce Climatography of the United States No. 85 Divisional Normals and Standard Deviations of Temperature, Precipitation, and Heating and Cooling Degree Days 1971–2000 (And Previous Normals Periods) Section 2: Precipitation. United States Department of Commerce, Washington, DC, USA, 2007.
27. Moody, D.W.; Chase, E.B.; Aronson, D.A. *National Water Summary 1985—Hydrologic Events and Surface-Water Resources*; United States Geological Survey Water-Supply Paper 2300: Chapter on Tennessee Surface-Water Resources; United States Geological Survey, Reston, VA, USA, 1986; pp. 425–429.
28. Abell, R.A.; Olson, D.M.; Dinerstein, E.; Hurley, P.T.; Diggs, J.T.; Eichbaum, W.; Walters, S.; Wettengel, W.; Allnutt, T.; Loucks, C.J.; *et al. Freshwater Ecoregions of North America: A Conservation Assessment.*; Island Press: Washington, DC, USA, 2000.
29. Olson, D.M.; Dinerstein, E. The Global 200: A Representation Approach to Conserving the Earth's Most Biologically Valuable Ecoregions. *Conserv. Biol.* **1998**, *12*, 502–515.
30. Etnier, D.A.; Starnes, W.C. *The fishes of Tennessee*; The University of Tennessee Press: Knoxville, TN, USA, 1993.

31. Master, L.L.; Flack, S.R.; Stein, B.A.; Conservancy, N. *Rivers of Life: Critical Watersheds for Protecting Freshwater Biodiversity*; Nature Conservancy: Arlington, VA, USA, 1998.
32. Thornton, P.E.; Thornton, M.M.; Mayer, B.W.; Wilhelmi, N.; Wei, Y.; Devarakonda, R.; Cook, R.B. *Daymet: Daily Surface Weather Data on a 1-km Grid for North America, Version 2*; Oak Ridge National Laboratory: Oak Ridge, TN, USA, 2014.
33. Monteith, J.L. Evaporation and Environment. In *The State and Movement of Water in Living Organism 19th Symposia of the Society Experimental Biology*; University Press: Cambridge, UK, 1965; pp. 205–234.
34. Walter, I.; Allen, R.; Elliott, R.; Itenfisu, D.; Brown, P.; Jensen, M.; Mecham, B.; Howell, T.; Snyder, R.; Eching, S.; *et al.* The ASCE Standardized Reference Evapotranspiration Equation. PREPARED BY Task Committee on Standardization of Reference Evapotranspiration of the Environmental and Water Resources Institute. Available online: http://kimberly.uidaho.edu/water/asceewri/ascestzdetmain2005.pdf (accessed on 3 April 2015).
35. Rotstayn, L.D.; Roderick, M.L.; Farquhar, G.D. A simple pan-evaporation model for analysis of climate simulations: Evaluation over Australia. *Geophys. Res. Lett.* **2006**, *33*, L17715.
36. Hobbins, M.; Wood, A.; Streubel, D.; Werner, K. What Drives the Variability of Evaporative Demand across the Conterminous United States? *J. Hydrometeorol.* **2012**, *13*, 1195–1214.
37. Rango, A.; Martinec, J. Revisiting the degree-day method for snowmelt computations. *J. Am. Water Resour. Assoc.* **1995**, *31*, 657–669.
38. Beven, K.; Freer, J. Equifinality, data assimilation, and uncertainty estimation in mechanistic modelling of complex environmental systems using the GLUE methodology. *J. Hydrol.* **2001**, *249*, 11–29.
39. Seibert, J. Multi-Criteria calibration of a conceptual runoff model using a genetic algorithm. *Hydrol. Earth Syst. Sci.* **2000**, *4*, 215–224.
40. Seibert, J. Regionalisation of parameters for a conceptual rainfall-runoff model. *Agric. For. Meteorol.* **1999**, *98–99*, 279–293.
41. Knight, R.R.; Brian Gregory, M.; Wales, A.K. Relating streamflow characteristics to specialized insectivores in the Tennessee River Valley: A regional approach. *Ecohydrology* **2008**, *1*, 394–407.
42. Thompson, J.; Archfield, S. *The EflowStats R package Introduction to EflowStats*; United States Geological Survey: Reston, VA, USA, 2014.
43. Castellarin, A.; Camorani, G.; Brath, A. Predicting annual and long-term flow-duration curves in ungauged basins. *Adv. Water Resour.* **2007**, *30*, 937–953.
44. McManamay, R.A. Quantifying and generalizing hydrologic responses to dam regulation using a statistical modeling approach. *J. Hydrol.* **2014**, *519*, 1278–1296.
45. Zhu, Y.; Day, R.L. Regression modeling of streamflow, baseflow, and runoff using geographic information systems. *J. Environ. Manag.* **2009**, *90*, 946–953.
46. Kennard, M.J.; Mackay, S.J.; Pusey, B.J.; Olden, J.D.; Marsh, N. Quantifying uncertainty in estimation of hydrologic metrics for ecohydrological studies. *River Res. Appl.* **2010**, *26*, 137–156.
47. Gupta, H.V.; Wagener, T.; Liu, Y. Reconciling theory with observations: Elements of a diagnostic approach to model evaluation. *Hydrol. Process.* **2008**, *22*, 3802–3813.

6

Effect of COD:SO_4^{2-} Ratio, HRT and Linoleic Acid Concentration on Mesophilic Sulfate Reduction: Reactor Performance and Microbial Population Dynamics

Chungman Moon [1,†], Rajesh Singh [1,‡], Sathyanarayan S. Veeravalli [1], Saravanan R. Shanmugam [1], Subba Rao Chaganti [2], Jerald A. Lalman [1,*] and Daniel D. Heath [2,3]

[1] Department of Civil and Environmental Engineering, University of Windsor, 401 Sunset Ave., Windsor, ON N9B 3P4, Canada; E-Mails: chungman.moon@gmail.com (C.M.); rajesh.singh@cug.ac.in (R.S.); sevilim@uwindsor.ca (S.S.V.); ramiah@uwindsor.ca (S.R.S.)

[2] Great Lakes Institute for Environmental Research, University of Windsor, 401 Sunset Ave., Windsor, ON N9B 3P4, Canada; E-Mails: chaganti@uwindsor.ca (S.R.C.); dheath@uwindsor.ca (D.D.H.)

[3] Department of Biological Sciences, University of Windsor, 401 Sunset Ave., Windsor, ON N9B 3P4, Canada

† Current address: Korea Institute of Energy Research, 152 Gajeong-ro, Yuseong-gu, Daejeon 305343, Korea.

‡ Current address: Central University of Gujarat, School of Environment and Sustainable Development, Sector-30, Gandhinagar 382030, India.

* Author to whom correspondence should be addressed; E-Mail: lalman@uwindsor.ca;

Academic Editor: Say-Leong Ong

Abstract: Biological sulfate (SO_4^{2-}) reduction was examined in anaerobic sequential batch reactors (ASBRs) operated under different hydraulic retention times (HRTs) ranging from 12 to 36 h and COD (Chemical Oxygen Demand)/SO_4^{2-} ratios of 2.4, 1.6 and 0.8. Competition between SO_4^{2-} reducing bacteria (SRBs), methane producing archaea (MPAs) and homoacetogens (HACs) was examined in controls and cultures treated with linoleic acid (LA). The ASBR performance was influenced by the COD/SO_4^{2-} ratio in control cultures with a SO_4^{2-} reduction of 87% at a COD/SO_4^{2-} ratio of 0.8. At a 12 h HRT, in both control and LA treated cultures, greater than 75% SO_4^{2-} removal was observed under all the conditions examined. In control reactors operating at a 36 h HRT, high levels of MPAs belonging to *Methanobacteriales* and *Methanosarcinales* were detected; however, in comparison, under low COD/SO_4^{2-} ratio and with decreasing HRT conditions, a relative increase in SRBs belonging to *Desulfovibrio* and *Desulfatibacillum* was observed. Adding 0.5 g·L^{-1} LA suppressed *Methanobacteriales,* while increasing the LA concentration to 1 g·L^{-1} completely suppressed MPAs with a relative increase in SRBs. HACs belonging to Bacteroidetes were observed in the control and in cultures operated at 12 h HRT with a COD/SO_4^{2-} ratio of 1.6 and fed 0.5 g·L^{-1} LA; however, with all other LA levels (0.5 and 1.0 g·L^{-1}) and HRTs (12, 24 and 36 h), HACs were not detected.

Keywords: sulfate reduction; sulfate reducing bacteria; methanogens; anaerobic sequencing batch reactor; COD/SO_4^{2-} ratio

1. Introduction

Sulfate (SO_4^{2-}), an abundant anion in the environment, is discharged in effluents from various industrial sectors including edible oil processors, tannery operations, food processors and pulp and paper mills. Typically these effluents contain chemical oxygen demand (COD) and SO_4^{2-} concentrations ranging from 0.5 to 50 g·L^{-1} and 0.3 to 7 g·L^{-1}, respectively [1,2]. In another sector such as mining, the management of sulfide ores is important as oxidation of ores exposed to precipitation results in the production of acid mine drainage which can cause severe environmental damage when discharged into receiving water bodies [1,3].

Many studies have employed up-flow anaerobic sludge blanket reactors (UASBRs) and continuous stirred tank reactors (CSTRs) to treat effluents containing high COD and SO_4^{2-} concentrations [4–6]. However, using anaerobic sequential batch reactors (ASBRs) is advantageous because the reactor configuration allows for treatment in a single tank without the need for a final clarifier. In addition, the important features of ASBRs include a lower cost reactor configuration combined with higher organic removal efficiency.

Advantages of employing mixed anaerobic cultures when compared to pure cultures include the ability of sourcing cultures from natural ecosystems and engineered bioreactors, ability to operate under non-sterile conditions, capability of adapting to fluctuating the operational conditions and the potential ability to utilize a variety of substrates. According to Colleran *et al.* [7] and Muyzer and Stams [8] using mixed anaerobic cultures is associated with a major challenge of enriching SO_4^{2-} reducing bacteria (SRBs) which compete with methane producing archaea (MPAs) and homoacetogens (HACs) for substrates such as H_2 and acetate. Evidence by Schonheit *et al.* [9] has shown that SRBs have a greater affinity for acetate when compared to MPAs. In comparison, studies by Isa *et al.* [10] using anaerobic filters continuously operating at a high organic loading concluded that 15% of the acetate added was utilized for SO_4^{2-} reduction while as much as 85% was converted into methane. Both the SRBs and MPAs are able to utilize acetate as the major source of organic carbon and energy source under anaerobic conditions [8]. A reaction sequence showing the microbial degradation of organic matter in an anoxic environment in the presence and absence of SO_4^{2-} is shown in Figure 1. Also shown are H_2 and acetate consumption reactions by SRBs, MPAs and HACs.

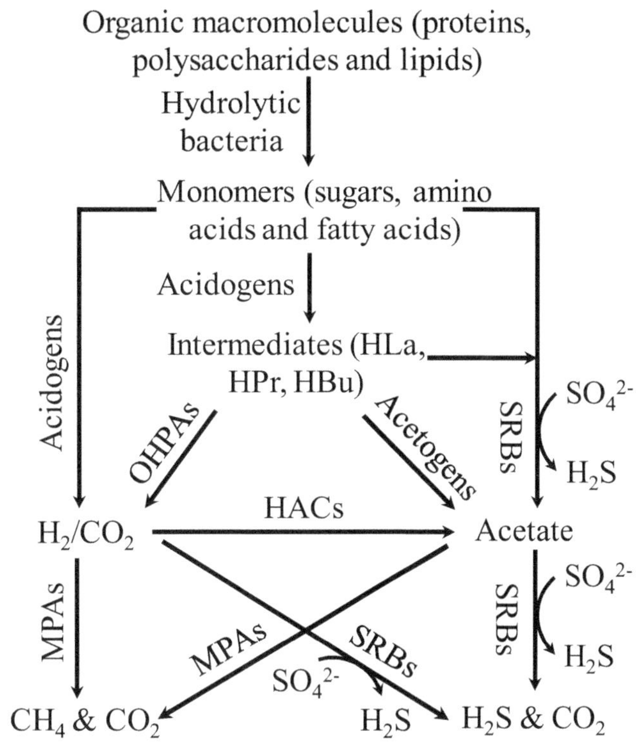

Figure 1. Pathways of organic compound degradation under methanogenic and sulphidogenic conditions (Colleran *et al.* [7]). Notes: VFAs = volatile fatty acids; H_2S = hydrogen sulfide; CH_4 = methane; HPr = propionic acid; HLa = lactic acid; HBu = butyric acid; HAC = homoacetogenic bacteria; MPA = methane-producing archaea; SRB = sulfate-reducing bacteria; OHPA = obligate hydrogen-producing acetogen.

Bacterial abundance and activity are affected by factors such as the presence of terminal electron acceptors [11]. For example, in the presence of HCO_3^-, MPAs are dominant over SRBs. At high SO_4^{2-} concentrations, SRBs belonging to *Desulfobacter* sp. and *Desulfobacca acetoxidans* are able to out-compete MPAs such as *Methanosaeta* sp. and *Methanosarcina* sp. for acetate [6,9]. Dar *et al.* [12] reported that in upflow anaerobic sludge blanket reactors (UASBRs) the detection of *Desulfovibrio-Desulfomicrobium* group indicated the presence of a large diversity of SRBs. In addition, they reported detecting members of other phylogenetic SRB groups, *i.e.*, *Desulfotomaculum, Desulfobulbus*, and *Desulfococcus-Desulfobacca-Desulfosarcina.* In comparison to work reported by Raskin *et al.* [13], *Desulfobacterium* and *Desulfobacter* were not detected by Dar *et al.* [12]. This variation in the SRB populations is likely due to the different electron donors used by these researchers.

Factors affecting competition between the co-existence of SRBs, MPAs and HACs include pH, temperature, substrate and reactor type, COD to SO_4^{2-} ratio (COD/SO_4^{2-} ratio), hydraulic retention time (HRT) as well as physical structure of microbial cultures [5,14,15]. White and Gadd [5] claimed that interaction between the COD/SO_4^{2-} ratio and HRT controlled SO_4^{2-} reduction and COD/SO_4^{2-} ratio 1.0 to 3.0 was preferred for SRBs. In comparison, Dar *et al.* [16] reported that at a low COD/SO_4^{2-} ratio of 0.34, SRBs out-competed MPAs and to some extent HACs. These authors reported that at a limiting SO_4^{2-} concentration (high COD/SO_4^{2-} ratio), the relative abundance of MPAs and HACs was approximately 45% of the total microbial consortia. These results indicate that further investigation on COD to SO_4^{2-} ratio is necessary to clearly understand the competition between MPAs, HACs and SRBs.

Competition between different microbial species can also be controlled by altering operational factors such as pH, temperature and adding methanogenic inhibitors. Sipma *et al.* [14] reported that both pH and temperature can impact MPAs more in comparison to HRT. Chaiprapat *et al.* [17] observed that for an ASBR operating at neutral pH, increasing SO_4^{2-} removal was observed in comparison to under low pH conditions. In the neutral pH range, SRBs and MPAs are active but suppressed under low pH conditions [18]. Inhibiting MPAs by utilizing chemical inhibitors could be of great significance in reducing methanogenesis by diverting electron fluxes to SRBs and subsequently improve SO_4^{2-} removal. Inhibiting methanogens with long chain fatty acids (LCFAs) have gained some attention over other inhibitory chemicals [19] because they are environmentally friendly, relatively abundant and biodegradable [20].

The presence of recalcitrant compounds, heavy metals, SO_4^{2-} and total dissolved solids can affect the anaerobic microbial dynamics and hence, the process efficiency. Sulfate-reducing anaerobic bioreactors have been treated as 'black boxes' without a thorough understanding of the microorganisms involved in SO_4^{2-} reduction [21]. The operation of these bioreactors is highly dependent on microbial activities and a better understanding of the role of microbial communities in these systems will assist in improving their design and performance [22]. Hence, further work is required to understand methanogenesis and SO_4^{2-} reduction from a more fundamental perspective. Hence, the objectives of this study were as follows: 1. To investigate the effect of SO_4^{2-} reduction at varying levels of COD/SO_4^{2-} ratio, HRT and LA (methanogenic inhibitor) concentration. 2. To investigate the microbial population dynamics at varying levels of these experimental variables. An integrated approach to characterize the microflora using terminal restriction fragment length polymorphism (T-RFLP) together with chemical analysis of the fermentation byproducts was used to analyze trends between the microbial community structure and changes in the operational parameters.

2. Materials and Methods

2.1. Inoculum Source

The anaerobic inoculum was procured from an UASBR located at a brewery wastewater treatment facility (Guelph, ON, Canada) (A) and at the municipal wastewater treatment plant (Chatham, ON, Canada) (B). Culture A and culture B were selected based on sources of MPAs and SRBs, respectively. The volatile suspended solid (VSS) of the culture A and B was 50 and 20 g $VSS{\cdot}L^{-1}$ respectively. The cultures (A and B) were diluted with basal medium to 25 and 12 g $VSS{\cdot}L^{-1}$ in 10 L reactors, respectively (designated as reactor A and B). The bioreactors were operated in accordance to procedures reported by Ray *et al.* [23]. The cultures in reactors A and B were maintained at 37 °C and at pH 7.0 ± 0.5 in sequencing batch mode with a 14 d (days) HRT and a feed concentration of 2000 mg $glucose{\cdot}L^{-1}$. In addition to glucose, reactor B was acclimated incrementally to increasing SO_4^{2-} levels from 250 to 2000 $mg{\cdot}L^{-1}$ over 2 months. During the acclimation period, the quantity of gas and VFAs were monitored to establish quasi-steady state conditions. Inoculum for the experiments under consideration was combined from reactors A (80%) and B (20%) and diluted with basal medium to 8 g VSS·L^{-1}. The basal medium composition used for dilution and feed was adapted from Wiegant and Lettinga [24]. All the chemicals (99% purity) for preparing the basal medium were procured from ACP Chemicals Inc. (Montreal, Quebec, Canada) and Sigma Aldrich, (Oakville, Ontario, Canada). Glucose (99% purity) and LA (99% purity) were procured from Spectrum Chemicals (Gardena, CA, USA) and TCI America (Portland, OR, USA).

2.2. Sulfate Reduction Studies

Two 7 L (total volume) reactors (New Brunswick Scientific, New Brunswick, NJ, USA) with a 5 L working volume were used to conduct the experiments. The reactors (R1 and R2) were operated as ASBRs at 37 ± 1 °C. Liquid samples were collected at the end of each cycle. Continuous mixing of the reactor contents during the reaction phase was conducted at 200 rpm using a stirring plate mixer. The pH (6.5 ± 0.1) was maintained using 1 M NaOH and 0.5 M HCl. The ASBRs (R1 and R2) were seeded with the inoculum from reactors A and B (8 g VSS·L^{-1}) and then purged with nitrogen (N_2) (99.99% purity, Praxair, Windsor, ON, Canada) for 5 min to maintain anaerobic conditions. A three factor three level Taguchi design was used for conducting the experimental run (Table 1) as described by Singh *et al.* [25]. The factors investigated in this study include LA concentration, COD/SO_4^{2-} ratio and HRT. The reactors (R1 and R2) were operated under the same conditions with a feed concentration of 2000 mg glucose·L^{-1} (2.134 g COD·L^{-1}) as a carbon source and SO_4^{2-} concentration varied according to the COD/SO_4^{2-} ratios shown in Table 1. The reactors were operated as follows: 40 min settling; 10 min decanting and 10 min fill. The reaction times maintained were 5, 11 and 17 h for HRT values of 12, 24 and 36 h, respectively. The volume decanted per cycle was constant at 2.5 L and the HRT was calculated using Equation (1):

$$\text{HRT} = \frac{\text{(Working volume of the reactor)}}{\text{(Volume decanted per cycle) (No. of cycles per day)}} \tag{1}$$

The reactors under each HRT were operated until they achieved quasi-steady state condition (constant SO_4^{2-} reduction with ±10% variation). Different LA levels (0, 0.5 and 1.0 g·L^{-1}) were fed to cultures according to experimental conditions shown in Table 1. During the inhibition studies, the cultures were incubated with LA for 24 h prior to initiating the experiment (adding SO_4^{2-} and glucose).

Table 1. Experimental outline to elucidate the effect of operational parameters on sulfate removal in ASBRs.

Experiment #s	LA Concentration (g·L^{-1})	COD/SO_4^{2-} Ratio	HRT (h)	Initial SO_4^{2-} Concentration (g·L^{-1})
1		0.8	12	2.67
2	0	1.6	36	1.34
3		2.4	24	0.89
4		0.8	24	2.67
5	0.5	1.6	12	1.34
6		2.4	36	0.89

Table 1. *Cont.*

Experiment #s	LA Concentration ($g{\cdot}L^{-1}$)	COD/SO_4^{2-} Ratio	HRT (h)	Initial SO_4^{2-} Concentration ($g{\cdot}L^{-1}$)
7		0.8	36	2.67
8	1.0	1.6	24	1.34
9		2.4	12	0.89

Notes: The pH and influent substrate concentration were 6.5 ± 0.1 and 2 $g{\cdot}glucose{\cdot}L^{-1}$, respectively; COD = Chemical oxygen demand; LA = linoleic acid; SO_4^{2-} = sulfate and ASBRs = anaerobic sequential batch reactors; The experiment design is based on the Taguchi model described by Singh *et al.* [25]; Cultures fed 0 $g{\cdot}L^{-1}$ LA is referred as control cultures; All the experiments were conducted in duplicate reactors designated as R1 and R2.

2.3. Analytical Methods

Biogas production was monitored using a tipping bucket gas meter [26] and the composition of biogas was quantified using a gas chromatograph [19]. 25 μL of each gas was injected and the detection limits for CH_4 and H_2 were 0.0032 kPa [0.5 mL/bottle (160 mL)] and H_2S was 0.0315 kPa [5 mL/bottle (160 mL)], respectively.

The liquid samples collected at the end of each cycle were analyzed for VFAs, dissolved sulfur compounds (SO_4^{2-} and sulfide) and alcohols. The sulfur compounds were analyzed according to methods described by Moon *et al.* [19] using an ion chromatograph (IC). The detection limits for the sulfide and SO_4^{2-} were 2.0 and 0.5 $mg{\cdot}L^{-1}$, respectively.

The IC and high performance liquid chromatograph (HPLC) methods used to analyze alcohol and VFAs in the effluent liquid samples were conducted using the methods described by Chowdhury *et al.* [27] and Moon *et al.* [19], respectively. The detection limit for alcohols (ethanol, *i*-propanol, *n*-propanol, *n*-butanol, and *i*-butanol) and VFAs (lactate, acetate, propionate, formate and butyrate) were 5 and 2 $mg{\cdot}L^{-1}$, respectively. The total suspended solids (TSS) and VSS were measured according to *Standard Methods* [28].

2.4. Microbial Methods

The microbial community diversity in the mixed culture subjected to different operational conditions was determined using nested polymerase chain reaction (PCR) of the 16S rRNA gene followed by terminal restriction fragment length polymorphism (T-RFLP) analysis. The cultures samples collected at the end of the experiment were used for microbial community analysis. Details for DNA isolation, PCR amplification and T-RFLP methods were previously reported by Chaganti *et al.* [29]. The data obtained from the T-RFLP analysis comprised the peaks reflecting the size of terminal restriction fragments (T-RFs) in base pairs (bp) together with the area of each peak measured in fluorescence units. For the T-RFs generated by the digestion of PCR-amplified 16S rRNA genes from culture samples in the current study, a phylogenetic assignment was performed using a modified database generated for T-RFs which were previously described by Chaganti *et al.* [29].

The relative abundances of the terminal restriction fragments (T-RFs) was used to detect phyla of samples observed under different operational conditions based on the taxonomy annotation using the modified database generated from Microbial Community Analysis (MiCA) plus experimentally determined TRFs for microorganisms identified from the 16S rRNA gene clone library analysis.

3. Results and Discussion

3.1. Comparison of Sulfate Removal and the Degradation Byproducts at Different Operating Conditions

The SO_4^{2-} removal based on an influent SO_4^{2-} concentration range of 0.89 to 2.67 g·L^{-1} varied from approximately 56% to 94% (Figure 2). The mean SO_4^{2-} removal showed a small improvement when 0.5 g·L^{-1} of LA was added; however, with a further increase in LA concentration to 1 g·L^{-1}, the SO_4^{2-} removal increased from 69% to 83%. In comparison, according to Moon *et al.* [19], increasing the LA concentration from 0.5 to 1.5 g·L^{-1} in batch reactors showed no significant effect on SO_4^{2-} reduction. Moon *et al.* [19] observed that adding LA improved SO_4^{2-} reduction by ≥15% in comparison to the control cultures operating at a similar COD/SO_4^{2-} ratio and pH.

In the current study, varying the COD/SO_4^{2-} ratio had less of an effect on the percent SO_4^{2-} removal while decreasing the HRT from 36 to 12 h increased the SO_4^{2-} removal efficiency from approximately 68% to 84% [25]. Studies using a lactate feed revealed that the HRT and the COD/SO_4^{2-} significantly influenced the SO_4^{2-} reduction with increasing SO_4^{2-} removal detected with decreasing HRT (from 20 to 10 h) and with increasing the COD/SO_4^{2-} ratio from 1 to 3 [5].

Dissolved sulfide (DS) was the major liquid degradation by-product from SO_4^{2-} reduction. In control cultures, the DS concentration varied between 100 and 250 mg·L^{-1} at elevated COD/SO_4^{2-} ratios ranging from 1.6 to 2.4 (Figure 2a). In comparison, for cultures fed 0.5 g·L^{-1} LA and operating at COD/SO_4^{2-} ratios of 1.6 to 2.4, the sulfide concentrations ranged from 100 to 400 mg·L^{-1} (Figure 2b). Increasing DS concentration ranging from 450 to 650 mg·L^{-1} were detected in control cultures at a low COD/SO_4^{2-} ratio of 0.8 while sulfide concentrations ranging from 250 to 400 mg·L^{-1} were detected in cultures fed 0.5 g·L^{-1} LA (Figure 2a,b). Alvarez *et al.* [30] reported a maximum sulfide concentration (300 to 500 mg·L^{-1}) could be obtained with maximum SO_4^{2-} removal (50% to 62%) with a COD/SO_4^{2-} ratio ranging from 0.7 to 1.5 for packed bed bench-scale biofilm reactors operating at a 100 h HRT, 20 °C and a pH at 7.5.

A maximum SO_4^{2-} removal efficiency of 90% was observed at a 36 h HRT and with a COD/SO_4^{2-} ratio of 0.8 in cultures fed 1 g·L^{-1} LA. Under these conditions, the sulfide concentration observed was approximately 550 mg·L^{-1} (Figure 2c). For a feed containing, 2500 mg·L^{-1} of SO_4^{2-}, Chaiprapat *et al.* [17] reported an elevated sulfide levels reaching 929 mg·L^{-1} in ASBRs fed a rubber skim wastewater and operating at a 10 d HRT with a pH at 7.0. Chaiprapat *et al.* [17] attributed the high sulfide concentrations at high COD/SO_4^{2-} ratios to a high rate of sulfidogenesis. In the current study, high DS concentrations reaching 641 mg·L^{-1} were observed under low COD/SO_4^{2-} ratio conditions with a feed SO_4^{2-} concentration of approximately 2670 mg·L^{-1} and a 12 h HRT (Figure 2). Similarly, Dar *et al.* [16] reported sulfide levels of approximately 250 mg·L^{-1} in reactors operating at a 50 h HRT, a high COD/SO_4^{2-} ratio of 1.94 and cultures fed lactate containing wastewater.

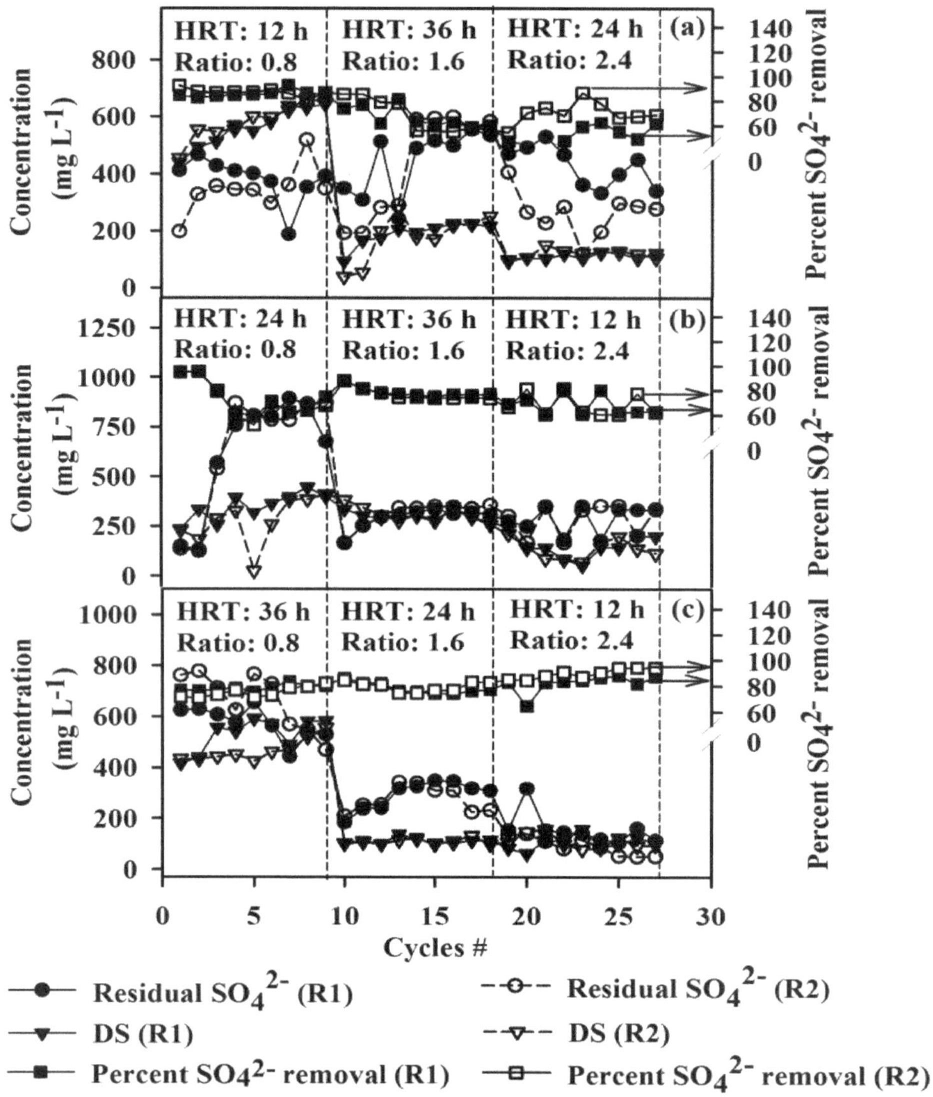

Figure 2. Effect of HRT and COD/SO_4^{2-} ratio on SO_4^{2-} reduction and sulfide formation with cultures fed **(a)** 0 $g{\cdot}L^{-1}$ LA; **(b)** 0.5 $g{\cdot}L^{-1}$ LA; **(c)** 1.0 $g{\cdot}L^{-1}$ LA. Notes: 1. HRT = hydraulic retention time; Ratio = COD/SO_4^{2-} ratio; SO_4^{2-} = sulfate and DS = dissolved sulfide; 2. Initial SO_4^{2-} concentration with respect to COD/SO_4^{2-} ratios of 0.8, 1.6 and 2.4 are 2667, 1333 and 889 $mg{\cdot}L^{-1}$ of sulfate, respectively; 3. Reactor 1 (R1) and Reactor 2 (R2) were operated in sequential batch mode under the conditions in Table 1.

3.2. Effect of COD/SO_4^{2-} Ratio and HRT on Biogas Production Using LA Treated and Untreated Cultures

Increasing CH_4 yield ($mol{\cdot}mol^{-1}$ glucose) with increasing COD/SO_4^{2-} ratios revealed that decreasing the SO_4^{2-} concentration showed increasing methanogenic activity. In the case of control cultures, the CH_4 yield ($mol{\cdot}mol^{-1}$ glucose) increased from approximately 0.09 to 0.5 when the COD/SO_4^{2-} ratio increased from 0.8 to 2.4 (Figure 3a). Sarti and Zaiat [31] reported decreasing CH_4 yields with decreasing COD/SO_4^{2-} ratio (3.5 to 2.5) for an ASBR operating with a 48 h HRT using butanol as a carbon source. In cultures fed 0.5 $g{\cdot}L^{-1}$ LA, increasing CH_4 yield (from 0.16 to 0.95) was detected in the reactor designated as R1 and from 0.46 to 1.1 in R2 when the COD/SO_4^{2-} ratio was increased from 0.8 to 1.6 (Figure 3b). In contrast, for cultures receiving 1 $g{\cdot}L^{-1}$ LA, increasing CH_4

yield was initially observed with increasing COD/SO_4^{2-} ratio from 0.8 to 1.6; however, the CH_4 yield decreased to approximately 0.07 mol·mol^{-1} glucose, when the COD to SO_4^{2-} ratio was increased to 2.4 (Figure 3c).

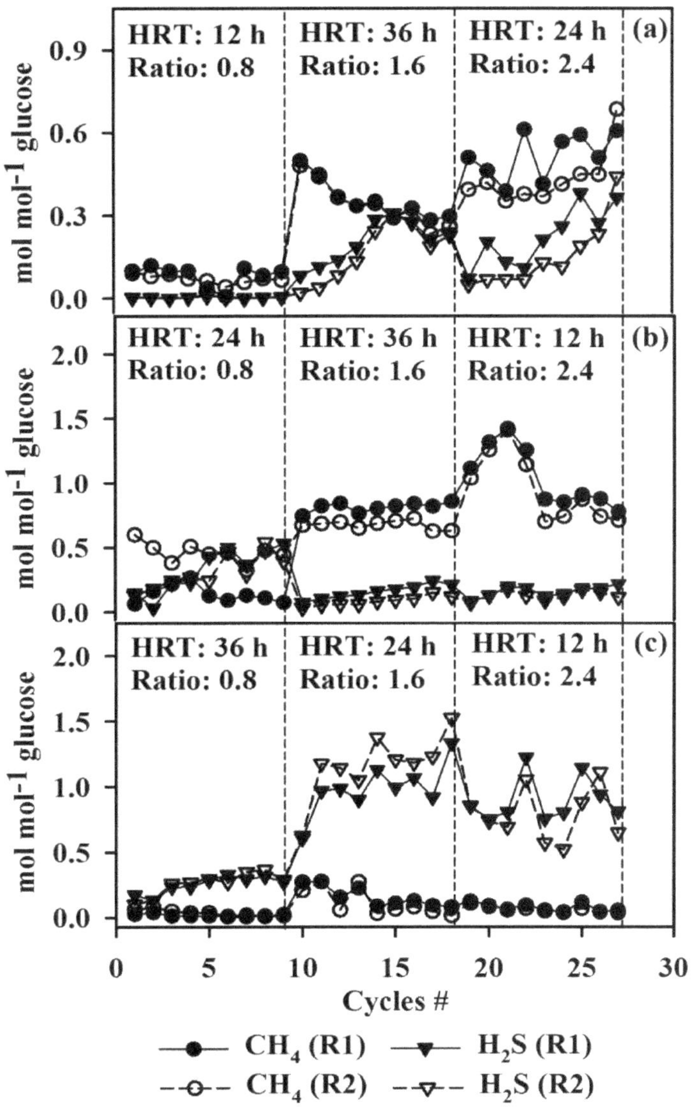

Figure 3. Effect of HRT and COD/SO_4^{2-} ratio on CH_4 and H_2S production with cultures fed (**a**) 0 g·L^{-1} LA (**b**) 0.5 g·L^{-1} LA (**c**) 1.0 g·L^{-1} LA. Notes: 1. HRT = hydraulic retention time; Ratio = COD/SO_4^{2-} ratio; CH_4 = methane; LA = linoleic acid and H_2S = hydrogen sulfide; 2. Hydrogen detected under the conditions examined is ≤0.03 mol·mol^{-1} glucose; 3. Reactor 1 (R1) and Reactor 2 (R2) were operated under the same conditions shown in Table 1.

In both control and LA fed cultures, no clear trend was observed in CH_4 yield with varying HRT conditions. Isa *et al.* [10] reported decreasing the HRT from 10 to 0.5 d decreased CH_4 production and increased H_2S production coupled with high SO_4^{2-} reduction. These authors also observed increasing competition between SRBs and MPAs at high substrate concentrations (*i.e.*, high COD/SO_4^{2-} ratio) using high-rate anaerobic reactors operating with a feed containing a mixture of acetate and ethanol. At high substrate concentrations (5 g COD·L^{-1}), Isa *et al.* [10] observed that the quantity of electron equivalents diverted towards CH_4 production reached 89% and at 0.5 g COD·L^{-1}, the quantity diverted to CH_4 production was 66%.

The trend observed for the H_2S yield (mol·mol^{-1} glucose) was similar to CH_4 yield. Increasing the COD/SO_4^{2-} ratio resulted in an increase in the H_2S yield except for cultures fed 0.5 g·L^{-1} LA. In control cultures and cultures fed 1 g·L^{-1} LA and operating at 1.6 COD/SO_4^{2-} ratio, the H_2S yield increased to approximately 0.30 and 1.1 mol·mol^{-1} glucose, respectively (Figure 3a,c). This indicates that an increase in LA concentration results in more electron diversion to sulfide (H_2S) production coupled with SO_4^{2-} reduction in comparison to the CH_4 production. Wei *et al.* [32] reported for a mesophilic fluidized bed reactor operating with SO_4^{2-} concentrations ranging from 1.0 to 3.0 g·L^{-1}, the quantity of H_2S produced ranged from 150 to 370 mg·L^{-1}. In the current study, for an influent SO_4^{2-} concentrations ranging from 0.8 to 2.6 g·L^{-1}, the corresponding H_2S concentration ranged from 50 to 225 mg·L^{-1}. In comparison, Wei *et al.* [32] reported elevated H_2S levels reaching 370 mg·L^{-1} in jet-loop anaerobic fluidized bed reactors operating at a 6 h HRT, 3000 mg·L^{-1} SO_4^{2-} and a pH at 5.8.

3.3. Volatile Fatty Acid Production

Acetate was the major VFA observed under all the experimental conditions (Figure 4). In control cultures, the acetate concentration ranged from 190 to 900 mg·L^{-1} within the range of HRT conditions (Figure 4a). In the control cultures maintained at high COD/SO_4^{2-} ratios, propionate and formate were also observed with concentrations ranging from 80 to 300 mg·L^{-1} and 150 to 450 mg·L^{-1}, respectively. Cultures fed 0.5 and 1.0 g·L^{-1} LA had lower acetate levels ranging from 200 to 650 mg·L^{-1}. In comparison to the control cultures, a possible reason for the low acetate levels could be due to the inhibition of methanogenesis and hence, acetate utilization by SRBs in LA treated cultures (Figure 4b). According to Visser *et al.* [33], in a heat treated culture to inhibit methanogenesis, acetate rapid consumption by SRBs resulted in increased SO_4^{2-} reduction.

In studies conducted by Dar *et al.* [16], increasing the COD/SO_4^{2-} ratio from 0.34 to 20.9 led to decreasing acetate production while increasing levels of propionate was observed. In the current study, except for control cultures, increasing COD/SO_4^{2-} ratio caused the accumulation of acetate while no trend was observed for propionate. Propionate and formate levels ranging from 70 to 150 mg·L^{-1} at COD/SO_4^{2-} ratios $\leq$1.6 was observed in cultures fed 0.5 g·L^{-1} LA while at a high COD/SO_4^{2-} ratio of 2.4, the quantity of propionate and formate was negligible (Figure 4b). McCartney and Oleszkiewicz [34] reported that at high COD/SO_4^{2-} ratios (non-SO_4^{2-} reducing pathway), propionate and acetate were the major end products while at low COD/SO_4^{2-} ratios (SO_4^{2-} reducing pathway), acetate was the only end product. Supporting evidence in the current study showed low propionate and formate levels were observed at low COD/SO_4^{2-} ratio (0.8) in comparison to ratios at 1.6 and 2.4 (Supplementary Figure S1).

In cultures fed 1 g·L^{-1} LA and operating at a COD/SO_4^{2-} ratio of 0.8, both acetate and butyrate were dominant while butyrate was not detected at a COD/SO_4^{2-} ratio of 1.6 (Figure 4c). Lactate levels ranging between 200 to 350 mg·L^{-1} as well as acetate (450 to 600 mg·L^{-1}) and butyrate (50 to 70 mg·L^{-1}) were observed in cultures operating at a COD/SO_4^{2-} ratio of 2.4 and fed 1 g·L^{-1} LA (Figure 4c). Increasing SO_4^{2-} reduction at a COD/SO_4^{2-} ratio of 2.4 for cultures fed 1 g·L^{-1} LA is likely due to preferred lactate consumption by SRBs when compared to propionate and butyrate (Figure 2c) [15,34].

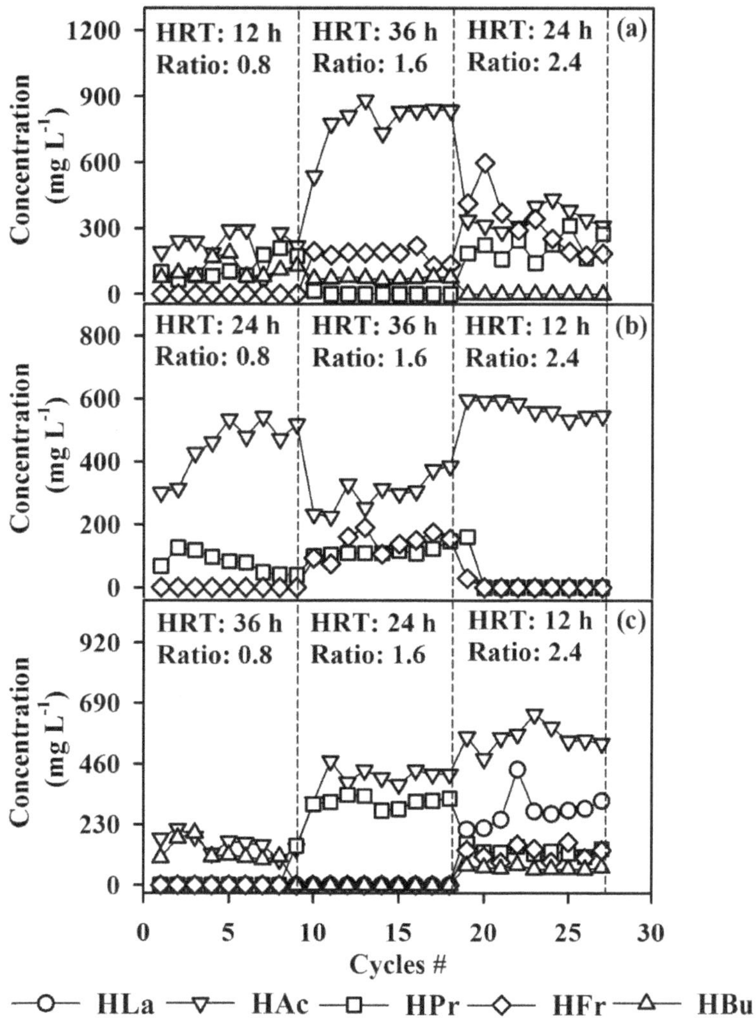

Figure 4. Effect of HRT and COD/SO_4^{2-} ratio on volatile fatty acids production with cultures fed (**a**) 0 g·L^{-1} LA; (**b**) 0.5 g·L^{-1} LA; (**c**) 1.0 g·L^{-1} LA. Notes: 1. HRT = hydraulic retention time; LA = linoleic acid; Ratio = COD/SO_4^{2-} ratio; HAc = acetic acid; HLa = lactic acid; HPr = propionic acid; HFr = formic acid; HBu = butyric acid; 2. Only the major volatile fatty acids produced under each condition are shown; 3. Data shown in the plot represent averages from reactors R1 and R2.

3.4. Competition and Coexistence of Sulfate-Reducing Bacteria, Homoacetogens and Methanogens

Microbial samples for quantifying various microorganisms were collected after completing the experiments. However, the discussion in this section is based on data for samples #1, #2, #4, #5, #6 and #9 because microbial amplification of the samples from experiments #3, #7 and #8 did not yield any quantification data after conducting the analyses twice.

Variation in the microbial community structure is based on the reactor operational conditions. The four major phyla associated with anaerobic SO_4^{2-} reduction in the ASBRs belonged to *Firmicutes*, *Proteobacteria*, *Euryarchaeota* and *Bacteroidetes*. The relative abundance of these phyla in samples obtained under different operating conditions is shown in Figure 5. In addition to the classified phylum, each sample containing T-RFs with a relative abundance of 22%–38% could not be assigned to known microbial phyla (shown as Uncultured in Figure 5).

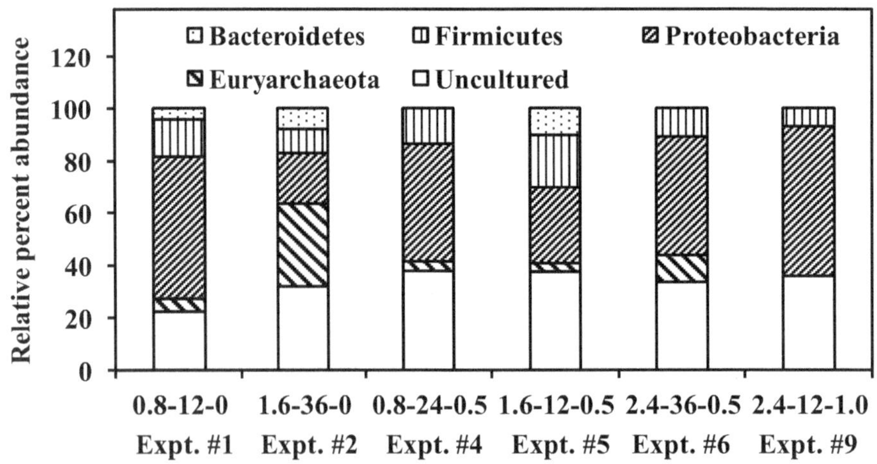

Figure 5. Microbial communities consisting of bacteria and Archaea in cultures operated under sulfate rich and limiting conditions at phylum level. The *x*-axis label describing the culture conditions is as follows: 1. COD/SO_4^{2-} ratio–HRT (h)–LA concentration (g L^{-1}); 2. The phyla with abundance greater than 1% in all samples are shown; 3. Sequences that could not be assigned at phylum level were marked as 'Uncultured'; 4. Refer Supplementary Table S1 for the organisms at genus level.

Species belonging to *Euryarchaeota* accounted for 32% and 5% in control cultures operating at 36 and 12 h HRT, respectively (Figure 5). The quantity of electron diverted to MPAs under these conditions corresponds to 4.8% and 1.4%, respectively (Table 2). The low CH_4 yields at high HRTs is likely due to the inactivity of HACs. In cultures operating at a 36 h HRT with a COD/SO_4^{2-} ratio of 1.6, HACs which belonging to *Bacteroidetes* accounted for approximately 8% of the population (Figure 5). Note acetate was the major byproduct observed under this condition (Expriment #2, Supplementary Figure S1). Similarly, work by Dar *et al.* [16] indicated that at high COD/SO_4^{2-} ratios and at a 50 h HRT, HACs were able to out-compete MPAs and SRBs at low SO_4^{2-} concentrations.

The quantity of *Proteobacteria* and *Firmicutes* in control cultures in ABSRs operating at 12 h HRT was approximately 55% and 14%, respectively (Figure 5). Members of this group consisted of species belonging to *Desulfovibrio*, *Desulfatibacillum* and *Desulfomicrobium* (Supplementary Table S1). Several studies have reported that species belonging to this group are able reduce SO_4^{2-} under high SO_4^{2-} concentrations (low COD/SO_4^{2-} ratios) [7,30,35]. Low HRTs and high SO_4^{2-} levels are major factors leading to the suppression of MPAs and HACs. Under these conditions, a relative increase in SRBs activity was associated with an 87% SO_4^{2-} reduction (Experiment #1, Table 2; Figure 2a and Supplementary Table S1).

In cultures fed 0.5 g·L^{-1} LA, *Euryarchaeota* was in the range from 3% to 10% of the species detected (Figure 5). Cultures operating at 36 h HRT reached a 10% maximum of the species belonging to *Methanosarcinales* [36] (Supplementary Table S1). The percent electrons diverted to MPAs accounted for approximately 13% under this condition (Experiment #6, Table 2). Results from the current study suggest that long HRTs and high COD/SO_4^{2-} ratios are favorable to MPAs when compared to SRBs at a 0.5 $g\cdot L^{-1}$ LA (Supplementary Figure S1 and Supplementary Table S1). Evidence by Isa *et al.* [10] suggest that at high COD/SO_4^{2-} ratios and long HRTs more electrons are diverted towards MPAs in comparison to that of SRBs. In addition, species belonging to methylotrophic methanogens and *Methanosarcinales* are able to out-compete SRBs at high acetate levels [7,37]. In the current study, high acetate levels were observed in the cultures fed 0.5 $g\cdot L^{-1}$ (Figure 4b and Supplementary Figure S1). In cultures fed 0.5 $g\cdot L^{-1}$ LA, *Methanococcoides methylutens* and *Methanosaeta concilii* belonging to methylotrophic methanogens and aceticlastic methanogens were detected (Supplementary Table S1) [36].

Table 2. Percent electron flow from substrate to SRBs and MPAs.

Experiment #s	LA Concentration ($g\cdot L^{-1}$)	COD/SO_4^{2-} Ratio	HRT (h)	COD Equivalent ($mg\cdot L^{-1}$) SO_4^{2-} Reduced	COD Equivalent ($mg\cdot L^{-1}$) Methane Produced	Percent Electron Flow to SRBs	Percent Electron Flow to MPAs
1		0.8	12	1281.4	28.7	60.0	1.4
2	0	1.6	36	543.6	101.7	25.4	4.8
3		2.4	24	351.8	194.7	16.4	9.1
4		0.8	24	971.3	91.9	45.4	4.3
5	0.5	1.6	12	663.8	265.4	31.0	12.4
6		2.4	36	394.8	288.8	18.5	13.5
7		0.8	36	1187.1	5.4	55.5	0.3
8	1.0	1.6	24	681.1	26.6	31.8	1.2
9		2.4	12	562.4	21.8	26.3	1.0

Notes: LA = linoleic acid; SO_4^{2-} = sulfate; SRB = sulfate reducing bacteria; MPA = methane producing archaea and HRT = hydraulic retention time; COD equivalent and percent electron flow values are average of duplicate reactors; COD equivalent of SO_4^{2-} reduced corresponds to the sulfide produced; 1 mol of sulfate reduced = 1 mol of H_2S produced = 2 mol of COD = 64 g of COD; 1 mol of CH_4 produced = 2 mol of COD = 64 g of COD; Electron flow to SRB = mol of SO_4^{2-} reduced × 64 g $COD\cdot mol^{-1}$ SO_4^{2-} = A g COD; Electron flow to MPA = mol of CH_4 produced × 64 $g\cdot COD\cdot mol^{-1}$ CH_4 = B g COD; Percent electron flow by SRB = [A/(COD influent)] × 100 and percent electron flow by MPA = [B/(COD influent)] × 100; Complete balance of COD is presented in supplementary Figure S1.

The amount of SO_4^{2-} reduction (63% to 75%) observed in cultures fed 0.5 $g\cdot L^{-1}$ LA could be due to the presence of SRBs belonging to *Proteobacteria* (29% to 45%) and *Firmicutes* (10% to 20%) (Figures 2b and 5). Under these conditions, *Deulfovibrio* sp. and *Desulfatibacillum* sp. were abundant (Supplementary Table S1). According to Sousa *et al.* [20] and Tan *et al.* [38], these microorganisms were detected in SO_4^{2-} rich or SO_4^{2-} limiting conditions and in the presence of electron donors such as fatty acids and alkanes. In cultures fed 0.5 $g\cdot L^{-1}$ LA, *Dethiosulfovibrio* sp. and *Clostridium* sp., organisms belonging to the *Firmicutes* phylum, were detected (Supplementary Table S1). The presence of *Clostridium* sp. and species belonging to *Bacteroidetes* is consistent with the findings from other studies, where these species were observed under SO_4^{2-} reducing conditions and capable of producing a mixture of VFAs from solid organic substrates and sugar beet molasses [39,40]. In the current study, acetate, propionate and formate which constituted approximately 30% to 40% of the initial COD fed to the reactors were observed in cultures fed 0.5 g·L^{-1} LA (Experiments #4–#6, Supplementary Figure S1).

In cultures fed 1 g·L^{-1} LA and operating at a 12 h HRT, the relative abundance of *Proteobacteria* and *Firmicutes* was 58% and 6%, respectively (Figure 5). The low HRT (12 h) in combination with 1 g·L^{-1} LA resulted in the suppression of MPAs belonging to *Euryarchaeota.* The SO_4^{2-} removal obtained under this condition corresponded to 90% with only 1% of the electrons diverted to MPAs (Figure 2c and Table 2, Expt. #9). Studies by Moon *et al.* [19] using batch reactors suggest that high LA concentrations inhibited methane production in the presence of SO_4^{2-}. Similarly, the suppression of MPAs was observed in SO_4^{2-} fed fluidized bed reactors operating at an HRT of approximately 16 h [41]. The results from this current study indicate that under low HRT conditions, SRBs were able to out-compete MPAs under SO_4^{2-} limiting conditions in the presence of 1 g·L^{-1} LA (Table 2 and Figure 5). Under SO_4^{2-} rich conditions, the SO_4^{2-} reduction efficiency was greater than 75% and the percent relative abundance of SRBs was approximately 50 (Figure 2 and Supplementary Table S1). In work related to this study, Weijma *et al.* [42] suggest that sulfidogenic and acetogenic cultures favors high SO_4^{2-} reduction under low HRT conditions by outcompeting MPAs. In addition to the SRBs belonging to *Desulfovibrio* sp. *Desulfomicrobium* sp. and *Desulfatibacillum* sp., *Geobacter sulfurreducens* was detected in cultures fed 1 g·L^{-1} LA and operating at a 12 h HRT.

Phylogenetically, *Syntrophus aciditrophicus,* along with SO_4^{2-} reducers and Geobacter species, are members of the delta subdivision of *Proteobacteria.* Levels of *S. aciditrophicus* ranging from 3% to 7.5% were observed under all the conditions examined. Studies by Jackson *et al.* [43] suggest that *S. aciditrophicus* were able to degrade fatty acids in syntrophic association with SRBs under SO_4^{2-} rich and limiting conditions. This relationship could enhance the SO_4^{2-} reduction in the process as short chain fatty acids such as propionate, butyrate could be degraded to acetate which is a more preferred carbon substrate for SRBs in comparison to propionate and butyrate [15]. Studies by Sousa *et al.* [20] also suggest that these syntrophic bacteria existed along with SRBs in LCFA degrading communities in the presence of SO_4^{2-}.

4. Conclusions

This study demonstrated that different levels of SO_4^{2-} reduction in ASBRs were observed under different operating conditions. A link between the microbial population dynamics and metabolic byproducts (from substrate and SO_4^{2-} reduction) to changes caused by HRT and COD/SO_4^{2-} ratio under control and LA treated conditions was established. Based on the factors under consideration in the study, the conclusions from this study are as follows:

1. Operating the ASBRs was suitable for SO_4^{2-} reduction with SO_4^{2-} removal efficiencies greater than 60%, under the conditions examined in this study.
2. ASBRs operating under low COD/SO_4^{2-} ratio (0.8) with low HRT (12 h) is preferred for SO_4^{2-} removal efficiencies greater than 75%.
3. When compared to the control cultures, in cultures fed 1 g·L^{-1} LA and at COD/SO_4^{2-} ratios 1.6 and 2.4, the SO_4^{2-} removal efficiencies improved by approximately 20% and 28%, respectively.
4. Dissolved sulfide and butyrate were associated with cultures fed with low COD/SO_4^{2-} ratios, while H_2S, acetate and propionate were linked with high COD/SO_4^{2-} ratios.

5. Methanogens belonging to *Methanobacteriales* and *Methanosarcinales* were abundant in control cultures operating at high HRTs while the addition of LA suppressed *Methanobacteriales* alone; however, with a reduction in HRT, washout of *Methanosarcinales* was observed.
6. In controls and cultures fed 0.5 $g \cdot L^{-1}$ LA, HACs belonging the *Bacteroidetes* phylum was abundant.
7. *Desulfovibrio* sp. and *Desulfatibacillum* sp., the major SRBs responsible for SO_4^{2-} reduction, were able to out-compete MPAs and HACs at low HRTs and low COD/SO_4^{2-} ratios.

Acknowledgments

Financial support for this work was provided by the Natural Sciences and Engineering Research Council of Canada (Grant No. 261797-2009), the Canada Research Chair program (Grant No. 950-203725) and the University of Windsor (Account No. 13320).

Author Contributions

The experimental work was conducted by Chungman Moon, Rajesh Singh, Sathyanarayanan S. Veeravalli and Saravanan R. Shanmugam. The manuscript was written by Sathyanarayanan S. Veeravalli and Jerald A. Lalman. Engineering data analysis was performed by Sathyanarayanan S. Veeravalli, Saravanan R. Shanmugam, Jerald A. Lalman. Microbial analysis was performed by Subba Rao Chaganti. Daniel D. Heath provided lab facility and funding for conducting the microbial analysis.

Conflicts of Interest

The authors declare no conflict of interest.

References

1. Lens, P.N.L.; Visser, A.; Janssen, A.J.H.; Pol, L.W.H.; Lettinga, G. Biotechnological treatment of sulfate-rich wastewaters. *Crit. Rev. Environ. Sci. Technol.* **1998**, *28*, 41–88.
2. Zitomer, D.H.; Shrout, J.D. High-sulfate, high-chemical oxygen demand wastewater treatment using aerated methanogenic fluidized beds. *Water Environ. Res.* **2000**, *72*, 90–97.
3. Johnson, D.B.; Hallberg, K.B. Acid mine drainage remediation options: A review. *Sci. Total Environ.* **2005**, *338*, 3–14.
4. Shayegan, J.; Ghavipanjeh, F.; Mirjafari, P. The effect of influent COD and upward flow velocity on the behaviour of sulphate-reducing bacteria. *Process Biochem.* **2005**, *40*, 2305–2310.
5. White, C.; Gadd, G.M. Mixed sulphate-reducing bacterial cultures for bioprecipitation of toxic metals: Factorial and response-surface analysis of the effects of dilution rate, sulphate and substrate concentration. *Microbiology* **1996**, *142*, 2197–2205.
6. Omil, F.; Lens, P.; Visser, A.; Pol, L.W.H.; Lettinga, G. Long-term competition between sulfate reducing and methanogenic bacteria in UASB reactors treating volatile fatty acids. *Biotechnol. Bioeng.* **1998**, *57*, 676–685.

7. Colleran, E.; Finnegan, S.; Lens, P. Anaerobic treatment of sulfate-containing waste streams. *Antonie Van Leeuwenhoek* **1995**, *67*, 29–46.
8. Muyzer, G.; Stams, A.J.M. The ecology and biotechnology of sulphate-reducing bacteria. *Nat. Rev. Microbiol.* **2008**, *6*, 441–454.
9. Schonheit, P.; Kristjansson, J.K.; Thauer, R.K. Kinetic mechanism for the ability of sulfate reducers to out-compete methanogens for acetate. *Arch. Microbiol.* **1982**, *132*, 285–288.
10. Isa, Z.; Grusenmeyer, S.; Verstraete, W. Sulfate reduction relative to methane production in high-rate anaerobic-digestion—Microbiological aspects. *Appl. Environ. Microbiol.* **1986**, *51*, 580–587.
11. Findlay, S.E.G.; Sinsabaugh, R.L.; Sobczak, W.V.; Hoostal, M. Metabolic and structural response of hyporheic microbial communities to variations in supply of dissolved organic matter. *Limnol. Oceanogr.* **2003**, *48*, 1608–1617.
12. Dar, S.A.; Yao, L.; van Dongen, U.; Kuenen, J.G.; Muyzer, G. Analysis of diversity and activity of sulfate-reducing bacterial communities in sulfidogenic bioreactors using 16S rRNA and dsrB genes as molecular markers. *Appl. Environ. Microbiol.* **2007**, *73*, 594–604.
13. Raskin, L.; Rittmann, B.E.; Stahl, D.A. Competition and coexistence of sulfate-reducing and methanogenic populations in anaerobic biofilms. *Appl. Environ. Microbiol.* **1996**, *62*, 3847–3857.
14. Sipma, J.; Lettinga, G.; Stams, A.J.M.; Lens, P.N.L. Hydrogenogenic CO conversion in a moderately thermophilic (55 °C) sulfate-fed gas lift reactor: Competition for CO-derived H_2. *Biotechnol. Progr.* **2006**, *22*, 1327–1334.
15. Liamleam, W.; Annachhatre, A.P. Electron donors for biological sulfate reduction. *Biotechnol. Adv.* **2007**, *25*, 452–463.
16. Dar, S.A.; Kleerebezem, R.; Stams, A.J.M.; Kuenen, J.G.; Muyzer, G. Competition and coexistence of sulfate-reducing bacteria, acetogens and methanogens in a lab-scale anaerobic bioreactor as affected by changing substrate to sulfate ratio. *Appl. Microbiol. Biotechnol.* **2008**, *78*, 1045–1055.
17. Chaiprapat, S.; Preechalertmit, P.; Boonsawang, P.; Karnchanawong, S. Sulfidogenesis in pretreatment of high-sulfate acidic wastewater using anaerobic sequencing batch reactor and upflow anaerobic sludge blanket reactor. *Environ. Eng. Sci.* **2011**, *28*, 597–604.
18. Gibson, G.R.; Cummings, J.H.; Macfarlane, G.T.; Allison, C.; Segal, I.; Vorster, H.H.; Walker, A.R.P. Alternative pathways for hydrogen disposal during fermentation in the human colon. *Gut* **1990**, *31*, 679–683.
19. Moon, C.; Singh, R.; Chaganti, S.R.; Lalman, J.A. Modeling sulfate removal by inhibited mesophilic mixed anaerobic communities using a statistical approach. *Water Res.* **2013**, *47*, 2341–2351.
20. Sousa, D.Z.; Alves, J.I.; Alves, M.M.; Smidt, H.; Stams, A.J.M. Effect of sulfate on methanogenic communities that degrade unsaturated and saturated long-chain fatty acids (LCFA). *Environ. Microbiol.* **2009**, *11*, 68–80.
21. Sheoran, A.S.; Sheoran, V.; Choudhary, R.P. Bioremediation of acid-rock drainage by sulphate-reducing prokaryotes: A review. *Miner. Eng.* **2010**, *23*, 1073–1100.
22. Kaksonen, A.H.; Plumb, J.J.; Franzmann, P.D.; Puhakka, J.A. Simple organic electron donors support diverse sulfate-reducing communities in fluidized-bed reactors treating acidic metal- and sulfate-containing wastewater. *FEMS Microbiol. Ecol.* **2004**, *47*, 279–289.

23. Ray, S.; Chowdhury, N.; Lalman, J.A.; Seth, R.; Biswas, N. Impact of initial pH and linoleic acid (C18:2) on hydrogen production by a mesophilic anaerobic mixed culture. *J. Environ. Eng. ASCE* **2008**, *134*, 110–117.
24. Wiegant, W.M.; Lettinga, G. Thermophilic anaerobic-digestion of sugars in upflow anaerobic sludge blanket reactors. *Biotechnol. Bioeng.* **1985**, *27*, 1603–1607.
25. Singh, R.; Moon, C.; Veeravalli, S.S.; Shanmugam, S.R.; Chaganti, S.R.; Lalman, J.A. Using a statistical model to examine the effect of COD: SO_4^{2-} Ratio, HRT and LA concentration on sulfate reduction in an anaerobic sequencing batch reactor. *Water* **2014**, *6*, 3478–3494.
26. Speece, R.E. Gas Flow Totalizer. US Patent 4064750 A, 27 December 1977.
27. Chowdhury, N.; Lalman, J.A.; Seth, R.; Ndegwa, P. Biohydrogen production by mesophilic anaerobic fermentation of glucose in the presence of linoleic acid. *J. Environ. Eng. ASCE* **2007**, *133*, 1145–1152.
28. American Publishers Health Association (APHA). *Standard Methods for the Examination of Water and Wastewater*, 20th ed.; APHA: Washington, DC, USA, 1999.
29. Chaganti, S.R.; Lalman, J.A.; Heath, D.D. 16S rRNA gene based analysis of the microbial diversity and hydrogen production in three mixed anaerobic cultures. *Int. J. Hydrogen Energy* **2012**, *37*, 9002–9017.
30. Alvarez, M.T.; Pozzo, T.; Mattiasson, B. Enhancement of sulphide production in anaerobic packed bed bench-scale biofilm reactors by sulphate reducing bacteria. *Biotechnol. Lett.* **2006**, *28*, 175–181.
31. Sarti, A.; Zaiat, M. Anaerobic treatment of sulfate-rich wastewater in an anaerobic sequential batch reactor (AnSBR) using butanol as the carbon source. *J. Environ. Manag.* **2011**, *92*, 1537–1541.
32. Wei, C.H.; Wang, W.X.; Deng, Z.Y.; Wu, C.F. Characteristics of high-sulfate wastewater treatment by two-phase anaerobic digestion process with jet-loop anaerobic fluidized bed. *J. Environ. Sci. China* **2007**, *19*, 264–270.
33. Visser, A.; Gao, Y.; Lettinga, G. Effects of short-term temperature increases on the mesophilic anaerobic breakdown of sulfate containing synthetic wastewater. *Water Res.* **1993**, *27*, 541–550.
34. McCartney, D.M.; Oleszkiewicz, J.A. Competition between methanogens and sulfate reducers: Effect of COD/Sulfate ratio and acclimation. *Water Environ. Res.* **1993**, *65*, 655–664.
35. Abed, R.M.M.; Musat, N.; Musat, F.; Mussmann, M. Structure of microbial communities and hydrocarbon-dependent sulfate reduction in the anoxic layer of a polluted microbial mat. *Mar. Pollut. Bull.* **2011**, *62*, 539–546.
36. Garcia, J.L.; Patel, B.K.C.; Ollivier, B. Taxonomic phylogenetic and ecological diversity of methanogenic Archaea. *Anaerobe* **2000**, *6*, 205–226.
37. Zinder, S.H. Physiological ecology of methanogens. In *Methanogens: Ecology, Physiology, Biochemistryt and Genetics*; Ferry, J.G., Ed.; Chapman and Hall: London, UK, 1993, pp. 128–206.
38. Tan, B.; Dong, X.L.; Sensen, C.W.; Foght, J. Metagenomic analysis of an anaerobic alkane-degrading microbial culture: Potential hydrocarbon-activating pathways and inferred roles of community members. *Genome* **2013**, *56*, 599–611.
39. Pereyra, L.P.; Hiibel, S.R.; Pruden, A.; Reardon, K.F. Comparison of microbial community composition and activity in sulfate-reducing batch systems remediating mine drainage. *Biotechnol. Bioeng.* **2008**, *101*, 702–713.

40. Ren, N.; Zhao, Y.; Wang, A.; Gao, C.; Shang, H.; Liu, Y.; Wan, C. The effect of decreasing alkalinity on microbial community dynamics in a sulfate-reducing bioreactor as analyzed by PCR-SSCP. *Sci. China Ser. C* **2006**, *49*, 370–378.
41. Bijmans, M.F.M.; Dopson, M.; Ennin, F.; Lens, P.N.L.; Buisman, C.J.N. Effect of sulfide removal on sulfate reduction at pH 5 in a hydrogen fed gas-lift bioreactor. *J. Microbiol. Biotechnol.* **2008**, *18*, 1809–1818.
42. Weijma, J.; Gubbels, F.; Pol, L.W.H.; Stams, A.J.M.; Lens, P.; Lettinga, G. Competition for H_2 between sulfate reducers, methanogens and homoacetogens in a gas-lift reactor. *Water Sci. Technol.* **2002**, *45*, 75–80.
43. Jackson, B.E.; Bhupathiraju, V.K.; Tanner, R.S.; Woese, C.R.; McInerney, M.J. *Syntrophus aciditrophicus* sp. nov., a new anaerobic bacterium that degrades fatty acids and benzoate in syntrophic association with hydrogen-using microorganisms. *Arch. Microbiol.* **1999**, *171*, 107–114.

7

Effect of Incident Rainfall Redistribution by Maize Canopy on Soil Moisture at the Crop Row Scale

Marco Martello [1], Nicola Dal Ferro [1,*], Lucia Bortolini [2] and Francesco Morari [1]

[1] Department of Agronomy, Food, Natural resources, Animals and Environment, Agripolis, University of Padova, Viale Dell'Università 16, Legnaro (Padova) 35020, Italy; E-Mails: marco.martello@unipd.it (M.M.); francesco.morari@unipd.it (F.M.)

[2] Department of Land, Environment, Agriculture and Forestry, Agripolis, University of Padova, Viale Dell'Università 16, Legnaro (Padova) 35020, Italy; E-Mail: lucia.bortolini@unipd.it

* Author to whom correspondence should be addressed; E-Mail: nicola.dalferro@unipd.it;

Academic Editor: Miklas Scholz

Abstract: The optimization of irrigation use in agriculture is a key challenge to increase farm profitability and reduce its ecological footprint. To this context, an understanding of more efficient irrigation systems includes the assessment of water redistribution at the microscale. This study aimed to investigate rainfall interception by maize canopy and to model the soil water dynamics at row scale as a result of rain and sprinkler irrigation with HYDRUS 2D/3D. On average, 78% of rainfall below the maize canopy was intercepted by the leaves and transferred along the stem (stemflow), while only 22% reached the ground directly (throughfall). In addition, redistribution of the water with respect to the amount (both rain and irrigation) showed that the stemflow/throughfall ratio decreased logarithmically at increasing values of incident rainfall, suggesting the plant capacity to confine the water close to the roots and diminish water stress conditions. This was also underlined by higher soil moisture values observed in the row than in the inter-row at decreasing rainfall events. Modelled data highlighted different behavior in terms of soil water dynamics between simulated irrigation water distributions, although they did not show significant changes in terms of crop water use efficiency. These results were most likely affected by the soil type (silty-loam) where the experiment was conducted, as it had unfavorable physical conditions for the rapid vertical water movement that would have increased infiltration and drainage.

Keywords: maize; irrigation; soil water content; stemflow; HYDRUS 2D/3D

1. Introduction

The competition for freshwater between agricultural, industrial and civil uses has intensified over the last few years, worsened by the negative effects of climate change and pollution. The sustainable use of water in agriculture is therefore a priority in order to minimize potential future inter-sectorial conflicts and meanwhile increase benefits in terms of farm profitability and ecological footprint.

Innovations are mainly required in the irrigation sector [1], which accounts for *ca.* 85% of the global freshwater consumption [2]. Advances in irrigation have focused on the assessment of water interception by the crop canopy [3,4] and its effect on water redistribution [5]. Indeed, incident rainfall is partly intercepted by the leaves and then flows down the stem (*i.e.*, stemflow) and partly falls directly to the ground without leaf interception (*i.e.*, throughfall). This results in microscale variability in infiltration, distribution and root water uptake [6]. Van Weesenbeck and Kachanoski [7] studied the soil moisture dynamics in maize, observing that in-row soil water infiltration was greater than in the inter-row, as a consequence of stemflow interception. These results were confirmed by Hupet and Vanclooster [6], who measured a wide variability of rainfall beneath the maize canopy, ranging from 78% to 189%, with some areas receiving up to 4.5-times more water than incident rainfall. Paltineanu and Starr [8] found an inverse third-order relationship between the stemflow/throughfall ratio and rainfall that highlighted a sharp increase of water interception localized on the stem at decreasing values of incident rainfall.

So far, traditional simulation models have assumed that in the "soil-plant-atmosphere" system, the incident rain that reaches the ground is uniform and evenly distributed on the soil surface [6]. However, representing the canopy interception effects in hydraulic models could be useful for simulating irrigation water dynamics at the microscale. This would be helpful for selecting the most appropriate irrigation systems and, in some cases, justifying the adoption of down canopy application systems, such as low-energy precision application (LEPA) irrigation [9,10].

Although several studies have measured the redistribution of rainfall by maize vegetation, only a few have conceptualized the dynamics of the plant interception at various rainfall intensities and its effects on water distribution in the soil [6,8]. Therefore, a key question is whether the maize canopy interception is a major factor for an optimal prediction of water use efficiency in irrigation systems. In order to investigate the role of canopy interception on water balance at the microscale, we aimed to measure and conceptualize the rainfall canopy redistribution below a fully-developed maize canopy. Its effects on water dynamics were then simulated by a numerical model with the final goal of comparing sprinkler irrigation with an alternative down canopy method.

2. Materials and Methods

2.1. Description of Experimental Site and Agronomic Management

The study was conducted at the Experimental Farm of the University of Padova, Legnaro, Italy (45°20′26″ N; 11°58′0″ E). The local climate is sub-humid, with mean annual rainfall of about 850 mm distributed fairly uniformly throughout the year. From December to February, the temperature rarely falls below zero, while maximum temperatures in summer vary from 25 to 30 °C during the day and 18 to 20 °C during the night. The summer period (from June to August) is generally characterized by a mean rainfall of *ca.* 210 mm, with a maximum of 100 mm observed in June (95%) and a minimum of 50 mm in August (95%). Rainfall intensity can reach up to 30 $mm{\cdot}h^{-1}$ with a recurrence interval of 10 years, while more frequently (recurrence interval of 1 year), rain falls with an intensity of 20 $mm{\cdot}h^{-1}$. Finally, the average duration of rain events is 2.5 h (standard error = 0.26).

The soil is a silt loam (sand, 16.7%; silt, 78.5%; clay, 4.8%) Fluvi-Calcaric Cambisol [11], whose infiltration rates vary from 0.1 to 10 $cm{\cdot}h^{-1}$ [12], with 6.8 $g{\cdot}kg^{-1}$ of organic carbon.

The experiment was conducted in 2011 from 25 June to 13 August on a field (9880 m^2) cultivated with maize (*Zea mays* L. cultivar "Pioneer PR31N25"-FAO 600) when the canopy was fully developed and uniformly distributed with a leaf area index (LAI) of 4. During experimentation, the mean daily temperature varied between 16.3 °C on 24 July and 27.1 °C on 12 July, while the daily air relative humidity was 66% (Figure 1), and the wind speed was always below 1.3 $m{\cdot}s^{-1}$ during irrigation events, following the standard provided by ISO 7749-2 (1990) [13].

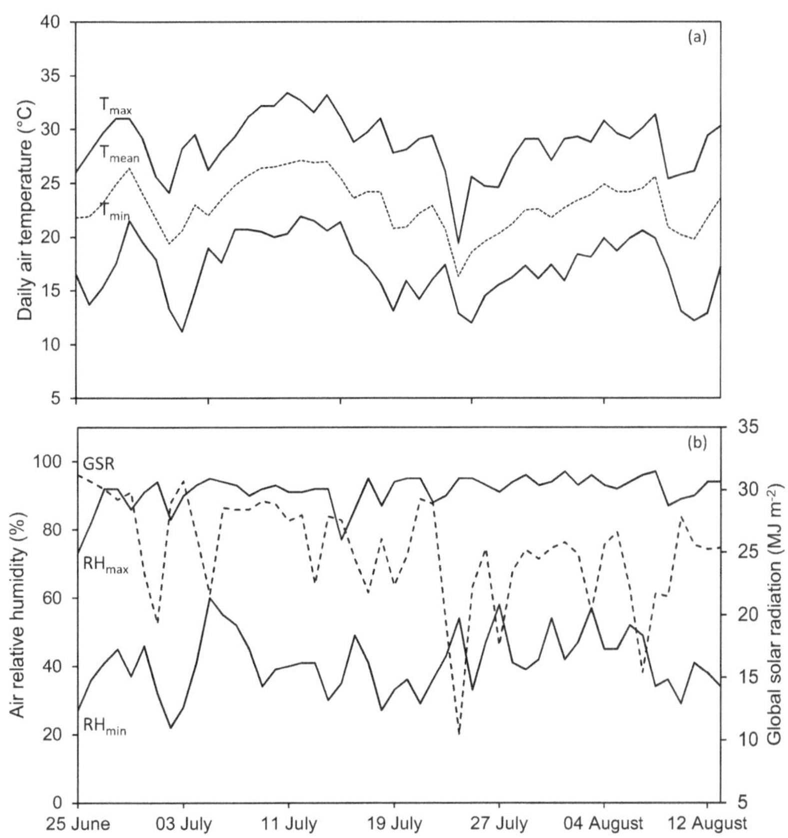

Figure 1. Weather conditions during the experimental season: (**a**) daily air temperature and (**b**) air relative humidity (RH) and global solar radiation (GSR).

The seedbed had been prepared in autumn with a 30-cm ploughing depth. The sowing was in April at a depth of 3 cm and a distance of 75 cm between the rows, 20 cm in the row. The nitrogen (N) fertilization consisted of three split-applications, 200 $kg{\cdot}N{\cdot}ha^{-1}$ as cattle slurry in autumn, 24 $kg{\cdot}N{\cdot}ha^{-1}$ and 138 $kg{\cdot}N{\cdot}ha^{-1}$ as mineral fertilizer at sowing and inter-row hoeing, respectively. A single travelling big gun sprinkler was used for irrigation twice at doses of 43 mm depth on 4 July and 28 mm depth on 2 August, with an average intensity of 15 $mm{\cdot}h^{-1}$.

2.2. Stemflow and Throughfall Water Collection

In order to quantify the rainfall redistribution within the field and its allocation between stemflow and throughfall, an experimental plot of 324 m^2 was equipped with a set of water collectors. Standard rain gauges were laid out in a grid of 3 m × 3 m to quantify the total water amount and distribution above the canopy from sprinkler irrigation and rainfall events (Figure 2). Rainfall redistribution between stemflow and throughfall was measured on 12 individual plants positioned in pairs across the plot, while soil water content (SWC) and the uniformity of water interception on the furrow (*i.e.*, of throughfall) were measured in three different areas of the plot labeled as A, B and C (Figure 2).

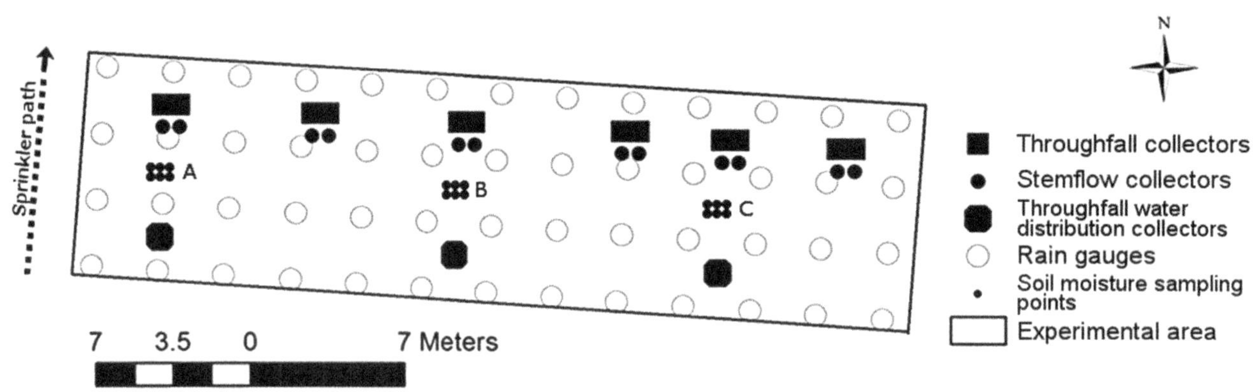

Figure 2. Plot setup with the relative position of throughfall and stemflow collectors, rain gauges and soil moisture sampling points.

The collectors for stemflow determination (Figure 3a), similar to those used by Lamm and Manges [3], were built at the Laboratory of Agricultural Mechanics of the University of Padova. They were designed as conical funnel-shaped polyvinyl chloride (PVC) cylinders, 18 cm tall with a 5-cm upper diameter; below this, a slot with a plastic pipe was connected to a can for draining the stemflow water. Each stemflow collector was fitted on a maize plant stem at about 20 cm above the ground. Throughfall collectors (Figure 3b) were 75 cm × 20 cm trays, with a central hole of 5 cm in diameter to avoid stemflow interception. A small plastic hose, glued into a hole at the corner of the base, drained the water into a tank in order to collect throughfall water. Each tray was positioned approximately 20 cm above the ground. Stemflow and throughfall collectors were placed on adjacent plants to avoid maize stress. All collected volumetric data were converted into depth (mm) considering the sowing density (0.15 $m^2{\cdot}seed^{-1}$).

Figure 3. Stemflow (**a**) and throughfall (**b**) collectors at the study site.

Throughfall water distribution was collected for each area (A, B and C) by a polystyrene board with 27 plastic glasses (7.5 cm upper diameter) laid out in an array of 3 × 9, placed side-by-side. Stemflow and throughfall amounts were measured a few hours after rainfall events.

2.3. Soil Moisture Monitoring

Soil moisture was monitored in order to quantify the effect of canopy interception on soil water distribution. Measurements were performed manually in the field following the rainfall events and gradually decreasing the sampling frequency from 4 times·day^{-1} after rainfall to once·day^{-1} during the following 8 days. A frequency domain reflectometry (FDR) probe (PR2, Delta-T Devices Ltd., Cambridge, UK) was used to measure the soil moisture profile at four different depths (10, 20, 30 and 40 cm). In each area (A, B and C), 6 probes were installed in an array of 2 × 3 at a distance of 40 cm × 37.5 cm, respectively, to monitor SWC in both the row and inter-row (Figure 2). FDR probes, calibrated in the laboratory, had an accuracy of ±4%.

2.4. Modelling Canopy Interception Effects with HYDRUS 2D/3D Model

Effects of canopy interception on water dynamics were simulated with the numerical model HYDRUS 2D/3D Version 1.11 [14]. The model was first calibrated and validated in the experimental conditions and then used to predict the performances of sprinkler and alternative irrigation method.

2.4.1. Calibration/Validation of HYDRUS 2D/3D

SWC data, monitored by the 24 sensors in each Site (A, B and C) at the four depths, were used to calibrate the soil hydraulic properties using the inverse solution approach [14]. Hydraulic properties were defined according to Van Genuchten and Mualem functions [15]. Inversion was applied to the first 22 days of experimental data, while the remaining 20 days were used for the validation.

The 2D domain (width 37.5 cm and depth 80 cm) encompasses the area between the row and the inter-row where the moisture probes were inserted (Figure 4), assuming the isotropy of the soil profile.

Variable flux conditions were assigned at the upper boundary for describing the spatial redistribution of the stemflow and throughfall. The stemflow flux was assigned to the top left corner in correspondence to the stem position (Figure 4, in green), while the throughfall flux was assigned to the upper boundary line (Figure 4, in pink).

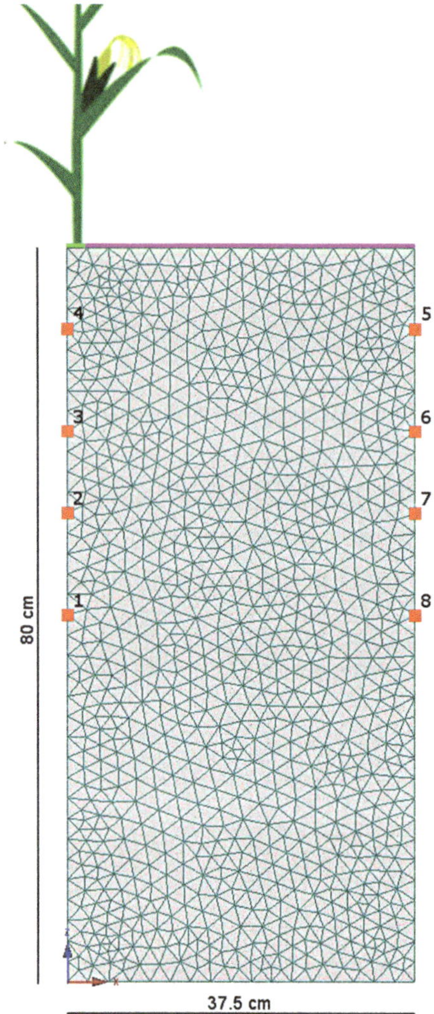

Figure 4. 2D domain geometry, mesh composition and node positions in HYDRUS 2D/3D for the inverse solution.

Potential evapotranspiration rates, calculated according to Allen *et al.* [16], were used as atmospheric boundary conditions. Root distribution was set according to Dal Ferro *et al.* [17], while the root water uptake followed the models proposed by Feddes *et al.* [18]. Initial water content conditions were specified using data measured by FDR, while free drainage was set as the lower boundary condition.

The goodness of validation was tested according to the linear regression between observed and simulated SWC and the residual deviation index (RPD) [19] as follows:

$$RPD = \frac{SD}{RMSE} \tag{1}$$

where *SD* is the standard deviation of observed data and *RMSE* is the root mean square error of fitted *versus* observed data.

2.4.2. Numerical Simulation of Irrigation Systems with HYDRUS 2D/3D

In order to compare sprinkler irrigation with an alternative down canopy method, 3D numerical simulations were performed at each field site (A, B and C). Three different upper boundary conditions distributions were considered: (1) rainfall application with an optimal uniform distribution down canopy (uniform rainfall, RU); (2) stemflow and throughfall fluxes, where the former was redistributed on the row and the latter on the inter-row according to the experimental data (ST); and (3) water application beneath the canopy as a low-energy precision application (LEPA) system into every other inter-row.

The lower boundary condition was free drainage, while the initial conditions of soil water content had a linear distribution between the top (0.25 $m^3 \cdot m^{-3}$) and the bottom (0.18 $m^3 \cdot m^{-3}$) of the soil domain. The 3D domain was 0.8-m depth, 0.75-m width and 1-m length; the root distribution was set according to Dal Ferro *et al.* [17], while the root water uptake followed the model proposed by Feddes *et al.* [18].

Simulations were performed during the 47-day experimental period. Two different water managements were simulated: the first one was according to the crop evapotranspiration (ET_c) compensation method, considering seven irrigation treatments as typical for sprinkler irrigation in Northeastern Italy. The second scenario was a deficit irrigation with 60% of ET_c adopting the same irrigation frequency. Rainfall intensity was 5 $mm \cdot h^{-1}$ for both simulations. The system performance was evaluated in terms of ET, *i.e.*, actual crop evapotranspiration (mm), and ET/ET_c, *i.e.*, the ratio between ET and the crop evapotranspiration under standard conditions [16] and drainage (mm).

3. Results

3.1. Rainfall Redistribution by Canopy Interception and Soil Water Dynamics

Total rainfall during the experimental period was 159 mm on average (±2 mm), as partly provided by rainfall events, which were homogeneous throughout the study area, and partly by sprinkler irrigation, which changed depending on the distance from the sprinkler. Rainfall events provided 87.5 mm (55%) of total incoming water, which is in line with the typical climatic conditions of the area [20], while 45% was provided by sprinkler irrigation. A total of five rain events (Figure 5) were used for the evaluation of water redistribution below the fully-developed maize canopy. On average, rainfall intensity was 3.7 $mm \cdot h^{-1}$, ranging from a minimum of 4.6 mm that fell intermittently during 1 July (maximum rainfall intensity = 2.2 $mm \cdot h^{-1}$) to a maximum observed on 24 July, when 36 mm fell in 19 h. This last rainfall event also showed the highest intensity during the experimentation (maximum rainfall intensity = 19.4 $mm \cdot h^{-1}$). Finally, during the night between 30 and 31 July, 17 mm fell in just 3 h, while between 6 and 7 August, 10 mm fell uniformly in 10 h.

Rainwater below the maize canopy was 78% intercepted by the leaves and transferred along the stem (*i.e.*, stemflow), while only 22% reached the ground directly (*i.e.*, throughfall). The maximum difference between stemflow and throughfall occurred during the most abundant event (13.5 mm, 24 July), while the minimum was observed on 1 July (2.9 mm). It is worth noting that in all cases, stemflow was greater than throughfall (Figure 5).

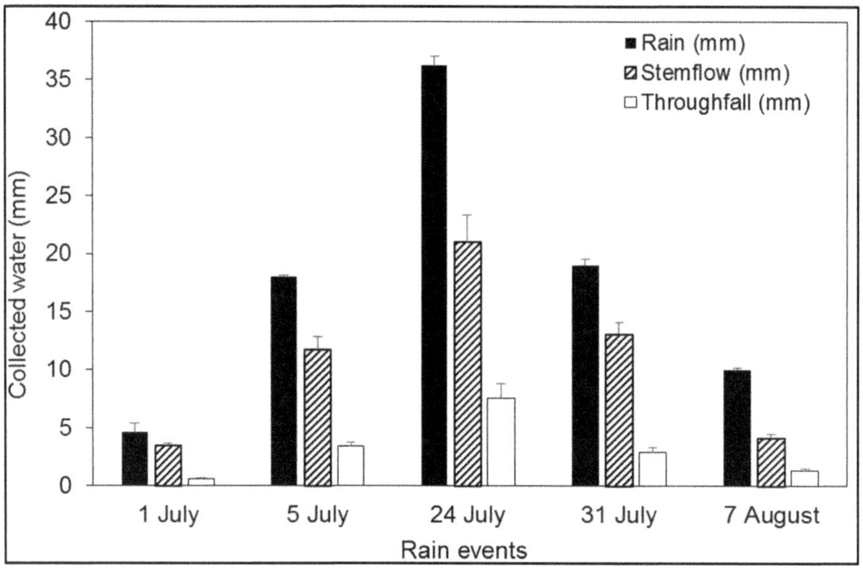

Figure 5. Rain redistribution as affected by stemflow and throughfall interception. Error bars indicate standard errors.

Sprinkler irrigation was applied on 4 July and 2 August, amounting to 43 mm (standard error, ±11 mm) and 28 mm (standard error, ±2 mm) throughout the study area, on average. Irrigation was not uniformly distributed in the study area, as it depended on the proximity to the sprinkler. Generally, a lower uniformity was observed during the first irrigation than the second one. In fact, the radius adjustment screw of the sprinkler was differently regulated to search an optimal setting, leading to a different spatial distribution between irrigations.

During the first irrigation event (Figure 6a), a linear decrease of water applied was observed, with a maximum of 57 mm, 6 m from the sprinkler, which reduced to a minimum of 10 mm at 37 m. As a result, water exceeded the mean value of 20 mm close to the sprinkler; *vice versa*, a difference of −21 mm between the amount applied and the mean value was found in the farthest sampling position. Water was mostly intercepted by the canopy as a stemflow (69.6%, on average), the distribution gradually increasing from 64% (6 m from the sprinkler) to 73% (37 m from the sprinkler).

During the second irrigation event (Figure 6b), the variability of the distribution (from 30 to 11 mm) was lower than that observed in the first one. Less variability was also observed when water interception by the canopy (*i.e.*, stemflow) was compared with throughfall. In fact, the major difference was observed at 13 m from the sprinkler, where stemflow reached 78% of water interception, while it was slightly reduced at 27 m, reaching values of 64%.

Throughfall was mainly concentrated in the middle of the inter-row (Figure 7), reaching a peak of 17% at 29 cm from the row and decreasing to *ca.* 7% in proximity to both sides. Nevertheless, a slight asymmetry in water distribution was found between the right and left side of the inter-row that may be affected by the directional impact of water drops due to atmospheric conditions (e.g., wind direction).

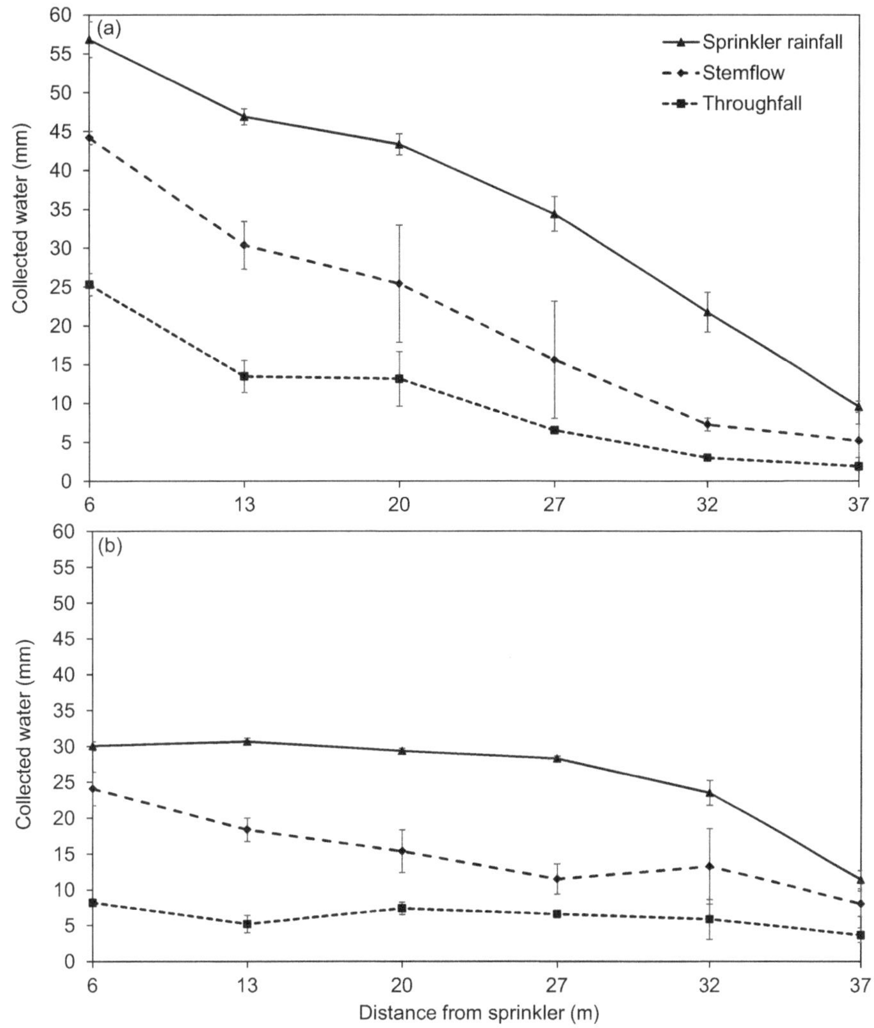

Figure 6. Spatial variability of sprinkler rainfall and its partitioning between stemflow and throughfall (**a**) during the first event (4 July) and (**b**) the second event (2 August). Error bars indicate standard errors.

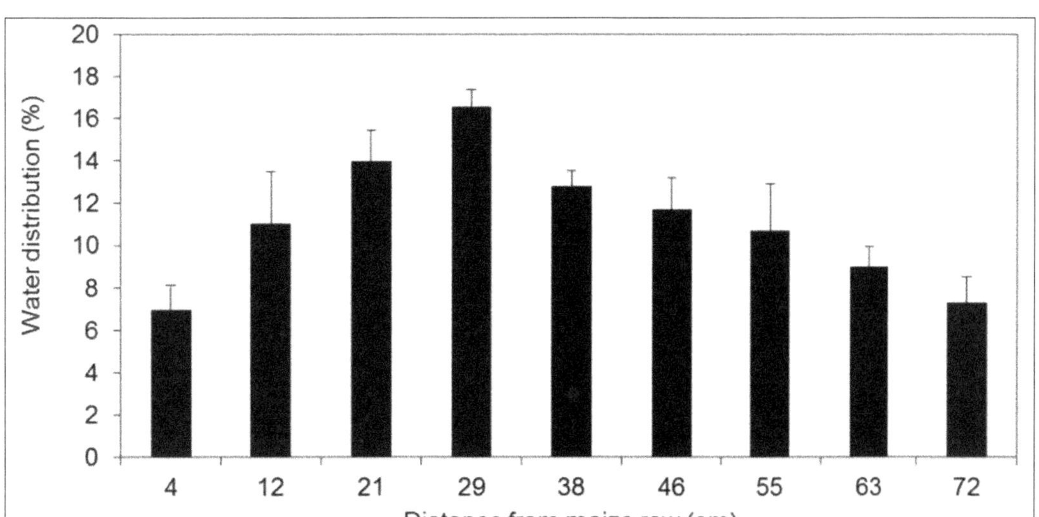

Figure 7. Spatial variability of the throughfall measurements collected with throughfall collectors. Mean values of the three trays are reported. Error bars indicate standard errors.

A general overview of water redistribution with respect to the amount (both rainfall and irrigation) showed that the stemflow/throughfall ratio decreased logarithmically (Figure 8) at increasing values of incident rainfall, underlining the plant maize capacity to intercept water during low rainfall events. Indeed, for incident rainfalls, <10 mm stemflow was four-times higher than throughfall, while for more abundant events (e.g., 35 mm), stemflow/throughfall was reduced to 2.6. By contrast, both the maximum and mean rainfall intensity data did not show a significant correlation with the stemflow/throughfall ratio.

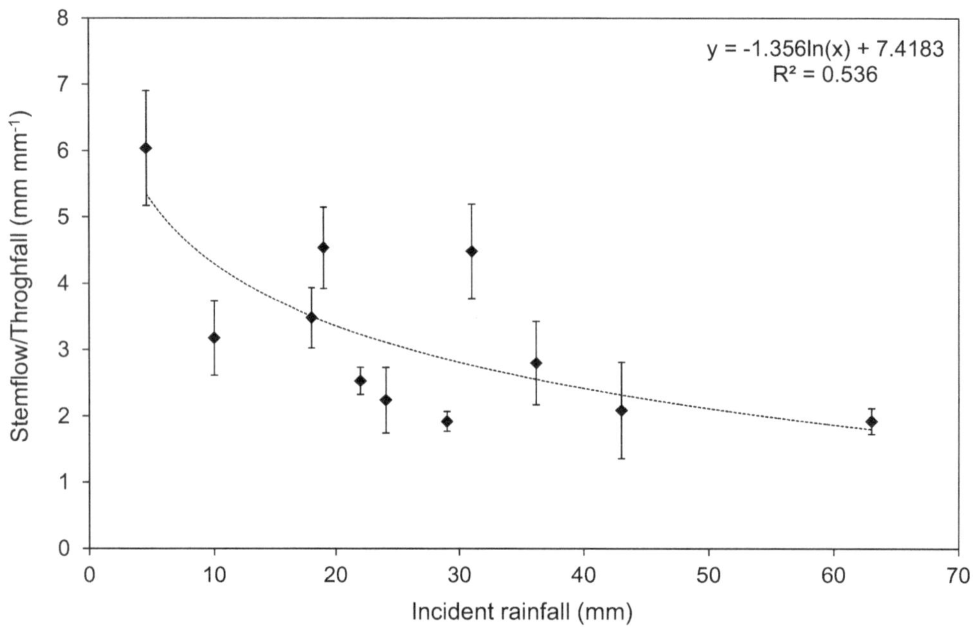

Figure 8. Relationship between total rainfall amounts and stemflow/throughfall ratio. Error bars indicate standard errors.

3.2. Inter-Row and Intra-Row Soil Water Dynamics

At the beginning of the experiment, soil water content (SWC) was similar in the three sampling areas, between the row and inter-row (Figure 9), with larger variations observed in the 0 to 20 cm soil profile ($-0.02\ m^3 \cdot m^{-3}$) than in the 20 to 40 cm on ($-0.03\ m^3 \cdot m^{-3}$). An increase in soil water content was generally observed from the top soil layer (10 cm depth) to a 40-cm depth, ranging from 0.18 to 0.27 $m^3 \cdot m^{-3}$, respectively.

Large variations were observed during the experiment in the top layer as affected by rainfall events, while they were gradually reduced in the deepest soil layers. Four different peaks in soil moisture were clearly visible at 10 cm (Figure 9), corresponding to the higher rainfall events. At 10 cm, the maximum soil water content was observed after the most abundant event (4 July, corresponding to sprinkler irrigation), reaching a peak of 0.26 $m^3 \cdot m^{-3}$. At greater depths (from 20 to 40 cm), all sampling sites showed reduced soil moisture variations as a result of soil water distribution after the rainfall events. Differences in terms of water storage between inter-row and row were negligible for all three study sites, as these were less than 1%, on average. However, higher soil moisture values were observed in the topsoil row position than in the inter-row as a consequence of low rainfall events (Figure 9), particularly after 19 and 20 July, when total rainfall was only 11 mm. In this case, SWC in the row exceeded 0.08 $m^3 \cdot m^{-3}$ of that in the inter-row.

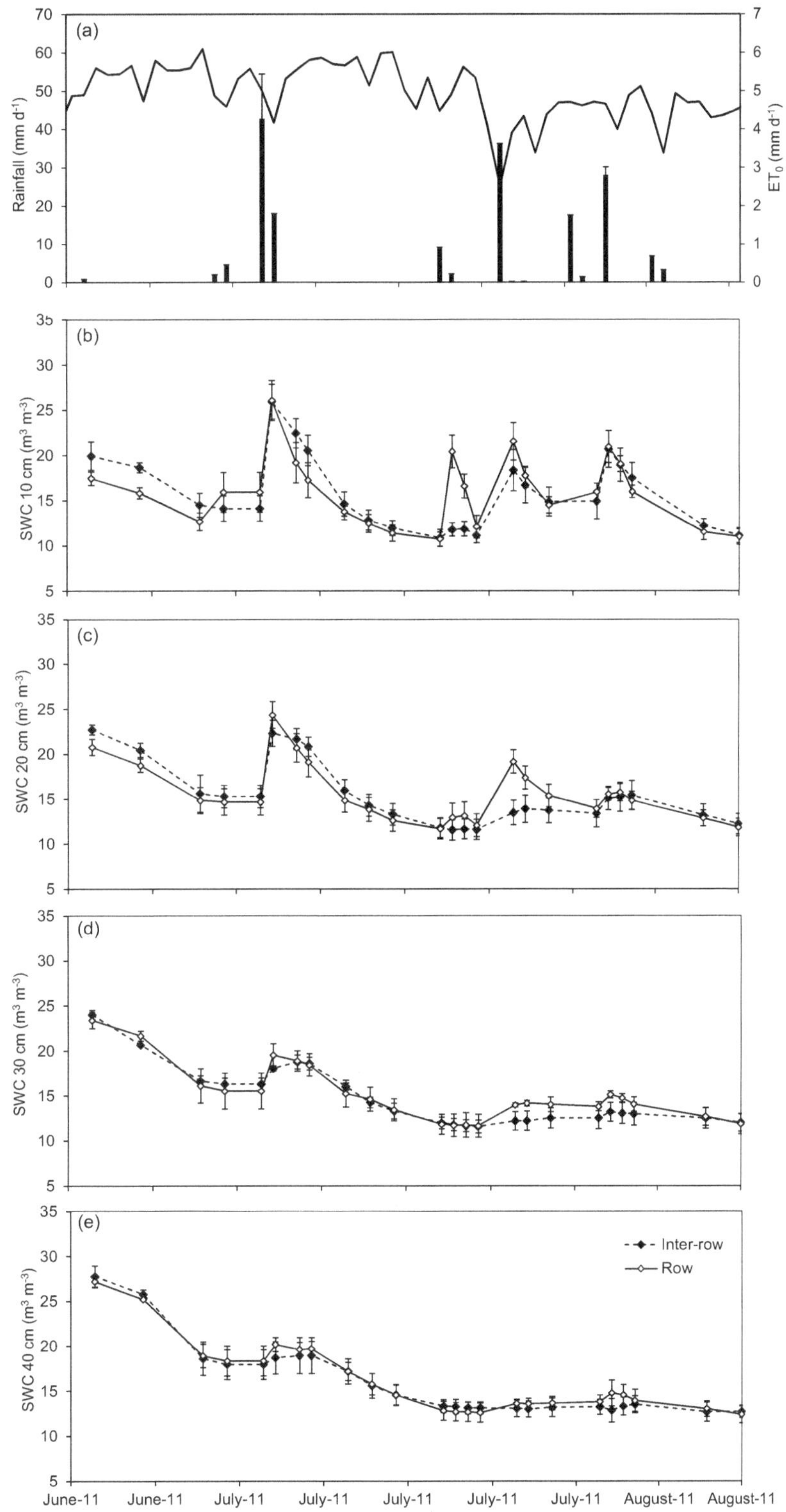

Figure 9. Daily evolution of (**a**) meteorology and profile soil water content (SWC) for: (**b**) 0 to 10 cm; (**c**) 10 to 20 cm; (**d**) 20 to 30 cm; (**e**) 30 to 40 cm. Error bars indicate standard errors.

3.3. Calibration/Validation of HYDRUS 2D/3D

Experimental SWC values matched the predicted ones well, indicating that HYDRUS 2D/3D was able to predict the effects of canopy interception on water dynamics. This was confirmed by the linear regression between observed and predicted data, both for the inversion solution (Figure 10a; $R^2 = 0.75$) and validation (Figure 10b; $R^2 = 0.66$), as well as by the model accuracy estimated by means of RPD (1.45) [21].

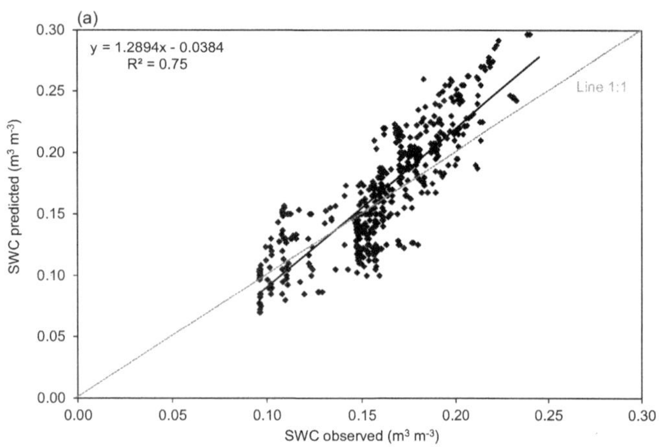

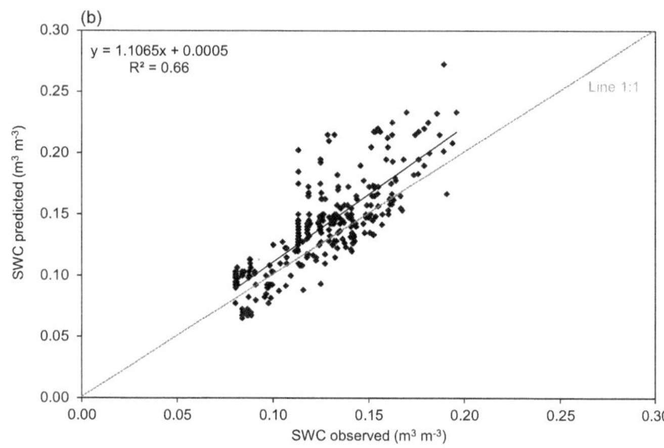

Figure 10. Linear regression between observed and predicted data as a result of (**a**) model inversion and (**b**) validation.

The Van Genuchten and Mualem models indicated that Sites A, B and C had similar soil structure characteristics (Table 1). Furthermore, the inversion process showed that the profiles were homogenous and properly described by a single soil material.

Table 1. Soil hydraulic parameters estimated by means of model inversion using HYDRUS 2D/3D for experimental Sites A, B and C.

Site	SWC_r [a] ($m^3 \cdot m^{-3}$)	SWC_s ($m^3 \cdot m^{-3}$)	α [b] (cm^{-1})	n [c]	K_s [d] ($cm \cdot h^{-1}$)	SWC ($m^3 \cdot m^{-3}$) −330 cm	SWC ($m^3 \cdot m^{-3}$) −15,000 cm
A	0.031	0.461	0.005	1.51	0.71	0.319	0.078
B	0.000	0.460	0.002	1.39	0.26	0.395	0.115
C	0.071	0.460	0.006	1.52	0.48	0.324	0.109

Notes: [a] SWC = residual (r) soil water content, at saturation (s) and at suctions of −330 and −15,000 cm; [b] α = parameter linked to the air entry suction; [c] n = parameter linked to pore size distribution; [d] K_s = saturated hydraulic conductivity.

3.4. Numerical Simulations

The first simulated water management (here called full irrigation) was based on crop evapotranspiration compensation with non-limiting water conditions for maize during ripening. Therefore, during simulations, total rainfall was 210 mm, corresponding to the crop evapotranspiration requirement (Table 2). Full irrigation management did not show water stress conditions for maize growth, since the ratio between actual evapotranspiration (ET) and crop ET_c (unstressed crop ET) was 0.92, on average. Generally, ET/ET$_c$ was higher in A than in B and C (0.97, 0.89 and 0.89, respectively) as a result of more favorable soil conditions for the crop. For example, plant growth was fostered in Site A compared to B and C at low moisture values, since it was characterized by a low water content at the wilting point (−15,000 cm; Table 1).

Table 2. Water mass balance estimated for two different irrigation management scenarios.

Site	Irrigation	Simulations [a]	RF (mm)	ET (mm)	ET$_c$ (mm)	ET/ET$_c$	Drainage (mm)	WC$_{in}$ [b] (mm)	WC$_{fin}$ [c] (mm)
A	Full	RU	210	261	270	0.97	0.90	172	122
		ST	210	261	270	0.97	0.91	172	122
		LEPA	210	261	270	0.97	1.00	172	120
	Deficit 60%	RU	126	217	270	0.80	0.90	172	82
		ST	126	220	270	0.81	0.90	172	81
		LEPA	126	213	270	0.79	0.90	172	86
B	Full	RU	210	241	270	0.89	0.19	172	142
		ST	210	241	270	0.89	0.19	172	142
		LEPA	210	243	270	0.90	0.19	172	138
	Deficit 60%	RU	126	190	270	0.70	0.19	172	108
		ST	126	193	270	0.71	0.19	172	107
		LEPA	126	186	270	0.69	0.19	172	113
C	Full	RU	210	240	270	0.89	0.34	172	144
		ST	210	241	270	0.89	0.34	172	144
		LEPA	210	241	270	0.89	0.35	172	143
	Deficit 60%	RU	126	193	270	0.71	0.34	172	108
		ST	126	196	270	0.72	0.34	172	106
		LEPA	126	189	270	0.70	0.34	172	112

Notes: [a] RU (uniform rainfall) = rainfall application with an optimal uniform distribution down canopy; ST = stemflow/throughfall water redistribution; LEPA = low-energy pressure application; [b] WC$_{in}$ = initial soil water content; [c] WC$_{fin}$ = final soil water content.

By comparing the water distributions, it was observed that the different irrigation simulations did not show any difference in terms of ET/ET$_c$, since it was always 0.97 in Site A and largely 0.89 in Sites B and C. Water losses due to drainage were negligible in all simulations and always less than 1 mm in Site A, 0.19 mm and 0.35 mm in B and C, respectively. The visual representation of soil water dynamics at increasing time steps (Figure 11) highlighted the effect of different water distributions. For example, after an irrigation event (at Hour 264), the ST model tended to increase the soil moisture below the maize row (central position of the soil domain) and gradually reduced it towards the inter-row (right and left areas of the domain). Irrigation only affected the first 40-cm depth of the soil profile, while the soil water content (0.15 to 0.19 $m^3 \cdot m^{-3}$) did not change in the layer below (40 to 80 cm). By contrast, with the LEPA irrigation system, water was mainly concentrated in *ca.* 40 cm around the area of application, influencing the soil water content both in the row and in the furrow of application (left side, inter-row). For example, SWC, at 266-h time step, ranged between 0.46 and 0.19 $m^3 \cdot m^{-3}$ within the soil profile.

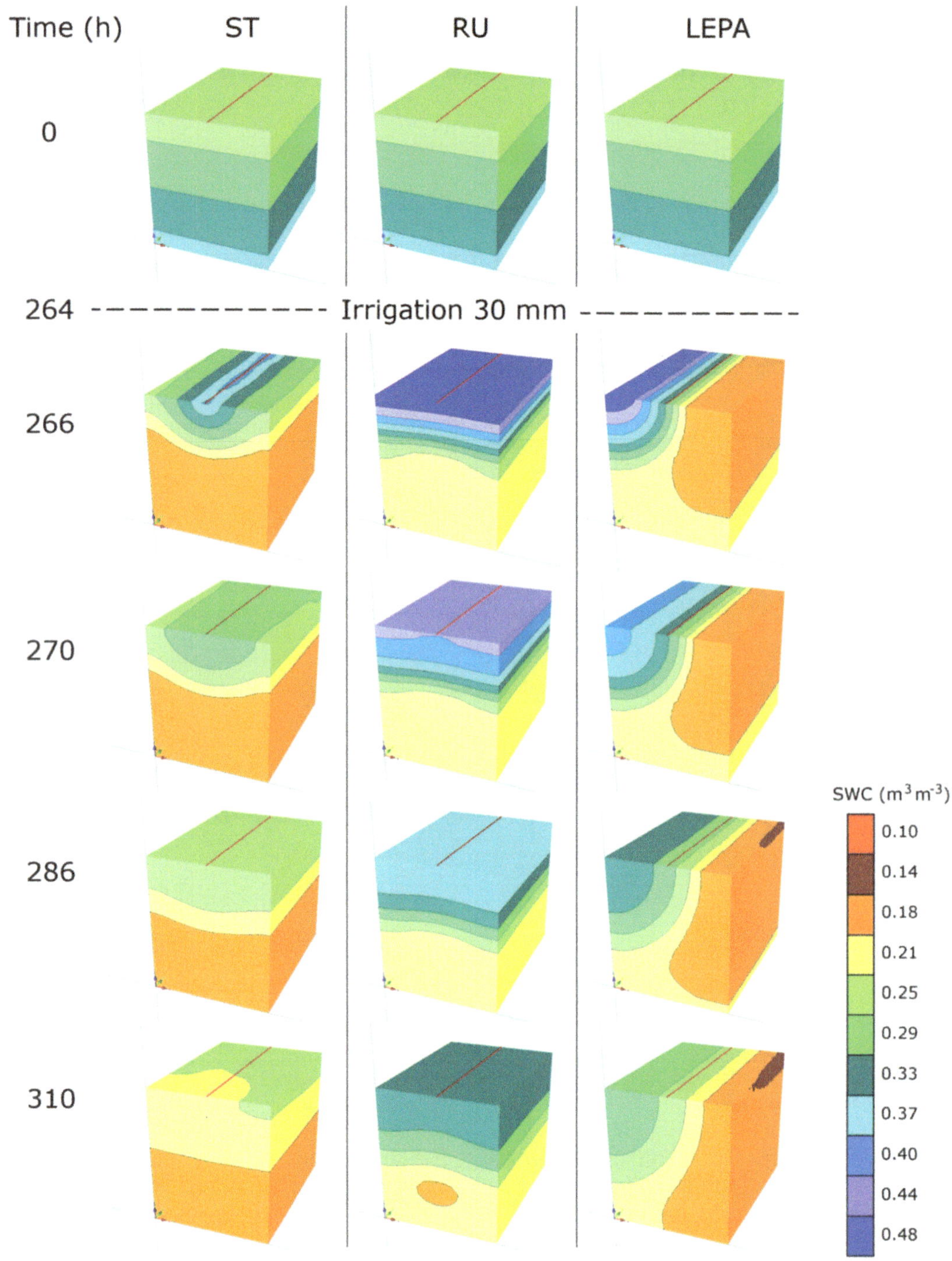

Figure 11. Visual 3D representation of soil water content (SWC, $m^3 \cdot m^{-3}$) in the 3D HYDRUS domain under full irrigation and characterized by: ST, stemflow/throughfall redistribution; RU, uniform rainfall application; LEPA, low-energy precision application system. The maize row is placed in the middle position of the soil domain (red line).

Generally, ET/ET_c with deficit irrigation (Figure 12) was lower than under full irrigation, as it was −20% on average, with a maximum observed in ST, Site A ($ET/ET_c = 0.81$), and a minimum in LEPA, Site B ($ET/ET_c = 0.69$), highlighting critical conditions for maize growth. In such cases, the higher root density below the row contributed to a reduction in soil water content as a result of increased water uptake.

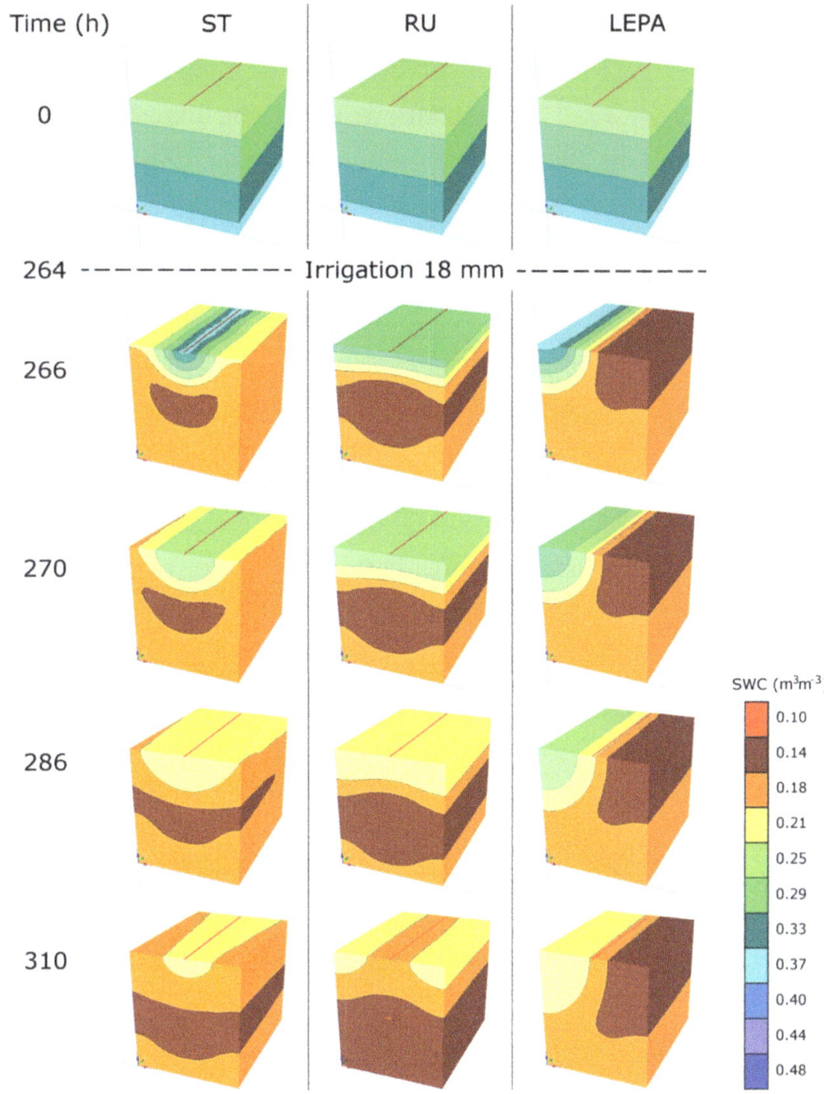

Figure 12. Visual 3D representation of soil water content (SWC, $m^3 \cdot m^{-3}$) in the 3D HYDRUS domain under deficit irrigation and characterized by: ST, stemflow/throughfall redistribution; RU, uniform rainfall application; LEPA, low-energy precision application system. The maize row is placed in the middle position of the soil domain (red line).

4. Discussion

Water partitioning between stemflow and throughfall was affected by the rainfall amount and described by a negative logarithmic equation (Figure 8). Nevertheless, the source of uncertainty could be due to the water evaporation from the maize canopy, although being estimated at less than 10% [22]. A similar stemflow/throughfall partitioning had already been observed in other studies [8], where the water redistribution between the row and the inter-row was measured in the field with conditions similar to ours (*i.e.*, closed maize canopy with LAI = 3.6 ± 0.2): the authors showed that the lowest rainfall events were associated with the highest stemflow/throughfall ratio, while with higher rainfall (*ca.* ≥20 mm), the ratio stabilized at values <10. Most likely, high rainfall events increase the probability of dripping in the inter-row, as raindrops overload and bend the leaves and accordingly decrease stemflow volumes. Conversely, Hupet and Vanclooster [6] reported that the stemflow/throughfall ratio increased with increasing rainfall, although their results were obtained with a low maize leaf coverage (2.5 < LAI < 3.6) and were not representative of a wide range of data, since the rainfall amounts were low (*i.e.*, <20 mm). Rainfall redistribution below the maize canopy was crucial to allocate the water in the soil, particularly after medium and low events, while during intense rainfall, the soil water was fairly uniform between stemflow and throughfall. As a result, the variation was large when rainfall was <40 mm (20 July), while a negligible effect was found after significant events (e.g., 4 July), as also observed by other authors [8]. According to Logsdon *et al.* [5], the high stemflow/throughfall ratio led to an abrupt increase of the row soil moisture that was followed by its decrease a few days after the rainfall event. This was probably due to the activity of maize roots that were mostly concentrated below the stem, leading to site-specific evapotranspiration rates and consequently decreasing the soil water content. Extensive literature has already reported that root water uptake within the maize rows is spatially variable, both laterally and vertically [6,7], and can be influenced by incident rainfall [23]. As a result, the soil moisture differences observed between stemflow and throughfall in a silty-loam soil were rapidly smoothed by root activity. Nevertheless, local soil conditions and moisture characteristics could have partly affected the heterogeneity in the root distribution and, as a consequence, the variation of water partitioning. Moreover, capillary forces increased the horizontal infiltration and consequently reduced the SWC difference induced by stemflow and throughfall. This was also supported by the simulations performed with diversified fluxes (stemflow and throughfall): in spite of the soil receiving higher amounts of water along the row according to ST redistribution (*i.e.*, in the middle of the soil domain; Figures 11 and 12), it was also the fastest to dry out due to a greater root water uptake.

Simulations under different water distributions provided significant indications of the effectiveness of irrigation management during maize ripening. In general, optimal crop growth conditions were observed with irrigation scheduling based on maize evapotranspiration compensation, responding to the ET_c demand and limiting the water loss by drainage (always less than 1% total rainfall). This emphasized that maize root water uptake, in our silty-loam soil, was weakly sensitive to irrigation systems. Similar results were also supported by Schneider and Howell [24], who measured maize growth under full irrigation with LEPA and a spray sprinkler (*i.e.*, irrigation below and above the canopy), reporting no significant yield differences between the systems.

Water use efficiency, estimated by means of the ratio ET/ET_c, did not change with RU and ST when compared with LEPA under full irrigation management as a consequence of soil water content variations between row and inter-row. Conversely, with deficit irrigation, SWC was a limiting factor for crop evapotranspiration, and ET/ET_c was slightly lower in LEPA than ST and RU. In particular, a sharp distinction in SWC was observed between alternate furrows under LEPA. In fact, the left side of the soil domain received rainfall that was gradually distributed in the row, while the right side of the furrow was only marginally affected by water input, limiting the optimal water uptake in proximity to a higher root density (*i.e.*, close to the soil surface).

5. Conclusions

Experimental data suggested the importance of considering rainfall canopy interception to improve water use efficiency and dynamics in the vadose zone, because the natural crop architecture was able to intercept water and direct it along the stem. This effect was particularly significant at low rainfall amounts, emphasizing maize's ability to mitigate water stress conditions as it localized water close to the roots. This was confirmed by soil moisture data, since the only differences observed between row and inter-row were found at rainfall amounts <15 mm.

Modelled data highlighted different behavior in terms of soil water dynamics between water distributions with both full and deficit irrigation management, although it did not affect the actual root water uptake and, consequently, the water use efficiency (ET/ET_c). Most likely, these results were affected by the soil type (silty-loam) where the experiment was conducted, as it showed unfavorable physical conditions for a rapid vertical water movement that would increase infiltration through the profile and, finally, drainage (e.g., low hydraulic conductivity). This suggests a need to test rainfall redistributions in further experiments considering a wider soil variability, in order to provide insights into water use efficiency under different irrigation managements.

Author Contributions

Lucia Bortolini and Francesco Morari conceived and designed the field experiments and modelling activities; Marco Martello and Lucia Bortolini performed the experiments; Marco Martello and Nicola Dal Ferro analyzed the data; all the authors equally contributed to write the paper.

Conflicts of Interest

The authors declare no conflict of interest.

References

1. Pereira, L.S.; Cai, L.G.; Hann, M.J. Farm water and soil management for improved water use in the North China Plain. *Irrig. Drain.* **2003**, *52*, 299–317.
2. Siebert, S.; Döll, P. Quantifying blue and green virtual water contents in global crop production as well as potential production losses without irrigation. *J. Hydrol.* **2010**, *384*, 198–217.
3. Lamm, F.R.; Manges, H.L. Partitioning of sprinkler irrigation water by a corn canopy. *Trans. ASAE* **2000**, *43*, 909–918.
4. Frasson, R.P.M.; Krajewski, W.F. Rainfall interception by maize canopy: Development and application of a processed-based model. *J. Hydrol.* **2013**, *489*, 246–255.
5. Logsdon, S.D.; Sauer, T.J.; Hernandez-Ramirez, G.; Hatfield, J.L.; Kaleita-Forbes, A.L. Effect of corn or soybean row position on soil water. *Soil Sci.* **2010**, *175*, 530–534.
6. Hupet, F.; Vanclooster, M. Micro-variability of hydrological processes at the maize row scale: Implications for soil water content measurements and evapotranspiration estimates. *J. Hydrol.* **2005**, *303*, 247–270.
7. Van Weesenbeck, I.; Kachanoski, R. Spatial and temporal distribution of soil water in the tilled layer under a corn crop. *Soil Sci. Soc. Am. J.* **1988**, *52*, 363–368.
8. Paltineanu, I.; Starr, J. Preferential water flow through corn canopy and soil water dynamics across rows. *Soil Sci. Soc. Am. J.* **2000**, *64*, 44–54.

9. Lyle, W.M.; Bordowsky, J.P. Low energy precision application (LEPA) irrigation system. *Trans. ASAE* **1981**, *24*, 1241–1245.
10. Schneider, A.D.; Buchleiter, G.W.; Kincaid, D.C. LEPA irrigation developments. In Proceedings of the 4th Decennial Symposium, National Irrigation Symposium, Phoenix, AZ, USA, 14–16 November 2000; pp. 89–96.
11. Food and Agriculture Organization (FAO); United Nations Educational, Scientific and Cultural Organization. *Soil Map of the World, Revised Legend*; FAO: Rome, Italy, 1990.
12. Borin, M.; Morari, F.; Bonaiti, G.; Paasch, M.; Skaggs, W. Analysis of DRAINMOD performances with different detaiul of soil input data inb the Veneto region of Italy. *Agric. Water Manag.* **2000**, *42*, 259–272.
13. Norme Internationale (ISO). Matériel D'irrigation Asperseurs Rotatifs Partie 2: Uniformité de la Distribution et Méthodes D'essai; ISO 7749-2; ISO: Lethbridge, AL, Canada, 1990. (In French).
14. Simunek, J.; Sejna, M.; van Genuchten, M.T. The HYDRUS-2D Software Package for Simulating Two-Dimensional Movement of Water, Heat, and Multiple Solutes in Variably Saturated Media; Version 2.0, IGWMC-TPS-53; International Ground Water Modeling Center, Colorado School of Mines: Golden, CO, USA, 1999.
15. Van Genuchten, M.T. A closed-form equation for predicting the hydraulic conductivity of unsaturated soils. *Soil Sci. Soc. Am. J.* **1980**, *44*, 892–898.
16. Allen, R.; Pereira, L.; Raes, D.; Smith, M. *Crop Evapotranspiration—Guidelines for Computing Crop Water Requirements*; Food Agriculture Organization Of the UN: Rome, Italy, 1998.
17. Dal Ferro, N.; Sartori, L.; Simonetti, G.; Berti, A.; Morari, F. Soil macro and microstructure as affected by different tillage systems and their effects on maize root growth. *Soil Till. Res.* **2014**, *140*, 55–65.
18. Feddes, R.A.; Kowalik, P.J.; Zaradny, H. *Simulation of Field Water Use and Crop Yield*; John Wiley and Sons: New York, NY, USA, 1978.
19. Chang, C.; Laird, D.A.; Mausbachc, M.J.; Hurburghd, C.R. Near-infrared reflectance spectroscopy–principal components regression analyses of soil properties. *Soil Sci. Soc. Am. J.* **2001**, *65*, 480–490.
20. Giardini, L. Productivity and Sustainability of Different Cropping Systems; Patron Editore: Bologna, Italy, 2004.
21. Viscarra Rossel, R.A.; Walvoort, D.J.J.; McBratney, A.B.; Janik, L.J.; Skjemstad, J.O. Visible, near infrared, mid infrared or combined diffuse reflectance spectroscopy for simultaneous assessment of various soil properties. *Geoderma* **2006**, *131*, 59–75.
22. Keller, J.; Bliesner, R.D. *Sprinkle and Trickle Irrigation*; Blackburn Press: Caldwell, NJ, USA, 2000.
23. Michot, D.; Benderitter, Y.; Dorigny, A.; Nicollaud, B.; King, D.; Tabbagh, A. Spatial and temporal monitoring of soil water content with irrigated corn crop cover using surface electrical resistivity tomography. *Water Resour. Res.* **2003**, *39*, doi:10.1029/2002WR001581.
24. Schneider, A.D.; Howell, T.A. LEPA and spray irrigation for grain crops. *J. Irrig. Drain. Eng.* **1999**, *125*, 167–172.

Spatiotemporal Analysis of Extreme Hourly Precipitation Patterns in Hainan Island, South China

Wenjie Chen [1], Chenghao Chen [2], Longbing Li [2], Litao Xing [2], Guoru Huang [1,3,*] and Chuanhao Wu [1]

1 School of Civil Engineering and Transportation, South China University of Technology, Guangzhou 510640, China; E-Mail: c.wenjie02@mail.scut.edu.cn
2 Hydrology and Water Resources Survey Bureau of Hainan, Hainan 571126, China
3 State Key Laboratory of Subtropical Building Science, South China University of Technology, Guangzhou 510640, China

* Author to whom correspondence should be addressed; E-Mail: huanggr@scut.edu.cn;

Academic Editor: Athanasios Loukas

Abstract: To analyze extreme precipitation patterns in Hainan Island, hourly precipitation datasets from 18 stations, for the period from 1967 to 2012, were investigated. Two precipitation concentration indices (PCI) and 11 extreme precipitation indices (EPI) were chosen. PCI1 indicated a moderate seasonality in yearly precipitation and PCI2 showed that at least 80% of the total precipitation fell in 20% of the rainiest hours. Furthermore, the spatial variations of PCI1 and PCI2 differed. Linear regression indicated increasing trends in 11 of the calculated EPI. Principal component analysis found that the first recalculated principal component represented the 11 EPI. The recalculated principal component revealed an increasing trend in precipitation extremes for the whole island (except the interior section). Trend stability analysis of several of EPI suggested that the southern parts of Hainan Island, and especially the city of Sanya, should receive more attention to establish the drainage facilities necessary to prevent waterlogging.

Keywords: extreme precipitation pattern; hourly datasets; extreme precipitation indices (EPI); principal component analysis (PCA); trend; stability

1. Introduction

Precipitation extremes are important hydrologic events, which receive a great deal of attention worldwide. The Intergovernmental Panel on Climate Change (IPCC) [1] indicated that the increasing frequency and intensity of extreme weather events is one of the major impacts of global warming. Keggenhoff [2] found that although the number of wet days in Georgia had decreased over the period of 1971 to 2010, the contribution of very heavy and extremely heavy precipitation to total precipitation had increased over the same period. Based on daily the precipitation records from 75 meteorological stations, Huang [3] indicated that both the frequency and intensity of the most extreme precipitation events showed an increasing trend in Hunan province, China. Results from the analyses of daily precipitation extremes in India by Sen Roy [4] were in general agreement with predictions from numerical models of increasing extreme precipitation events related to increasing concentrations of greenhouse gases in the atmosphere. Although much research has focused on extreme precipitation events in recent decades, short-duration rainstorms, which are one of the main factors leading to waterlogging, cannot be reliably described using daily precipitation datasets.

Characterized by a tropical monsoon climate, Hainan Island has an average annual precipitation of up to 1600 mm. Despite the very large amount of precipitation, there is extreme heterogeneity in monthly precipitation. Most of the rainfall occurs during the flood season. Furthermore, due to economic limitations, urban public facilities, especially drainage, tend to be underdeveloped in Hainan Island. The majority of Hainan residents are employed in the agricultural sector, with cereal crops covering the largest, most widely distributed area and being the most productive crops. Soil waterlogging causes this region to suffer from massive property and income losses. According to news reports, a rainstorm in 2010 affected 16 cities and caused economic losses of more than one billion yuan. Because of the urbanization and human activities on the island, precipitation extremes of short duration and high intensity, which can be learned from researching on hourly precipitation data rather than daily precipitation data, have an impact on urban hydrology and storm-water management [5]. Additionally, Beck *et al.* [5] also found that rainfall events exhibited different spatiotemporal characteristics in the early morning and in the afternoon, which cannot be described by daily precipitation data. Furthermore, precipitation is diverse temporally and spatially, so hourly data can reflect its variation more accuracy. However, almost no studies to date have focused on precipitation extremes in Hainan Island, based on hourly precipitation datasets. Therefore, we studied the spatial and temporal precipitation patterns in Hainan Island using hourly precipitation datasets, which could be used to inform prevention of waterlogging in the province.

2. Study Area and Datasets

2.1. Study Area

"Hainan", which can literally be translated to "south of the sea", is an island in the South China Sea (Figure 1). It is located between 18°10′ N and 20°10′ N, 108°37′ E and 111°03′ E, and is separated from Mainland China by a narrow strait. It covers an area of 35.4 thousand km^2. Characterized by a tropical monsoon and tropical oceanic climate, Hainan Island has an annual mean temperature of between 22 and 27 °C, and an annual mean precipitation of 1600 mm. The period from May to October is considered to be the flood season in Hainan Island. The precipitation in the flood season can account for about 70% to 90% of annual precipitation. Hainan Island can be divided into three basins; the Nandu River basin, Changhua River basin and Wanquan River basin. These basins cover 47% of the total area of Hainan Island.

Figure 1. Locations of the precipitation stations in the study area.

2.2. Datasets

This study is based on the hourly datasets acquired from 18 precipitation stations over the period from 1967 to 2012. The locations of the 18 stations are shown in Figure 1, with additional details provided in Table 1. The tipping-bucket rain recorders and the siphon rainfall recorders, which could record the precipitation traces on the recording papers automatically, were the most commonly used in the measuring regions. Then, the precipitation traces were scanned into the computers and the hourly precipitation data could be extracted by extraction software. The dividing line for a rainy day was set at 8:00 P.M., and only the precipitation greater than 0.1 mm per hour could be recorded. Quality and homogeneity control was applied to the datasets using RClimDex software (Climate Research Division, Environment Canada, Ottawa, ON, Canada) developed in "R" and presented by Wang and Feng [6]. The software can pick out the outliers and then we deal with these outliers using our judgment. The Illogical data, such as negative values, were treated as missing value. Less than 1% of the data was found to be missing for any station. The missing value was put to zero because it occupied less than 1% of the total data and would not have any effect on the calculation indices.

Table 1. Details of the precipitation stations in Hainan Island.

Station Name	Abbreviation	Longitude (°)	Latitude (°)	Elevation (m)
Baisha	BS	19.23	109.43	215.6
Baoting	BT	18.65	109.7	68.6
Changjiang	CJ	19.27	109.05	98.1
Chengmai	CM	19.73	110	31.4
Danzhou	DZ	19.52	109.58	169
Dingan	DA	19.7	110.33	24.2
Dongfang	DF	19.1	108.62	7.6
Ledong	LD	18.75	109.17	155
Lingao	LG	19.9	109.68	31
Lingshui	LS	18.5	110.03	13.9
Qionghai	QH	19.23	110.47	24
Qiongzhong	QZ	19.03	109.83	250.9
Sanya	SY	18.22	109.58	419.4
Tunchang	TC	19.37	110.1	118.3
Wanning	WN	18.8	110.33	39.9
Wenchang	WC	19.62	110.75	21.7
Wuzhishan	WZS	18.77	109.52	328.5
Haikou	HK	20	110.25	63.5

To detect any clustering in the distribution of the selected stations, nearest neighbor analysis was undertaken. The nearest neighbor ratio was found to be 1.818, indicating that the distribution of stations displays a dispersed pattern at the 0.01 level of statistical significance.

3. Methods

3.1. Precipitation Concentration Indices (PCI)

In this study, two type of PCI are used to explore the changing patterns of precipitation concentration in Hainan Island. PCI1 is used to reveal monthly rainfall heterogeneity [7], while PCI2 is proposed to show hourly rainfall heterogeneity.

PCI1 is calculated as follows:

$$PCI1 = \frac{\sum_{i=1}^{12} P_i^2}{\left(\sum_{i=1}^{12} P_i^2\right)^2} \times 100 \quad (1)$$

where P_i is the monthly precipitation in month *i*. According to Equation (1), the minimum value of PCI1 is 8.3, which occurs when the same amount of precipitation falls in each month. Oliver [7] suggested that monthly rainfall heterogeneity could be divided into four categories according to the PCI1 values: uniform precipitation distribution (PCI1 < 10), moderate precipitation distribution (PCI1: 10–15), irregular precipitation distribution (PCI1: 15–20) and highly irregular precipitation distribution (PCI1 > 20).

PCI2 has been used to study daily rainfall heterogeneity [8], based on the fact that the precipitation frequency can generally be described by a negative exponential distribution (Equation (2)). In this study,

PCI2 is used to describe hourly rainfall heterogeneity, which is defined as the similarity between hourly and daily precipitation frequency distribution. We can find out that the relative between the accumulated percentages of precipitation Y and the accumulated percentages of hours X can be described by a negative exponential distribution (Figure 2).

$$Y = aX \exp(bX) \tag{2}$$

where a and b are constants.

According to Martin-Wide [8], PCI2 could be calculated as follows:

$$PCI2 = 1 - A'/5000 \tag{3}$$

where A' is the definite integral of the exponential curve between 0 and 100.

$$A' = \left[\frac{a}{b}e^{bx}\left(x - \frac{1}{b}\right)\right]_0^{100} \tag{4}$$

$$\ln a = \frac{\sum X_i^2 \sum \ln Y_i + \sum X_i \sum (X_i \ln X_i) - \sum X_i^2 \sum \ln X_i - \sum X_i \sum (X_i \ln Y_i)}{N \sum X_i^2 - \left(\sum X_i\right)^2} \tag{5}$$

$$b = \frac{N \sum (X_i \ln Y_i) + \sum X_i \sum \ln X_i - N \sum (X_i \ln X_i) - \sum X_i \sum \ln Y_i}{N \sum X_i^2 - \left(\sum X_i\right)^2} \tag{6}$$

where N is the number of (X_i, Y_i).

We can find out that the more heterogeneous in hourly precipitation, the smaller area A' will be. Thus, larger PCI2 value indicates more precipitation falls in a same period, showing more heterogeneous in hourly precipitation.

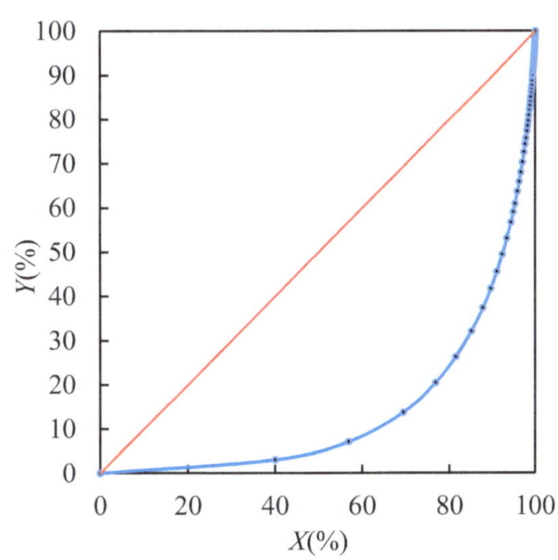

Figure 2. Accumulated percentages of precipitation Y—accumulated percentages of hours X curve (Blue curve represents the heterogeneous precipitation. Red curve represents the perfectly homogeneous precipitation.).

3.2. Extreme Precipitation Indices

A review of the methods used to test trends in extreme precipitation patterns regionally suggests that certain threshold values can be used to identify extreme values. For example, Haylock *et al.* [9] utilized two indices of extreme rainfall, which were calculated from daily precipitation data during the period from 1958 to 2000 in Europe. In another study by Sen Roy [10], nine extreme precipitation variables based on hourly datasets were utilized. Given the variety of indices that have been used to determine trends in extreme hourly precipitation values in combination with local conditions, the following indices have been used in this study (Table 2).

Table 2. Extreme precipitation indices (EPI) used in the current study. All precipitation events greater than 0.1 mm occurring throughout the entire study period were sorted in ascending order for each station separately to determine the percentile threshold values.

Index	Definition	Unit
FQ 90	Frequency of precipitation greater than the 90th percentile threshold value during the flood season	h
FQ 95	Frequency of precipitation greater than the 95th percentile threshold value during the flood season	h
FQ 97.5	Frequency of precipitation greater than the 97.5th percentile threshold value during the flood season	h
TP 90	Total amounts of precipitation greater than the 90th percentile threshold value during the flood season	mm
TP 95	Total amounts of precipitation greater than the 95th percentile threshold value during the flood season	mm
TP 97.5	Total amounts of precipitation greater than the 97.5th percentile threshold value during the flood season	mm
Max 1 h	Largest hourly precipitation during the flood season	mm
Max 3 h	Largest continuous 3-h precipitation during the flood season	mm
Max 6 h	Largest continuous 5-h precipitation during the flood season	mm
FQ 20 mm	Frequency of hourly precipitation during the flood season greater than 20 mm	h
FQ 30 mm	Frequency of hourly precipitation during the flood season greater than 30mm	h

3.3. Principal Component Analysis (PCA)

For each station, a matrix of 46 rows corresponding to each year from 1967 to 2012, and 12 columns containing the year and each of the 11 EPI described above was compiled. Linear regression analysis was then carried out for each station. A matrix of 18 rows, one for each station, and 12 columns, containing the year and the standardized regression coefficients was then compiled for each index. The standardized regression coefficients, calculated as the unstandardized regression coefficients multiplied by (S_x/S_y), where S_x is the standard deviation of the independent variable and S_y is the standard deviation of the dependent variable, represent the strength of the trend.

Table 3 presents the correlation coefficients for the 11 EPI. As shown in Table 3, all of the coefficients are greater than 0.5, indicating that they are suitable for inclusion in PCA. PCA was applied to the new 18-row by 12-column matrix. The first component calculated through PCA explained 82% of the variation for the 11 EPI. The loading matrix of standardized regression coefficients is shown for the first component in Table 4. All variables have loadings higher than 0.7, suggesting that the first component reflects both the frequency and the intensity of extreme precipitation. Finally, we recalculated [4] the component scores by setting their standard deviation to 0.14 (the mean standard deviation of the 11 EPI) and their mean to 0.23 (the average mean of the 11 EPI).

Table 3. Correlation coefficients of the extreme precipitation indices (EPI).

Correlation Coefficient	FQ 90	FQ 95	FQ 97.5	TP 90	TP 95	TP 97.5	Max 1 h	Max 3 h	Max 6 h	FQ 20 mm	FQ 30 mm
FQ 90	1.000	--	--	--	--	--	--	--	--	--	--
FQ 95	0.942	1.000	--	--	--	--	--	--	--	--	--
FQ 97.5	0.857	0.858	1.000	--	--	--	--	--	--	--	--
TP 90	0.946	0.913	0.930	1.000	--	--	--	--	--	--	--
TP 95	0.908	0.923	0.934	0.988	1.000	--	--	--	--	--	--
TP 97.5	0.810	0.788	0.949	0.948	0.956	1.000	--	--	--	--	--
Max 1 h	0.732	0.668	0.702	0.829	0.806	0.805	1.000	--	--	--	--
Max 3 h	0.728	0.629	0.806	0.822	0.790	0.852	0.817	1.000	--	--	--
Max 6 h	0.698	0.580	0.750	0.772	0.732	0.791	0.650	0.923	1.000	--	--
FQ 20 mm	0.816	0.806	0.922	0.918	0.927	0.944	0.697	0.826	0.777	1.000	--
FQ 30 mm	0.580	0.558	0.608	0.762	0.783	0.808	0.748	0.698	0.661	0.788	1.000

Table 4. Loadings of the standardized regression coefficients.

EPI	FQ 90	FQ 95	FQ 97.5	TP 90	TP 95	TP 97.5	Max 1 h	Max 3 h	Max 6 h	FQ 20 mm	FQ 30 mm
Loading	0.907	0.873	0.938	0.987	0.979	0.969	0.845	0.889	0.833	0.946	0.798

3.4. Stability Test

The Mann-Kendall (M-K) trend test [11,12], recommended by the World Meteorological Organization (WMO), is an effective tool to assess the significance of monotonic trends in hydrometeorological time series. This method has the advantages of not requiring a given data distribution and not being affected by outlying data, and has been used entensively to test trends in hydrometeorological data, including temperature, rainfall and runoff time series [13–16]. In this study, the M-K method was used to test the trend stability of extreme precipitation time series.

In general, the trend of a time series is detected for a specific time period, and test results are strongly influenced by the selection of the latter. For the same variable, it is likely that different trends will result from the analysis of different time periods, which reflects the instability of the trend. It is, therefore, difficult to determine the real trend of a time series. To access the stability characteristics of trends evaluated in this study, we determined the stability of extreme temperature and precipitation trends using the statistical significance of the trends tested by the M-K method, following the approach of Lupikasza [17] and Hidalgo-Muñoz *et al.* [18]. In this method, trend stability was defined as follows:

$$S = \frac{N_{0.1}}{K} \tag{7}$$

where K is the number of 30-year moving periods for one station (17 movable 30-year periods during the 1967–2012 period) and No.1 is the number of 30-year moving periods for each trend direction that passes the M-K test at the 90% confidence level. Trends are categorized as unstable ($S < 0.25$), stable ($0.25 \leq S < 0.6$), or highly stable ($S \geq 0.6$).

4. Results and Discussion

4.1. Spatial Pattern of PCI for 1967–2012

As is well known, Hainan Island is characterized by a tropical monsoon climate, which has distinct wet and dry seasons. To understand precipitation patterns in Hainan Island, we calculated of two PCI, PCI1 and PCI2.

4.1.1. PCI1

Figure 3 shows the spatial distributions of both calculated PCI. Throughout the island, PCI1 values range from 11.8 to 14.7. According to Oliver [7], such PCI1 values indicate a moderate precipitation distribution in Hainan Island. The maximum PCI1 value was found for the Western Changjiang County, while the minimum value was recorded for the Northeastern Wenchang County. Moreover, the seasonal precipitation concentration presents a decreasing trend from west to east.

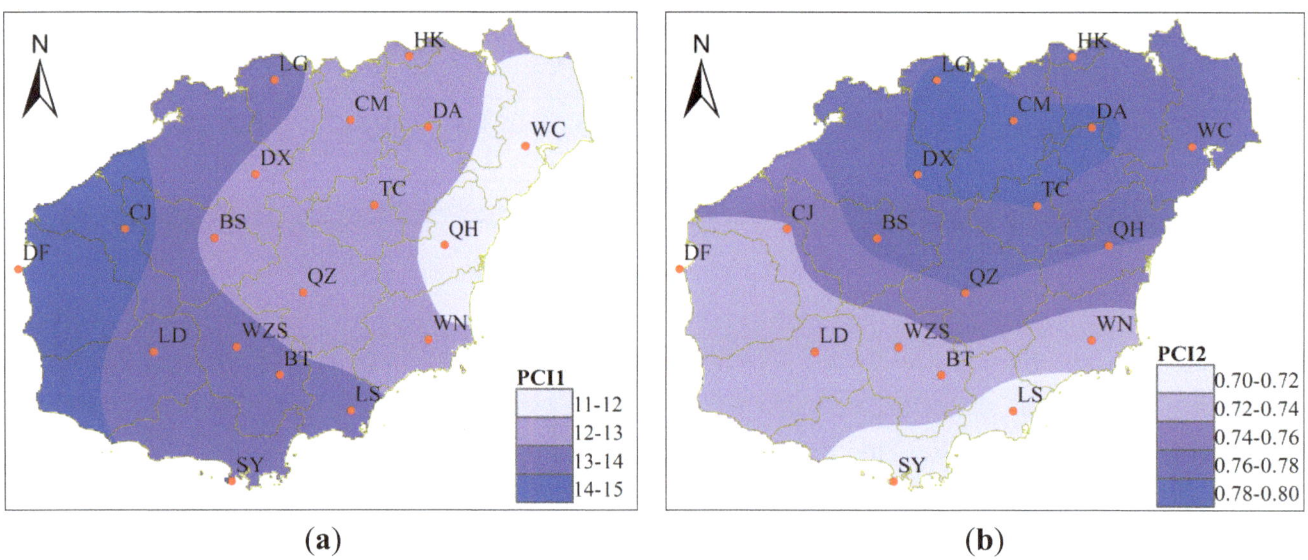

Figure 3. The spatial distribution of (**a**) PCI1 and (**b**) PCI2 from 1967 to 2012.

4.1.2. PCI2

PCI2 is applied to explore the contribution of precipitation extremes of certain lengths of time to total precipitation. Zhang *et al.* [19] found the lowest PCI2 values in the southwestern and northeastern regions of the Pearl River basin and the highest values in the northwestern and southern regions over the period 1960–2005. However, PCI2 values in Hainan Island are between 0.70 and 0.79, and display an obvious increasing trend from south to north, which is different from that of PCI1. The north-central regions of Hainan Island are characterized by the maximum PCI2 values, which are all above 0.78. On the contrary, the southern regions are characterized by the lowest values, below 0.72. Furthermore,

PCI2 was proposed to reflect the contribution of hours with the highest rainfall to the total amount of precipitation. In this study area, the PCI2 value of 0.70 corresponds with 80% of total precipitation falling in the 20% rainiest hours.

Due to the temporal heterogeneity of precipitation, the rest of this article focuses on analyzing precipitation extremes in the flood season (May to October).

4.2. Extreme Precipitation Indices (EPI) Trends

This section is devoted to the trends and significances thereof of the EPI calculated for the study region. The magnitude of the trends was calculated using linear regression analysis, which reveals the trend direction and the magnitude of variation of precipitation extremes. The M-K test was used to evaluate the trend significance. In recent years, several studies have undertaken similar work. For example, Chi *et al.* [20] used the generalized least squares method to analyze the linear regression trends, and found an increase in the frequency of extreme precipitation and a reduction in the frequency of light rain days in eastern forest regions of China in recent 50 years. Gocic *et al.* [21] analyzed precipitation and standardized precipitation index trends using linear regression in Serbia.

4.2.1. Statistical Trends

Regional precipitation indices, which are plotted in Figure 4, exhibit increasing trends. This result aligns with the statement of IPCC [1], that global warming has led to an increasing frequency and intensity of extreme precipitation events. The frequency indices investigated here, including FQ 90, FQ 95, FQ 97.5, FQ 20 mm and FQ 30 mm (Figure 4), show increasing trends for the whole island. The increasing rates of the five indices listed above are 2.5 h/decade, 1.4 h/decade, 1.1 h/decade, 1.2 h/decade and 0.6 h/decade, respectively. For TP 90, TP 95 and TP 97.5, rapidly increasing rates of 5.2 mm/year, 4.3 mm/year and 3.7 mm/year, respectively, can be observed. The Max 1 h, Max 3 h and Max 6 h display increasing rates of 1.9 mm/decade, 3.1 mm/decade and 4.6 mm/decade, respectively. In other words, both the frequency and the intensity of extreme precipitation events exhibit increasing trends. Furthermore, it also can be seen that almost all of the indices present the rapidly increase after 1987. It is more likely an oscillation in the long term variations instead of a tipping point. Shinjiro [22] found that the conclusion recent heavy precipitation in the 1990s was unprecedentedly intense and frequent in Tokyo was misunderstanding. He digitalized hourly precipitation since 1890 recorded at the Tokyo observatory and concluded that the local frequency minimum in the longer time series is just located in the 1970s. The 1990s was not the tipping point instead of a long-term cycle. In addition to Tokyo, Zhu [23] researched on summer monsoon rainfall in China and found an 80-year scale variability. Minobe [24] found a 50–70 year climatic oscillation over the North Pacific and North America. We can observe a 20-year oscillation in the precipitation extremes through the five-year moving average lines.

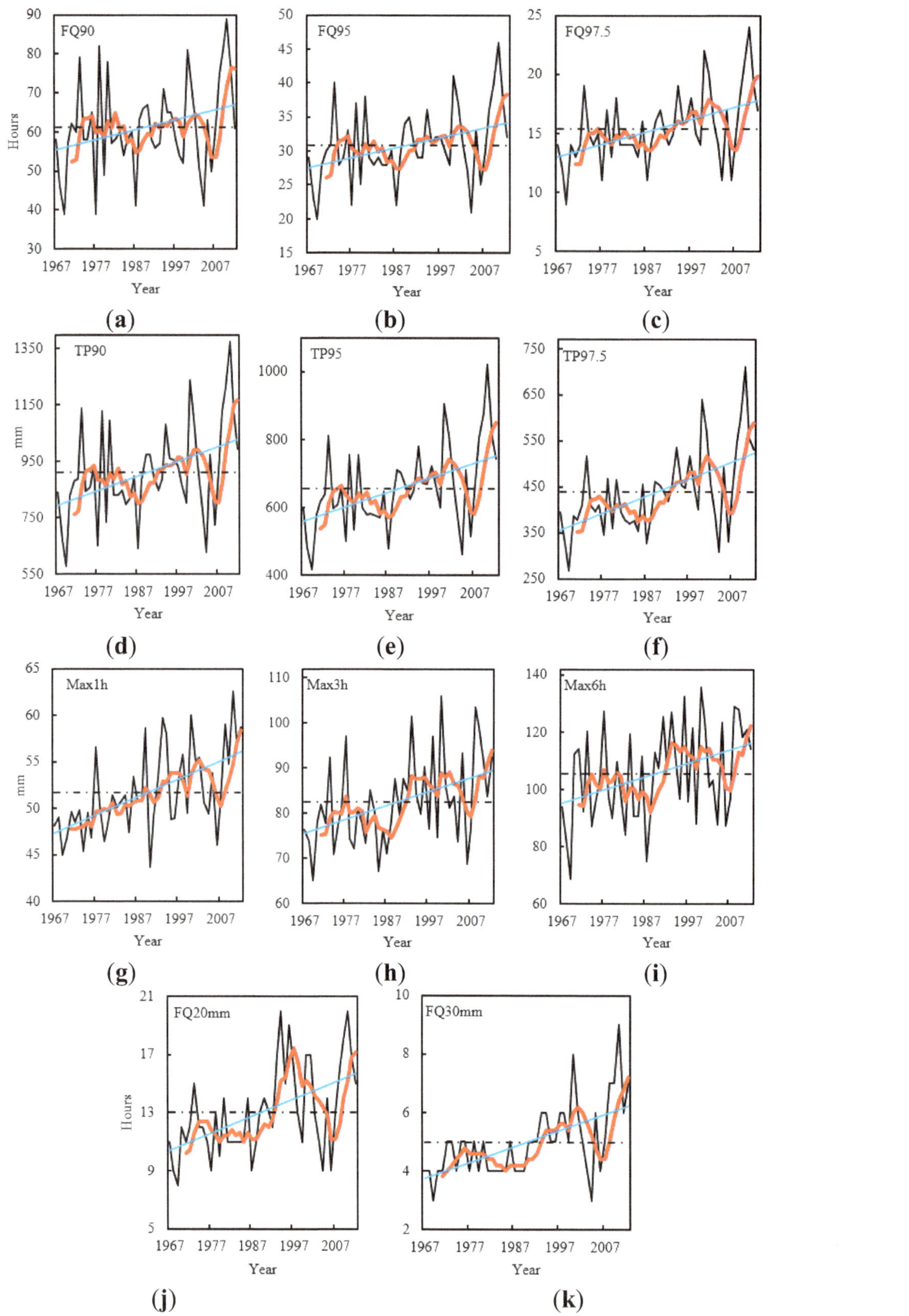

Figure 4. Regional time series of EPI for the flood season. (Black curves represent the arithmetic average of all station. Red curves represent the 5-year moving average. Blue lines are the trend lines of the indices. Black dotted lines show the 1967–2012 average.) (**a**) FQ 90; (**b**) FQ 95; (**c**) FQ 97.5; (**d**) TP 90; (**e**) TP 95; (**f**) TP 97.5; (**g**) Max 1 h; (**h**) Max 3 h; (**i**) Max 6 h; (**j**) FQ 20 mm; (**k**) FQ 30 mm.

4.2.2. Trend Significance

Table 5 shows the results of the trends significant tests for the regional EPI. All trends are significant at the 0.1 level, indicating that all the EPI display significant increasing trend. Only the FQ 90 is significant at the 0.1 level, whereas FQ 95 and Max 6 h are significant at the 0.05 level, and all others are significant at 0.01 level.

Table 5. Results of the trend significance tests for the regional EPI for the period 1967–2012.

EPI	FQ 90	FQ 95	FQ 97.5	TP 90	TP 95	TP 97.5	Max 1 h	Max 3 h	Max 6 h	FQ 20 mm
Z value	1.79 *	2.49 **	3.31 ***	2.84 ***	3.26 ***	3.60 ***	4.07 ***	2.78 ***	2.36 **	3.42 ***
Statistical trends	Sut	Sut	Sut	Sut	Sut	Sut	Sut	Sut	Sut	Sut

Notes: *Z* value is calculated by the M-K test. Sut, significant upward trends. * Significant at the 0.1 level; ** Significant at the 0.05 level; *** Significant at the 0.01 level.

4.3. Spatial Patterns of the Trends

The first component reflects most of the variation of EPI. Sen Roy [4] calculated the first component using seven EPI as input to describe the trends of precipitation extremes in India. To determine the spatial patterns of trends in the Hainan Island, the recalculated first component is plotted (Figure 5). The recalculated first component, ranging from −0.101 to 0.458, shows that the regression coefficients are generally positive in the northern and southern regions of Hainan Island, while the interior regions experience negative trends. Negative trends are also observed in LG and DF stations, which are located in the northern and eastern regions, respectively. CM, WC and SY stations exhibit the most positive trends. On the contrary, QZ station displays the most negative trend.

The maximum amount of precipitation occuring within a specified time period is one of the most common methods used to define extreme precipitation events [25]. Accordingly, to display the patterns of extreme precipitation in Hainan Island, we have also mapped three other EPIs, FQ 20 mm, TP 90 and Max 3 h (Figure 5). Spatial patterns of these three indices are similar, and show positive trends in the northern and southern regions, and negative trends in the interior regions. However, small differences exist for some stations. For example, LG station shows positive trends for FQ 20 mm and Max 3 h, and a negative trend for TP 90.

Moreover, the decreasing trends are observed to concentrate predominantly in interior section, where the elevation is higher than the surrounding coastal areas. Many people researched on the relationships between elevation and trends of EPI. Zhang *et al.* [26] researched precipitation extremes over the Hengduan mountains region and found that trends of some precipitation indices displayed statistically negative correlation with elevation while trends of some other indices showed positive correlation with elevation. However, in the paper of Chen *et al.* [27], they found that each EPI showed a different fluctuation in trend with elevation and they concluded that no significant correlations between trends of EPI and elevation. The elevation of the interior stations, such as BS, QZ and WZS, is usually higher than 200 m while the elevation of the coastal areas (except SY station) is lower than 200 m. It may therefore be concluded that high elevations have decreasing trends in precipitation extremes. However, this is not the case for all stations. For example, SY station is a coastal station with an elevation greater than 200 m, but shows an increasing trend. Furthermore, trends of EPI are associated with not only the elevation but also trends of annual precipitation. We then find that annual precipitation in QZ station exhibits decreasing trend from 1967 to 2012 while others exhibit increasing trends. The decreasing trend in annual precipitation is the main cause leading to the decreasing trend in precipitation extremes in QZ. We may venture to conclude that there is no significant correlation between elevation and trends of EPI.

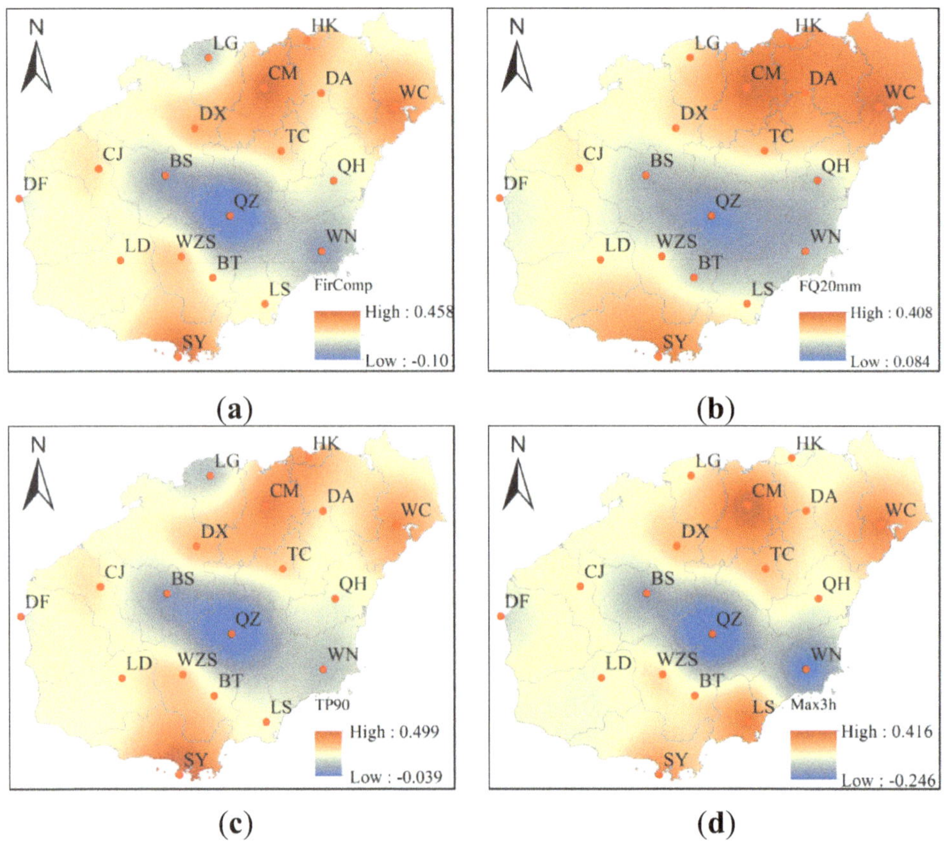

Figure 5. Flood season spatial patterns of trends in (**a**) the PCA first component; (**b**) FQ 20 mm; (**c**) TP 90 and (**d**) Max 3 h.

4.4. Trend Stability

To better illustrate the spatial patterns of extreme precipitation, the trend stability was also analyzed. Distinct from their trend distributions, the stabilities of FQ 20 mm, Max 3 h and TP 90 are dissimilar in Hainan Island (Figure 6). However, instabilities are present in most regions. The stability coefficients of FQ 20 mm in the northeastern and southwestern regions are between 0.25 and 0.6. As described by Wu [28], stability coefficients ranging from 0.25 to 0.6 indicate a stable trend. Most regions of Hainan Island show an unstable trend for Max 3 h. In the southern region, BT, LS and SY stations, show stable trends. For TP 90, stations associated with stable trends are located in the northern and southern regions of the study area. SY station exhibits strong stability for TP 90. Furthermore, combined with the increasing trend revealed in the previous section, the SY station shows a strongly stable increasing trend for 1967–2012. The stability in FQ 20 mm exhibits a similar phenomenon to TP 90, but the area for which stable trends are calculated is larger. Overall, most regions of Hainan display unstable EPI trends, while some parts of the southern and northern regions exhibit stable or strongly stable trends.

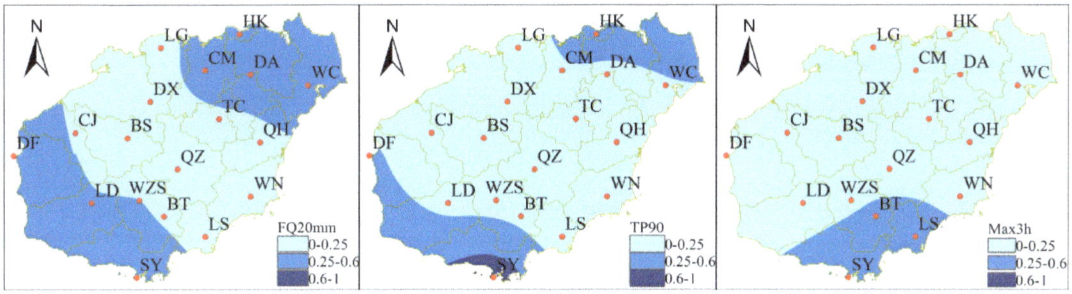

Figure 6. Trend stability for (**a**) FQ 20 mm; (**b**) TP 90 and (**c**) Max 3 h over the study period.

5. Conclusions

The precipitation heterogeneity and the spatial and temporal distribution of trend directions in precipitation extremes in Hainan Island during the period 1967–2012 were analyzed using two PCI and 11 EPI. In line with IPCC findings [1], the frequency and intensity of extreme precipitation events was found to exhibit increasing trends corresponding with observed global warming. According to Wu *et al.* [29], extreme events contribute significantly to variations in the extreme precipitation amount, as well as in the total annual precipitation. During the period from 1962 to 2005, both the number of days and the total precipitation associated with extreme events slightly increased in Hainan Island. Additionally, we also found a possible conclusion that a 20-year oscillation existed in precipitation extremes. Further investigation in the precipitation extremes oscillation is indispensable.

PCI1 in Hainan Island ranges from 11.8 to 14.7, indicating that yearly precipitation displays a moderate seasonality. PCI1 presents a decreasing trend from Western to Eastern Hainan. Precipitation in the western region of Hainan is known to be characterized by severe seasonality, while the seasonality of the southern region is known to be less severe. However, different trends appear in PCI2, which is calculated to reveal the concentration of hourly precipitation. Ranging from 0.70 to 0.79, PCI2 generally showed an increasing trend from south to north. In other words, more precipitation falls in a shorter period in the north, suggesting that more short-duration storms occur in this region. Wu *et al.* [29] also found that the heaviest 1% of daily precipitation contributes 23% of the annual precipitation in Hainan Island. In our study, at least 80% of the total precipitation is found to fall in the 20% rainiest hours. Overall, through the analysis of PCI1 and PCI2, it can be concluded that precipitation in the western region is more seasonal than in the eastern region, and that the northern hourly precipitation is more concentrated than that of the southern region.

Overall, for the entire island, it was observed that precipitation extremes show significant increasing trends. However, similar to the spatial patterns of precipitation extremes in Hainan Island, the southern and northern regions exhibit increasing trends, while the interior exhibits a decreasing trend. In other words, in the southern and northern Hainan Island, the frequency, amount, and intensity of rainstorms showed increasing trends for the period 1967–2012, which is the opposite of the interior region. Additionally, there is no significant correlation between elevation and trends of EPI.

Furthermore, an increasing trend of precipitation extremes can be predicted through the trend uncovered for the past decades and related trend stability. The northern and southern regions should receive more attention to prevent potential damage inflicted by the precipitation extremes according to the combination of statistical trends and stabilities. A strongly stable increasing trend in TP 90 for SY station indicates that the amount of precipitation in rainstorms can be expected to increase steadily, which may cause waterlogging in Sanya county.

Acknowledgments

This study has been financially supported by the Ministry of Water Resources Special Funds for Scientific Research Projects of Public Welfare Industry (201301093 and 201401048).

Author Contributions

Authors collaborated together for the completion of this work. Wenjie Chen designed this study, processed data and wrote the manuscript. Chenghao Chen, Longbing Li and Litao Xing provided the precipitation data, processed data and contributed to the EPI calculation. Guoru Huang designed the study, reviewed and revised the manuscript. Chuanhao Wu contributed to the EPI calculation, reviewed and revised the manuscript.

Conflicts of Interest

The authors declare no conflict of interest.

References

1. Working Group I Contribution to the Fifth Assessment Report of the Intergovernmental Panel on Climate Change. Available online: http://www.ipcc.ch/pdf/assessment-report/ar5/wg1/WG1AR5_Frontmatter_FINAL.pdf (accessed on 12 May 2015).
2. Keggenhoff, I.; Elizbarashvili, M.; Amiri-Farahani, A.; King, L. Trends in daily temperature and precipitation extremes over Georgia, 1971–2010. *Weather Clim. Extrem.* **2014**, *4*, 75–85.
3. Huang, J.; Sun, S.L.; Xue, Y.; Zhang, J.C. Spatial and temporal variability of precipitation indices during 1961–2010 in Hunan province, Central South China. *Theor. Appl. Climatol.* **2014**, *118*, 581–595.
4. Sen Roy, S.; Balling, R.C. Trends in extreme daily precipitation indices in India. *Int. J. Climatol.* **2004**, *24*, 457–466.
5. Beck, F.; Bárdossy, A.; Seidel, J.; Müller, T.; Fernandez Sanchis, E.; Hauser, A. Statistical analysis of sub-daily precipitation extremes in Singapore. *J. Hydrol. Reg. Stud.* **2015**, *3*, 337–358.
6. RClimDex. Available online: http://etccdi.pacificclimate.org/software.shtml (accessed on 17 November 2013).
7. Oliver, J.E. Monthly precipitation distribution: A comparative index. *Prof. Geogr.* **1980**, *32*, 300–309.
8. Martin-Vide, J. Spatial distribution of a daily precipitation concentration index in peninsular Spain. *Int. J. Climatol.* **2004**, *24*, 959–971.
9. Haylock, M.R.; Goodess, C.M. Interannual variability of European extreme winter rainfall and links with mean large-scale circulation. *Int. J. Climatol.* **2004**, *24*, 759–776.
10. Sen Roy, S.; Rouault, M. Spatial patterns of seasonal scale trends in extreme hourly precipitation in South Africa. *Appl. Geogr.* **2013**, *39*, 151–157.
11. Mann, H.B. Non-parametric tests against trend. *Econometrica* **1945**, *13*, 245–259.
12. Kendall, M.G. *Rank Correlation Methods*, 4th ed.; Charles Griffin: London, UK, 1975.

13. Wan, L.; Zhang, X.P.; Ma, Q.; Zhang, J.J.; Ma, T.Y.; Sun, Y.P. Spatiotemporal characteristics of precipitation and extreme events on the loess plateau of china between 1957 and 2009. *Hydrol. Process.* **2014**, *28*, 4971–4983.
14. Bocchiola, D. Long term (1921–2011) hydrological regime of alpine catchments in northern Italy. *Adv. Water Resour.* **2014**, *70*, 51–64.
15. Taxak, A.K.; Murumkar, A.R.; Arya, D.S. Long term spatial and temporal rainfall trends and homogeneity analysis in wainganga basin, Central India. *Weather Clim. Extrem.* **2014**, *4*, 50–61.
16. Anghileri, D.; Pianosi, F.; Soncini-Sessa, R. Trend detection in seasonal data: From hydrology to water resources. *J. Hydrol.* **2014**, *511*, 171–179.
17. Lupikasza, E. Spatial and temporal variability of extreme precipitation in Poland in the period 1951–2006. *Int. J. Climatol.* **2010**, *30*, 991–1007.
18. Hidalgo-Munoz, J.M.; Argueso, D.; Gamiz-Fortis, S.R.; Esteban-Parra, M.J.; Castro-Diez, Y. Trends of extreme precipitation and associated synoptic patterns over the Southern Iberian Peninsula. *J. Hydrol.* **2011**, *409*, 497–511.
19. Zhang, Q.; Xu, C.Y.; Gemmer, M.; Chen, Y.D.; Liu, C.L. Changing properties of precipitation concentration in the pearl river basin, China. *Stoch. Environ. Res. Risk A* **2009**, *23*, 377–385.
20. Chi, Y.; Zhang, C.; Liang, C.; Wu, H. The precipitation change in eastern forest regions of China in recent 50 years. *Acta Ecol. Sin.* **2013**, *33*, 217–226.
21. Gocic, M.; Trajkovic, S. Analysis of precipitation and drought data in Serbia over the period 1980–2010. *J. Hydrol.* **2013**, *494*, 32–42.
22. Shinjiro Kanae, T.O.; Akira, K. Changes in hourly heavy precipitation at Tokyo from 1890 to 1999. *J. Meteorol. Soc. Jpn.* **2004**, *82*, 241–247.
23. Zhu, J. 80 Year oscillation of summer rainfall over North China and East Asian summer monsoon. *Geophys. Res. Lett.* **2002**, *29*, doi:10.1029/2001GL013997.
24. Minobe, S. A 50–70 year climatic oscillation over the north pacific and North America. *Geophys. Res. Lett.* **1997**, *24*, 683–686.
25. Sen Roy, S. A spatial analysis of extreme hourly precipitation patterns in India. *Int. J. Climatol.* **2009**, *29*, 345–355.
26. Zhang, K.; Pan, S.; Cao, L.; Wang, Y.; Zhao, Y.; Zhang, W. Spatial distribution and temporal trends in precipitation extremes over the hengduan mountains region, china, from 1961 to 2012. *Quat. Int.* **2014**, *349*, 346–356.
27. Chen, F.; Chen, H.; Yang, Y. Annual and seasonal changes in means and extreme events of precipitation and their connection to elevation over Yunnan province, China. *Quat. Int.* **2015**, in press.
28. Wu, C.H.; Huang, G.R.; Yu, H.J.; Chen, Z.Q.; Ma, J.G. Spatial and temporal distributions of trends in climate extremes of the feilaixia catchment in the upstream area of the Beijiang river basin, South China. *Int. J. Climatol.* **2014**, *34*, 3161–3178.
29. Wu, Y.J.; Wu, S.G.; Zhai, P.M. The impact of tropical cyclones on Hainan Island's extreme and total precipitation. *Int. J. Climatol.* **2007**, *27*, 1059–1064.

9

Helminth Egg Removal Capacity of UASB Reactors under Subtropical Conditions

Rosa-Elena Yaya-Beas [1,2,*], Christian Ayala-Limaylla [1], Katarzyna Kujawa-Roeleveld [2,3], Jules B. van Lier [4,5] and Grietje Zeeman [2,3]

[1] Environmental Engineering Faculty, National University of Engineering, Lima 25, Peru; E-Mail: bgbcayalal@gmail.com

[2] Sub-Department of Environmental Technology, University of Wageningen, Wageningen 6700 AA, The Netherlands; E-Mails: katarzyna.kujawa@wur.nl (K.K.-R.); grietje.zeeman@wur.nl (G.Z.)

[3] Lettinga Associates Foundation (LeAF), P.O. Box 500, Wageningen 6700 AM, The Netherlands

[4] Section Sanitary Engineering, Department of Water Management, Faculty of Civil Engineering and Geosciences, Delft University of Technology, Delft 2600 GA, The Netherlands; E-Mail: J.B.vanLier@tudelft.nl

[5] Unesco-IHE Institute for Water Education, Westvest 7, Delft 2611 AX, The Netherlands

* Author to whom correspondence should be addressed; E-Mail: rosa.yayabeas@wur.nl or Ryaya@uni.edu.pe;

Academic Editor: Marc Henry

Abstract: This research was conducted to study the anaerobic sludge filtration capacity regarding helminth egg removal in upflow anaerobic sludge blanket (UASB) reactors. Two 25 L lab-scale UASB reactors were operated at an ambient temperature which varied between 17.1 and 28.6 °C. *Ascaris suum* egg was selected as the model egg considering its similarity in terms of size and morphology to *Ascaris lumbricoides*, a human pathogen. *Ascaris suum* eggs were obtained from female parasites of infected pigs. The anaerobic sludge filtration capacity was performed applying upflow velocities between 0.09 and 0.68 $m \cdot h^{-1}$. Three sludge bed heights in the range of 0.30–0.40 m, 0.50–0.60 m and 0.60–0.70 m were applied. These sludge bed heights corresponded to 19%–25%, 31%–38% and 38%–44% of the total reactor height, respectively. Under the mentioned conditions, the average helminth egg removal efficiency was reciprocally correlated to the imposed upflow velocity. The studied lab-scale reactors reported an average helminth egg removal between 34%–100%, 30%–91% and 34%–56%, when the sludge bed in the UASB reactor was 19%–25%, 31%–38% and 38%–44% of the total reactor height, respectively. The decreased filtration capacity at increasing sludge bed heights might be likely related to biogas production and channeling formation. The average helminth egg removal efficiency in the control experiments performed without any sludge bed, by plain sedimentation, varied between 44% and 66%.

Keywords: helminth eggs; *Ascaris suum*; pathogens; UASB reactor; sludge bed filtration capacity

---

1. Introduction

When treated wastewater is intended to be used for agricultural purposes, the presence of pathogens may limit its application potential [1,2]. Due to their shell resistance, helminth eggs are the most persistent pathogens to inactivation [3,4]. Particularly in developing countries, high concentrations of helminth eggs are present in domestic wastewater, which cause parasitic diseases like ascariasis, taeniasis and trichuriasis [5–7]. The prevailing symptoms caused by these diseases include diarrhea, effects on mental development, and anemia [8–10]. Within the group of pathogenic organisms, helminth eggs are infective agents which range in size from 10 μm to more than 100 μm [1,11,12].

Most literature regarding removal of helminth eggs is related to inactivation of helminth eggs contained in excess sludge [4,13–16] and physical removal from wastewater [1,17,18]. Technologies to inactivate helminth eggs in sludge are aimed at destroying the structure of the egg (mainly damages in its lipid layer) which prevents further development and survival of the eggs [4,19,20]. The best technologies for inactivation of helminth eggs present in sludge are thermal treatment at 108 °C [16], irradiation at 3500 Gy [14,15], pasteurization at 70 °C [13,21] or chemical treatment using sulfuric, hydrochloric, propionic, acetic or peracetic acid [19]. Processes like alkaline pre- and post-stabilization, by, for example, adding lime or other alkaline compound to the sludge, and thermophilic anaerobic digestion have shown high residual concentrations of worm eggs, *i.e.*, more than 1 egg·g^{-1} TS, and 0.99–1.1 egg·g^{-1} TS in the sludge, respectively [3,4]. These values are higher than the restrictive limit in developing countries, where the use of treated waste and wastewater in (irrigated) agriculture is commonly applied. [1,8,18,22]. Maya *et al.* [4] reported that four genera of helminth eggs, *i.e.*, *Ascaris lumbricoides*, *Ascaris suum*, *Toxocara canis and Trichuris trichiura*, are sensitive to environmental conditions in the larval state in the sludge. Furthermore, a proper combination of pH, dryness and contact time with temperatures above 60 °C can be applied to inactivate the eggs, efficiently [4,23]. Unfortunately, external energy and chemical-dependent technologies are in general not feasible for developing countries because they are complex, not sustainable and expensive in terms of investment and operation and maintenance costs [4,17,18,24].

Within the group of technologies applied to physical helminth egg removal (not inactivation) from wastewater, land-based post-treatment technologies such as sand filtration, wetlands and polishing ponds are reported to achieve helminth egg removal of 90%–99%, 100% and 100%, respectively [1,25,26]. In addition, Jimenez [19] reported that grit removal followed by a coagulation flocculation process in what is known as advanced primary treatment (APT), combined with an upflow sand filtration, reduced the amount of helminth eggs from 1.2 to 0.2 egg·g^{-1}. Additionally, a study using APT followed by a sand filter combined with a synthetic medium reduced the amount of helminth eggs in average from 26 to 1.2 egg·g^{-1}. Furthermore, APT followed by a multimedia filter and inclined parallel plates reduced the concentration from 27.0 to 1.2 egg·g^{-1} [19].

Limited research is executed on physical helminth egg removal in UASB reactors [1,3,25]. Filtration and sedimentation has been considered the main mechanism of helminth egg removal in UASB reactors [1,17,25]. During filtration and sedimentation, helminth eggs are respectively accumulated in the sludge bed and on the bottom of the reactor [1,27].

The removal of helminth eggs in UASB reactors has been reported to amount to 60%–90% [1]. UASB reactor technology is relatively cheap and compact and could contribute to domestic wastewater treatment in a sustainable way to improve environmental protection, resource recovery and public health protection [18,26,28–31]. However, the effect of different operational conditions of UASB reactors on helminth egg removal has not been evaluated thus far. Helminth egg removal through sedimentation and filtration would give an added value to UASB reactors. Mahmoud *et al.* [32] described the sludge bed filtration of UASB reactors as a mechanism for solids removal in domestic wastewater. Similar processes might affect the removal of helminth eggs in UASB reactors.

Pig helminths like *Ascaris suum*, *Trichuris suis* and *Oesophagostomum spp.* are often used in research as model organisms for human intestinal parasites, because they are very similar in morphology and size to the corresponding human parasite eggs and are relative easy to obtain in high numbers from infected pigs [33]. Maya *et al.* [4] reported that no significant differences were found between *Ascaris lumbricoides* and *Ascaris suum* regarding the inactivation conditions. In addition, *Ascaris* eggs were found to be the most resistant helminth egg genus to inactivation, combining unfavorable pH, dryness and temperature conditions, in comparison with *Taenia sp.* and *Toxocara sp.*, *Trichuris sp.* and *Hymennolepis* [4]. In previous work [34,35], it has been shown that in the municipal wastewater in Peru, *Ascaris lumbricoides* was the predominant specie. Therefore, this research was conducted using *Ascaris suum* as helminth eggs as surrogate for the human parasite.

Mature *Ascaris sp.* eggs have an ovoid shape with average sizes of 40–70 μm [1,36]. This helminth egg is very resistant to inactivation under different environmental conditions [1,4]. This resistance is related to their four-layered shell composed of a lipid layer with a total thickness of about 4.5 μm, a mechanically rigid chitinous layer, a vitelline membrane and an external coat [3,36,37]. The shell is sensitive to lipid solvents and shows reduced surfaces and ridges. This mammillated layer is bile-stained to a golden brown color, and its high hydration makes it limp in the natural environment [4,37]. Microorganisms present in anaerobic sludge may play a role in degrading nematode eggs, though limited research results are available. For example, it has been reported that *Duddingtonia flagrans* and *Angiostrongylus Cantonensis* (nematofagous fungi) feed on free-living nematodes at the larval stage at 27 °C [38–40]. These fungi could survive in the digestive tract of different animal species and kill parasite larvae as they develop in the feces. Evidence exists that they are able to degrade the eggshell enzymatically and infect the helminth eggs [41,42].

Sludge bed density, extracellular polymeric substances [43–46], stability [47] and methanogenic conversion capacity [47,48] are some of the parameters that may impact the sludge bed filtration capacity for helminth eggs. Depending on the applied solids retention time (SRT) and the concentration of helminth eggs in the influent, long-term filtration may lead to saturation of the sludge bed, possibly lowering the filtration capacity. According to reviewed literature [1,18,19,49–53], no studies have been done thus far to characterize the sludge bed capacity for helminth egg filtration. Therefore, the main aim of this research was to study the sludge bed filtration capacity of UASB reactors with respect to the physical retention of helminth eggs under different upflow velocities at the prevailing subtropical temperatures. Filtration capacity is defined in this research as the physical process to retain helminth eggs using anaerobic sludge as a filtration medium.

2. Materials and Methods

2.1. Influent

The research was carried using raw wastewater from two urban villages called El Angel and El Milagro located in Lima (Peru). This wastewater was fed into a pilot plant located at the Research Center for Wastewater Treatment and Hazardous Wastes (CITRAR) at the campus of the National University of Engineering (Lima, Peru). The main characteristics of the wastewater are shown in Table 1.

Table 1. Influent wastewater characteristics from two urban villages called El Angel and El Milagro located in Lima (Peru), used for this research.

Parameter	Units	Average	*n*
Chemical Oxygen Demand	$mg \cdot L^{-1}$	723.2 ± 320.3	90
Suspended Solids	$mg \cdot L^{-1}$	126.5 ± 28.5	36
Oils and Grease	$mg \cdot L^{-1}$	30.8 ± 14.1	36
Total Phosphorous—P	$mg \cdot L^{-1}$	6.6 ± 2	35
Total Kjeldahl nitrogen—TKN	$mg \cdot L^{-1}$	16.2 ± 6.5	36
Dissolved Oxygen	$mg \cdot L^{-1}$	6.8 ± 0.4	36
Temperature	°C	22.8 ± 4.1	233
pH	--	7.1 ± 0.3	233
Fecal Coliforms	MPN/100 mL	$9.67 \times 10^8 \pm 1.89 \times 10^8$	36
Helminth eggs	$egg \cdot g^{-1}$	2.4 ± 1.4	90

Note: Where *n* is a number of grab analyzed samples.

The wastewater was pumped daily into a 200 L tank. The tank was filled with fresh wastewater every morning for all cases except when the upflow velocity of 0.68 $m \cdot h^{-1}$ was tested. For the latter situation, it was filled again in the afternoon when the remaining volume of the wastewater was 20 L. After filling the tank, the wastewater was mixed using a mechanical stirrer (18 RPM) with a stock solution containing *Ascaris suum*. The helminth egg concentration in the tank varied between 20–50 $egg \cdot g^{-1}$. The tank was kept at ambient temperatures and its content was used to continuously feed the UASB reactors. The pH and temperature of the wastewater was measured daily at 9:00, 12:00 and 16:00. The setup of the experiments is shown in Figure 1.

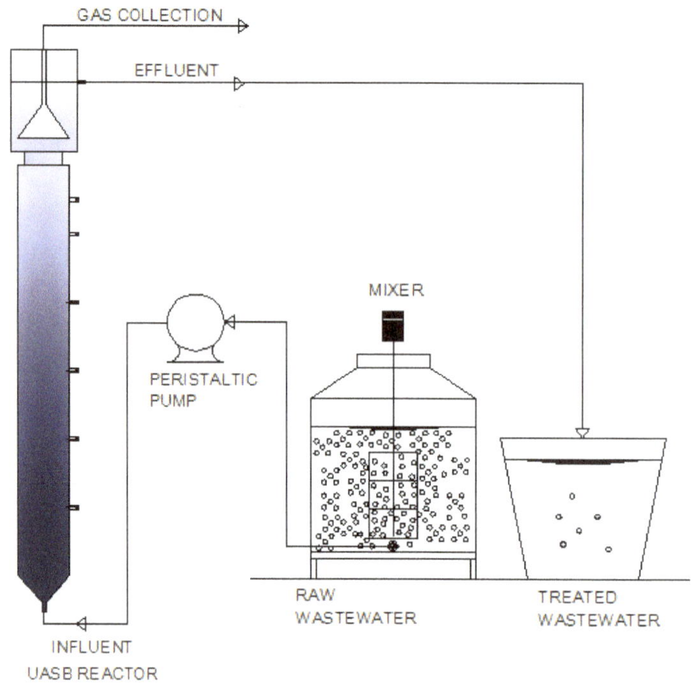

Figure 1. Set up of the filtration experiments in UASB (upflow anaerobic sludge blanket) reactors using wastewater inoculated with *Ascaris suum* eggs.

2.2. Upflow Anaerobic Sludge Blanket (UASB) Reactors

Two 25 L identical acrylic cylindrical lab-scale UASB reactors with a total height of 1.60 m and a diameter of 0.15 m were used separately in parallel. They were located at CITRAR. The experiments were performed from January 2010 to August 2013.

2.3. Inoculum

The inoculum was anaerobic flocculent sludge sampled from the 536 m^3 pilot-scale UASB reactor located at CITRAR. The inoculum was taken at a height of 1.5 m from the bottom of this reactor (total height of the reactor was 6.0 m). The total solids and volatile solids concentration of inoculum was 163 ± 37 and 106 ± 44 $g \cdot L^{-1}$ respectively.

2.4. Helminth Eggs

The experiment was conducted using *Ascaris suum* as helminth egg surrogate for the human parasite. The *Ascaris suum* helminth eggs were collected from female parasites of infected pigs (*Sus crofa domesticus*). In order to collect helminth eggs, dissections of the female parasite were performed according to Diawara *et al.* [54] by means of a longitudinal incision to obtain the reproductive system (womb and ovary). The womb and ovary were placed in 50 mL of physiological whey solution where they were opened to extract the helminth eggs. The optimal morphology and viability of the eggs of *Ascaris suum* were verified by microscopic observation according to Johnson *et al.* [55] and by using the staining procedure applied by de Victorica and Galván [56], respectively. Helminth eggs were added to the 200 L wastewater tank, which was fed to the UASB reactors.

2.5. Helminth Egg Counting

A multi-step methodology using local materials was developed from the modified Bailenger method [57,58]. This method was chosen due to its simplicity and the low cost of materials, in addition to the fact that it allows recovery of a wide range of helminths from the sample. The detailed methodology consists of collection of a 1 L sample, followed by settling for 24 h in a 1 L clear borosilicate glass bottle with graduations to concentrate the helminth eggs and to remove 90% of the supernatant (900 mL) by using a siphon. Then, 60 mL of the sediment are transferred to six centrifuge tubes of 10 mL each. Afterwards, the tubes are centrifuged at 1000 *g* for 15 min, and 70% of the supernatant (7 mL) is removed without shaking the tubes to avoid mixing the pellet with the supernatant. The remaining 40 mL of sediment is distributed over the same centrifuge tubes until the tubes are filled with 10 mL. Next, the bottle is rinsed two or more times with 10 mL of distilled water until it is completely clean. The corresponding rinse water is spread over the same centrifuge tubes or in new tubes. Distilled water is used to complete the remaining volume to fill 10 mL of water in each centrifuge tubes. Again, the centrifugation step is repeated. Subsequently, 2 mL of saturated sodium chloride solution with a specific gravity of 1.18 is added as flotation solution and, the tubes are shaken vigorously laterally. Afterwards, it is controlled whether all solids are located in the liquid phase. After 10 min, two phases are distinguished in the tubes. Finally, the top phase (1.5 mL) formed in the tubes is transferred to glass slides to be observed under the microscope (objectives lens 4× and 10×) and to count the eggs.

2.6. Physicochemical and Bacteriological Analysis

Total chemical oxygen demand (COD), suspended solid, volatile solids, oil/grease, pH, temperature, biochemical oxygen demand [59] and fecal coliform analysis were determined following standard methods [60]. Gravimetric and extractive-gravimetric methods carried out with hexane as a solvent were executed for solids and oil/grease determination, respectively. COD analysis was executed using high range Hach's COD digestion vials as well as a digester reactor DR 200, and program 17 from colorimeter DR 890. Dissolved oxygen, total nitrogen Kjeldahl and total phosphorous were measured according to Method HACH 10360, 8038 and 8048, correspondingly [61]. Microscopic views were performed with an optical microscope ZEISS Primo Star Serial number 3122001719.

2.7. UASB Operational Conditions

In order to study the influence of different upflow velocities and sludge bed heights in the UASB reactors, four experiments were carried out as indicated in Table 2. Each experiment was performed in duplicate (two reactors).

In order to facilitate the statistical interpretation of the results, it is assumed that at an upflow velocity near to zero, all helminth eggs are removed in the UASB reactor in experiment 1, 2 and 3. This assumption is in line with the results described in previous research [34,35].

Table 2. Setup of experiments in lab-scale UASB reactors to test the sludge filtration capacity to remove helminth eggs.

Experiment	SB Height Variation (m)	SBp (%)	Upflow Velocities ($m \cdot h^{-1}$)
1	0.30 to 0.40	19 to 25	0.09, 0.17, 0.23, 0.34 and 0.68
2	0.50 to 0.60	31 to 38	0.09, 0.11, 0.17, 0.23, 0.34 and 0.68
3	0.60 to 0.70	38 to 44	0.09, 0.14, 0.17, 0.23, 0.34, 0.45 and 0.68
4 (blank experiment)	0	0	0.09, 0.11, 0.14, 0.17, 0.23, 0.34, and 0.68

Notes: Where SB means sludge bed and SBp is the sludge bed expressed as a percentage of the total reactor height. The upflow velocities of 0.09, 0.11, 0.14, 0.17, 0.23, 0.34, 0.45 and 0.68 $m \cdot h^{-1}$ correspond to an hydraulic retention time (HRT) of 15, 12, 10, 8, 6, 4, 3 and 2 h, respectively. Each experiment was repeated three times.

The startup of the UASB reactors was performed at an upflow velocity of 0.34 $m \cdot h^{-1}$ and hydraulic retention time (HRT) of 4 h. Each upflow velocity for every experiment was applied during seven days and samples were taken on the last day. The samples were taken after an elapsed time equivalent to one HRT, after introducing a known wastewater corresponding in the influent tank. The effluent of UASB reactors was collected separately from each reactor in order to be able to take separate samples. For every upflow velocity, six samples were analyzed for COD and helminth egg content and temperature in the influent and effluent. A total of six samples were performed per upflow velocity, which were collected respectively from three measurements in each UASB reactor.

Experiment 1 was performed after 85 days of the start of the UASB reactor. Before starting experiment 2, reactors were operated for approximately 30 days and continuously fed with domestic wastewater containing an average helminth egg concentration of 2.4 $egg \cdot g^{-1}$ and an HRT of 4 h. The two reactors were fed with exactly the same influent using two peristaltic pumps (2 Masterflex, Oldham, UK). Some samples from the effluent in experiment 2 were taken for each upflow velocity in order to do microscopic observations. Experiment 3 started immediately after finishing experiment 2. Experiment 4 (control experiment) was performed without sludge in the acrylic UASB reactor 7 days after all experiments were finished. All experiments were performed at ambient temperatures. Sludge was removed in each UASB reactor in order to maintain the established sludge bed height variation according to Table 2.

3. Results and Discussion

A summary of the results of experiments 1, 2, 3 and 4 is listed in Table 3. The sludge filtration capacity at ambient temperatures and sludge bed heights in the ranges of 0.30–0.40 m and 0.50–0.60 m, showed a reciprocal correlation between the average helminth egg removal efficiency and upflow velocity with a coefficient of determination of 0.94 and 0.91, respectively. When the sludge bed height increased to 0.60–0.70 m, the reciprocal correlation is still present but the coefficient of determination decreased to 0.57.

Table 3. Results of helminth egg removal and chemical oxygen demand (COD) removal efficiencies at applied upflow velocities and wastewater temperatures. Each upflow velocity was applied three times in each UASB reactor. Then a total of six samples per upflow velocity was analyzed.

Experiment	Upflow Velocity ($m·h^{-1}$)	Temperature (°C)	Helminth Egg removal (%)	COD (%)
Experiment 1	0.09	24.6 ± 2.4	93 ± 5	71.9 ± 7.1
	0.17	28.6 ± 2	77 ± 4	66.4 ± 8.2
	0.23	25.6 ± 3.5	61 ± 7	63.1 ± 8.6
	0.34	23 ± 3.3	52 ± 9	60.3 ± 6.4
	0.68	26.5 ± 2	26 ± 7	45.4 ± 6.3
Experiment 2	0.09	22 ± 6	91 ± 3	71.6 ± 10.3
	0.11	22.5 ± 5.3	75 ± 10	71.6 ± 2.2
	0.17	24.2 ± 1.5	71 ± 11	66.2 ± 12.7
	0.23	23.2 ± 3.1	61 ± 10	65 ± 7.4
	0.34	26.1 ± 0.5	51 ± 7	63.7 ± 15.1
	0.68	25.5 ± 3	30 ± 15	63 ± 19.1
Experiment 3	0.09	23.3 ± 0.9	55 ± 1	80.3 ± 2.4
	0.14	21.4 ± 2.9	53 ± 5	80.2 ± 15.5
	0.17	27.1 ± 0.5	56 ± 7	80.2 ± 8.1
	0.23	23.3 ± 5.7	56 ± 8	79.3 ± 0.8
	0.34	22.1 ± 4.2	55 ± 11	69.5 ± 14.8
	0.45	28.5 ± 2	46 ± 8	60.5 ± 0.4
	0.68	26.2 ± 2.3	34 ± 8	45.3 ± 3.4
Experiment 4	0.09	16.9 ± 1	66 ± 3	77.6 ± 2.4
	0.11	16.9 ± 0.5	48 ± 3	44.8 ± 9.8
	0.14	17.3 ± 1	57 ± 3	84 ± 1.9
	0.17	17.3 ± 2	44 ± 3	50.7 ± 5.7
	0.23	16.9 ± 0.8	53 ± 3	64.9 ± 5
	0.34	18.1 ± 1	52 ± 10	71.1 ± 3.7
	0.68	17.7 ± 1	54 ± 8	55.2 ± 7.3

3.1. Experiment 1: Upflow Velocity between 0.09 and 0.68 $m·h^{-1}$ and Sludge Bed Height between 0.30 and 0.40 m (19% to 25% of the Total Reactor Height)

The efficiencies of helminth egg removal as a function of the upflow velocity, applying a sludge bed height between 0.30 and 0.40 m in two reactors, are shown in Figure 2, both operated at five upflow velocities of 0.09, 0.17, 0.23, 0.34 and 0.68 $m·h^{-1}$. Results show a decreasing trend for helminth egg removal efficiency at an increasing upflow velocity. A reciprocal linear relationship was observed between upflow velocity and helminth egg removal with a high coefficient of determination (R^2 = 0.92). The current results of the experiment applying a low sludge bed height of 19%–25% show that an increment of the upflow velocity leads to a decrease of the sludge filtration capacity. The latter statement could be explained because as soon as the wastewater upflow velocity increases, the associated sludge viscosity probably decreases [44]. Analogous to the removal of helminth eggs, the COD removal efficiency is decreasing at an increasing upflow velocity (Figure 3). A reciprocal linear relationship was observed between upflow velocity and COD removal with a high coefficient of determination ($R^2 = 0.99$). Average ambient temperature varied between 23 and 28.6 °C in both UASB reactors.

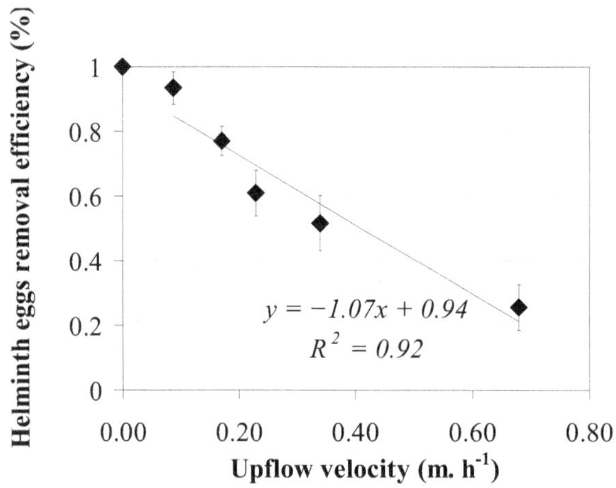

Figure 2. Helminth egg removal efficiencies (♦) *versus* upflow velocity at a sludge bed height between 0.30 and 0.40 m.

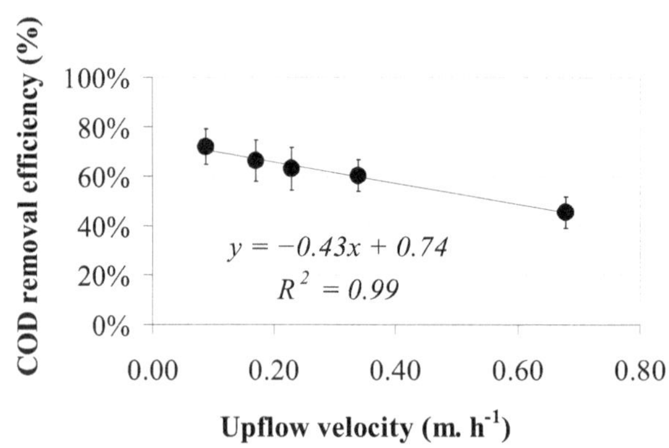

Figure 3. Total COD removal efficiencies in two UASB reactors, characterized by a sludge bed height between 0.30 and 0.40 m.

3.2. Experiment 2: Upflow Velocity between 0.09 and 0.68 m·h^{-1} and Sludge Bed Height between 0.50 and 0.60 m (31% to 38% of the Total Reactor Height)

The efficiencies of helminth egg removal as a function of the upflow velocity, applying a sludge bed height between 0.50 and 0.60 m in two reactors, are shown in Figure 4, both operated six different upflow velocities: 0.09, 0.11, 0.17, 0.23, 0.34 and 0.68 m·h^{-1}. Results show a decreasing trend for helminth egg removal efficiency at an increasing upflow velocity (Figure 4). A reciprocal linear relationship was observed between upflow velocity and helminth egg removal with a high coefficient of determination (R^2 = 0.91). The observed results applying a sludge bed height of 31%–38% were similar to those at a sludge bed height of 19%–25%.

In contrast, the COD removal efficiency did not show a clear trend at an increasing upflow velocity (Figure 5) when applying a sludge bed height of 0.50–0.60 m. Average ambient temperature varied from 22.0 to 26.1 °C.

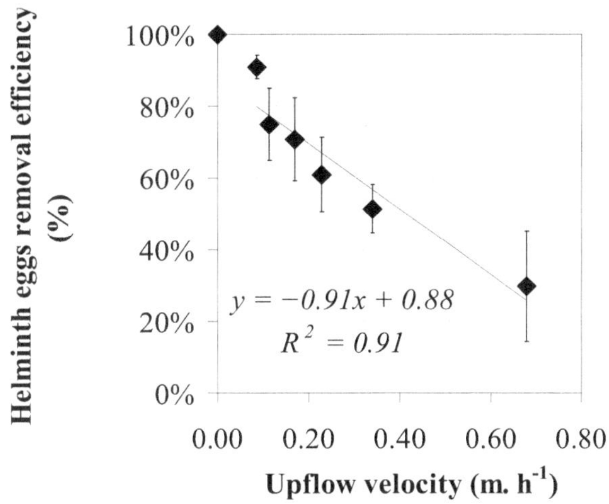

Figure 4. Helminth egg removal efficiencies (♦) *versus* upflow velocity using a sludge bed height between 0.50 and 0.60 m.

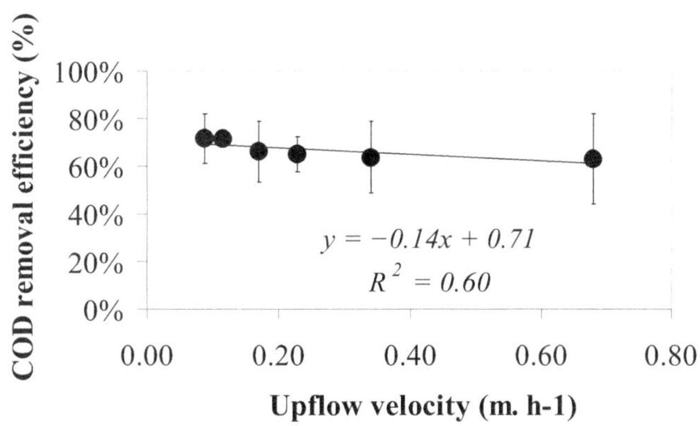

Figure 5. Total COD removal efficiencies *versus* upflow velocity using a sludge bed height between 0.50 and 0.60 m.

3.3. Experiment 3: Upflow Velocity between 0.09 and 0.68 m·h^{-1} and Sludge Bed Height between 0.60 and 0.70 m (38% to 44% of the Total Reactor Height)

Results of the helminth egg removal as a function of the upflow velocity at a sludge bed height between 0.60 and 0.70 m, in two UASB reactors operated at 0.09, 0.14, 0.17, 0.23, 0.34, 0.45 and 0.68 m·h^{-1}, are shown in Figure 6. Though a slightly decreasing trend is shown with increasing upflow velocity, the coefficient of determination is low (R^2 = 0.83). Moreover, standard deviations are large.

Results on COD removal efficiency show a decreasing trend at an increasing upflow velocity (Figure 7). A reciprocal linear relationship was observed between upflow velocity and COD removal with a high coefficient of determination (R^2 = 0.97). Average ambient temperature varied from 21.4 to 28.5 °C.

Although counterintuitive, the decreasing trend in helminth egg removal efficiency at an increasing sludge bed height in the studied lab-scale reactor might be explained by an increase in turbulence, created by the biogas production and formation of channels through the sludge bed [62,63] during all studied velocities. The possible saturation with helminth eggs during previous experiments could also have influenced the stability of the system with respect to helminth egg removal. Since none of the eggs are spherical [1,3,36,37], it is likely that they settle with different, but unknown orientations [64].

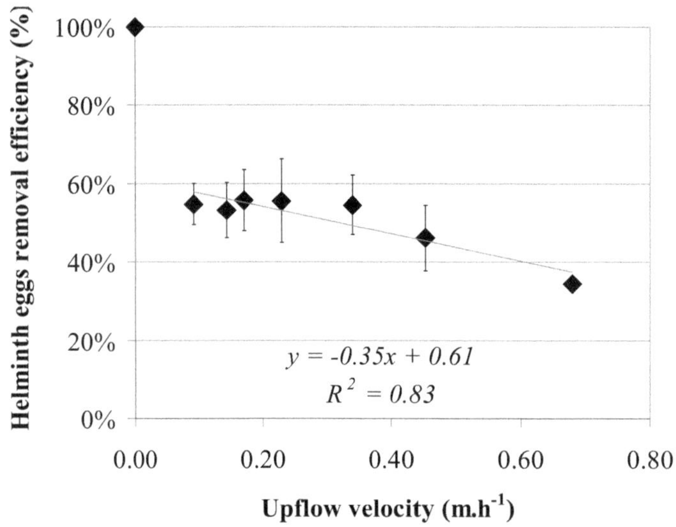

Figure 6. Helminth egg removal efficiencies (♦) *versus* upflow velocity using a sludge bed height between 0.60 and 0.70 m.

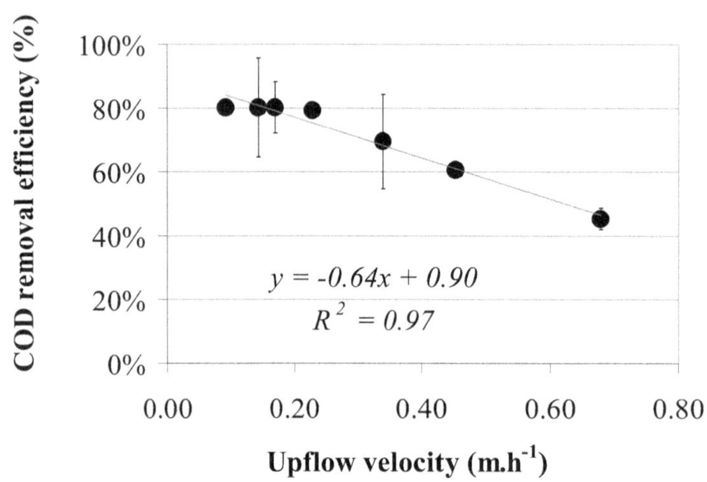

Figure 7. Total COD removal efficiencies *versus* upflow velocity using a sludge bed height between 0.60 and 0.70 m. Results show the average of two reactors and standard deviation and three measurements per reactor.

3.4. Experiment 4: Blank Experiment Using Upflow Velocity between 0.09 and 0.68 $m\cdot h^{-1}$ and No Sludge Bed

The effect of the upflow velocity on the settling of helminth eggs is demonstrated by the results of the control experiment, applying a UASB reactor without sludge. Results for seven different upflow velocities, 0.09, 0.11, 0.14, 0.17, 0.23, 0.34, and 0.68 $m\cdot h^{-1}$, are shown in Figures 8 and 9. Each point is the average of three samples in two reactors. Figure 8 indicates that the best efficiency for helminth egg removal was obtained at 0.09 $m\cdot h^{-1}$, when the removal efficiency reached 66% ± 3%. For upflow velocities higher than 0.09 $m\cdot h^{-1}$, the helminth egg efficiency removal was lower but always exceeded 44% ± 3%. It should be noted that standard deviations were large, so no significant differences were observed for helminth egg removal at an increasing upflow velocity. Figure 9 shows the trend for TSS and VSS removal. The best TSS and VSS removal efficiency (about 80%) was obtained at the lowest upflow velocities. For the higher upflow velocities the TSS and VSS removal efficiencies dropped to 40%–50% and 30%–40% for TSS and VSS, respectively.

The control experiment was executed in winter at a relatively low temperature and average ambient temperature varied from 16.9 to 18.1 °C. Latter temperatures were colder compared to previous experiments since experiment 4 was carried out coincidentally during winter.

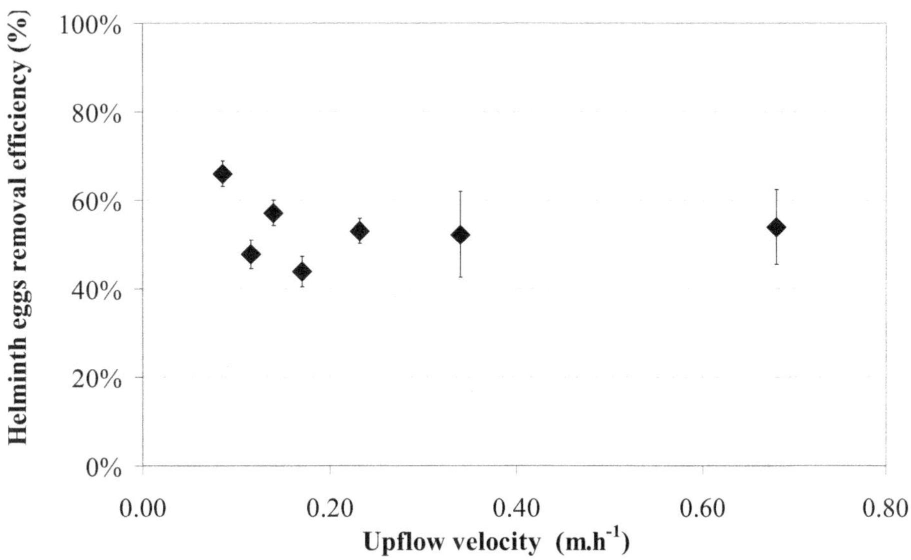

Figure 8. Helminth egg removal efficiencies (♦) *versus* upflow velocity without sludge in the control experiment. Results shows the average of three samples and standard deviation; no sludge bed.

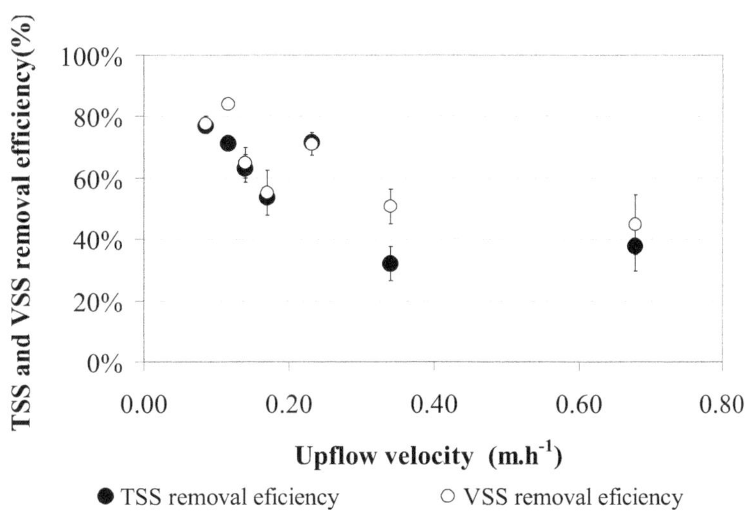

Figure 9. TSS (•) and VSS (○) removal efficiencies in the control experiment—No sludge bed. Results show the average of three samples and standard deviation.

Results of experiments 1, 2, 3 and 4 show that the sludge bed in a UASB reactor is an inappropriate and unreliable filter medium for helminth eggs. Therefore, for achieving a complete helminth egg removal, a UASB reactor must be followed by an adequate post-treatment unit like land-based settling units or a post-filtration step [18,26,53]. For the control experiment (without sludge), average helminth egg removal efficiency varied between 44% and 66% at upflow velocities between 0.09 $m \cdot h^{-1}$ and

0.68 $m \cdot h^{-1}$. Unexpectedly, these values exceed the removal efficiencies of the reactors filled with high volumes of sludge, particularly at the high upflow velocities. Previous research [34,35] showed that viscosity of the flocculent anaerobic sludge is approximately more than 50 times higher than the viscosity of the liquid water, thereby theoretically leading to a better retention of helminth eggs. The explanation in the control experiments for why the levels of helminth eggs removal were so high is not very clear. Likely, the better retention might be associated to the absence of biogas production and thus turbulence. Therefore, in the absence of turbulence, the wastewater flow is more homogeneous [65–68], and helminth eggs settling follow a discrete settling pattern. Another remarkable observation is that even at the lowest upflow velocity (0.09 $m \cdot h^{-1}$) helminth eggs do not settle completely nor are retained completely by the sludge bed. The wash out of helminth eggs under these conditions may indicate that either the egg density is much less than expected, or the flow distribution is far from laminar [47,68,69]. A higher degree of channeling, which is expected at higher volumes of sludge [47,62,70], will aggravate the extremes in the flow distribution patterns. The latter will certainly lead to poorer filtration performances, as was also observed in the conducted experiments.

COD removal efficiencies showed a similar trend to the helminth egg removal efficiency. An increased removal of COD with decreasing upflow velocity was shown by Mahmoud [67,71]. Though completely different in nature, helminth eggs are also particles that could be expected to behave similarly.

3.5. Microscopic Observations in the Effluent

Microscopic observations were performed only for experiment 2. The presence of helminth eggs was detected in the sludge samples. In addition, several damages have been microscopically observed in the morphology of helminth eggs in the effluent (Figure 10a,c,f) with respect to the influent (Figure 10b–e,g,h) of the UASB reactor. The observed damages in the internal morphological structure of the eggs might be related to a possible loss in egg viability. These damages could be possibly caused by the retention of the eggs in the sludge bed for the applied HRT prior to their washout. The indicated hypotheses need to be confirmed in further research. Figure 10c,d present some microscopic views of helminth eggs, respectively in the influent and effluent of the UASB at an applied upflow velocity of 0.09 $m \cdot h^{-1}$. There are some changes in the internal morphology like probable larval development but without progression to the next stage (Figure 10c–e). Figure 10g,h shows some observed damages in the structure of helminth eggs.

The percentage of damaged helminth eggs present in the effluent of the UASB reactor that operated under different conditions was not determined, but visually they only were present in the effluent and not in the influent.

Following microscopic observations of the sludge sampled at different upflow velocities in experiment 2, it is shown that damages of helminth eggs occurred in all applied upflow velocities. It is shown in Figure 10h that some helminth eggs formed clusters of eggs. The mechanisms behind this phenomenon are not known.

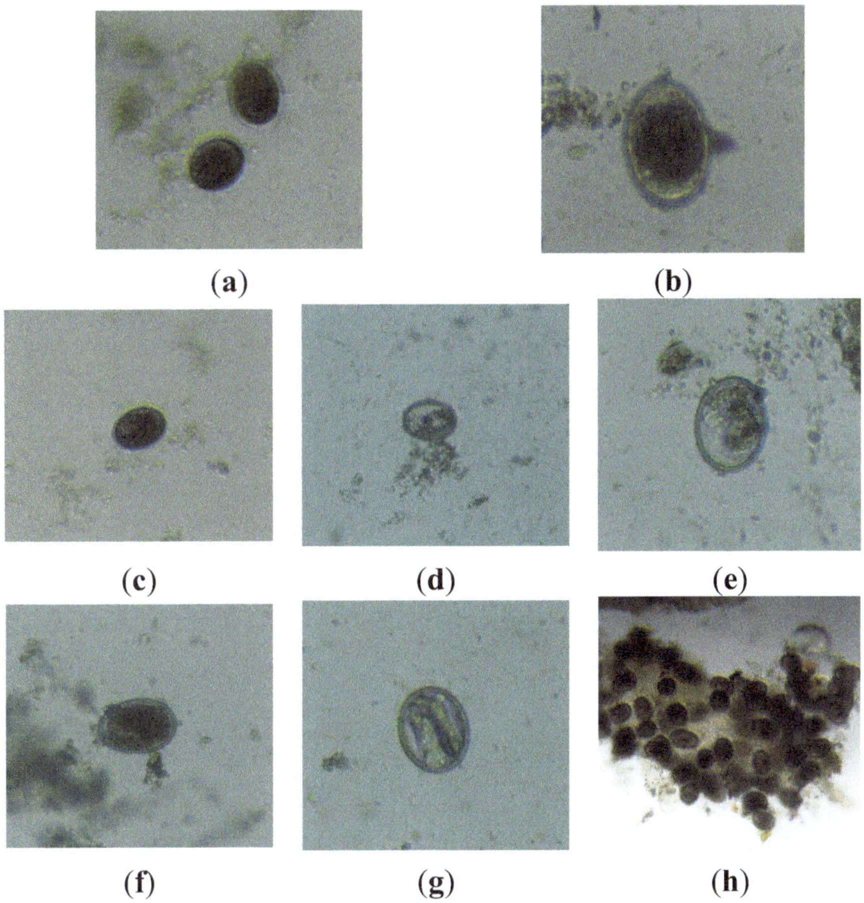

Figure 10. Helminth eggs in the influent (**a**) and in the effluent (**b**) for an applied upflow velocity of 0.09 $m \cdot h^{-1}$. Helminth eggs in the influent (**c**) and in the effluent (**d**) and (**e**) for an upflow velocity of 0.34 $m \cdot h^{-1}$. Helminth eggs from the effluent show an internal morphology likely affected by the experimental conditions. Helminth eggs in the influent (**f**) and in the effluent (**g**) and (**h**) of lab-scale UASB reactor for an upflow velocity of 0.68 $m \cdot h^{-1}$. Helminth eggs from the effluent (**g**) show an apparently deteriorated semi-crystalline internal morphology (possible larval development but interrupted by the conditions of the experiment) and (**h**) group of attached helminth eggs.

The observed damages on *Ascaris suum* might be attributed to the prevailing physicochemical conditions in the direct vicinity of the eggs or to other microorganisms, which could be present in the anaerobic sludge like nematofagous fungi [38–40]. The relatively low helminth egg removal in the experiment with the highest sludge bed might also be related to a high percentage of damaged helminth eggs as a result of a long retention of helminth eggs in the sludge bed. Damaged helminth eggs might have a decreased density and thus a lowered settleability. The latter hypothesis could also explain why the removal of helminth eggs in the blank experiment (without sludge bed) is relatively high compared to the sludge bed reactors. This hypothesis needs to be verified in future research.

The results showed that helminth egg removal will not be sufficient for UASB systems operated with conventionally collected domestic wastewater where relatively high upflow velocities need to be applied as a result of the low COD concentration [72,73]. News trends in domestic wastewater collection and transport like uncoupling rainwater [74] and source separation [75–77] increase wastewater concentration and therefore reduce applied upflow velocities. The observed increased helminth egg removal at reduced upflow velocity might imply that application of UASB reactors, with similar loading rates in source separated domestic wastewater, leads to improved helminth egg removal.

4. Conclusions

This study demonstrated that with an increased sludge bed height there is a reduction in the sludge filtration capacity for helminth egg removal. If treated wastewater is used for irrigation purposes, the UASB reactor must be followed by an adequate post-treatment unit. The sludge filtration capacity at ambient temperatures and sludge bed height in the range of 0.30–0.40 m and 0.50–0.60 m, which agrees with 19%–25% and 31%–38% of the total height reactor, respectively, showed a reciprocal correlation between the average helminth egg removal efficiency and upflow velocity. This study reported an average helminth egg removal between 34%–100%, 30%–91% and 34%–56% when the sludge bed height was 19%–25%, 31%–38% and 38%–44%, respectively, of the total height in the UASB reactor at upflow velocities varying between 0.09 and 0.68 $m \cdot h^{-1}$. Several damages were observed during microscope observations in the morphology of helminth eggs present in the sludge and the effluent of UASB reactors at upflow velocities between 0.09 and 0.68 $m \cdot h^{-1}$.

Acknowledgments

This work was coordinated by the Sub-Department of Environmental Technology of Wageningen University.

Author Contributions

Rosa Elena Yaya Beas developed the experimental strategy, performed the experiments, was the principal author of the the paper, and contributed to the paper revisions. Christian Ayala Limaylla contributed to the biological part of experiments concerning helminth eggs counting. Katarazyna Kujawa-Roeleveld and Grietje Zeeman contributed to the paper revisions and experimental strategy. Jules van Lier provided advice on the experimental set-up and draft manuscript; all authors have reviewed the manuscript.

Conflicts of Interest

The authors declare no conflict of interest.

References

1. Jimenez, B. Helminth ova removal from wastewater for agriculture and aquaculture reuse. *Water Sci. Technol.* **2007**, *55*, 485–493.
2. Navarro, I.; Jiménez, B. Evaluation of the who helminth eggs criteria using a qmra approach for the safe reuse of wastewater and sludge in developing countries. *Water Sci. Technol.* **2011**, *63*, 1499–1505.

3. Jimenez, B. Helminthes (worms) eggs control in wastewater and sludge. In Proceedings of the International Symposium on New Directions in Urban Water Management, Paris, France, 12–14 September 2007; pp. 12–14.
4. Maya, C.; Torner-Morales, F.; Lucario, E.; Hernández, E.; Jiménez, B. Viability of six species of larval and non-larval helminth eggs for different conditions of temperature, ph and dryness. *Water Res.* **2012**, *46*, 4770–4782.
5. Blumenthal, U.J.; Cifuentes, E.; Bennett, S.; Quigley, M.; Ruiz-Palacios, G. The risk of enteric infections associated with wastewater reuse: The effect of season and degree of storage of wastewater. *Trans. R. Soc. Trop. Med. Hyg.* **2001**, *95*, 131–137.
6. Cooper, P.J.; Chico, M.E.; Sandoval, C.; Espinel, I.; Guevara, A.; Kennedy, M.W.; Urban, J.F.; Griffin, G.E.; Nutman, T.B. Human infection with ascaris lumbricoides is associated with a polarized cytokine response. *J. Infect. Dis.* **2000**, *182*, 1207–1213.
7. Cruz Toribio, L.I.F. *Gastrointestinal Helminthiasis in Livestock Herding Dogs Puno Communities*; Universidad Nacional Mayor de San Marcos: Lima-Peru, Spain, 2010.
8. World Health Organization (WHO). *Who Guidelines for the Safe Use of Wastewater, Excreta and Greywater Organization*; World Health Organization: Geneva, Switzerland, 2006; Volume 2
9. De Bonilla, L.C. El problema de las parasitosis intestinales en venezuela. *Investig. Clín.* **1990**, *31*, 1–2.
10. Santiso, R. Effects of chronic parasitosis on women's health. *Int. J. Gynecol. Obstet.* **1997**, *58*, 129–136.
11. Tchobanoglous, G.; Burton, F.; Stensel, H.D. *Wastewater Engineering: Treatment and Reuse*; McGraw Hill: New York, NY, USA, 2003.
12. Qadir, M.; Wichelns, D.; Raschid-Sally, L.; McCornick, P.; Drechsel, P.; Bahri, A.; Minhas, P. The challenges of wastewater irrigation in developing countries. *Agric. Water Manag.* **2010**, *97*, 561–568.
13. Keller, R.; Passamani-Franca, R.; Cassini, S.; Gonçalves, F. Disinfection of sludge using lime stabilisation and pasteurisation in a small wastewater treatment plant. *Water Sci. Technol.* **2004**, *50*, 13–17.
14. Borrely, S.; Cruz, A.; del Mastro, N.; Sampa, M.; Somessari, E. Radiation processing of sewage and sludge. A review. *Prog. Nucl. Energy* **1998**, *33*, 3–21.
15. De Souza, G.S.; Rodrigues, L.A.; de Oliveira, W.J.; Chernicharo, C.A.; Guimarães, M.P.; Massara, C.L.; Grossi, P.A. Disinfection of domestic effluents by gamma radiation: Effects on the inactivation of ascaris lumbricoides eggs. *Water Res.* **2011**, *45*, 5523–5528.
16. Gantzer, C.; Gaspard, P.; Galvez, L.; Huyard, A.; Dumouthier, N.; Schwartzbrod, J. Monitoring of bacterial and parasitological contamination during various treatment of sludge. *Water Res.* **2001**, *35*, 3763–3770.
17. Mara, D. *Domestic Wastewater Treatment in Developing Countries*; Earthscan: London, UK, 2003.
18. Von Sperling, M.; Chernicharo, C.; Andreoli, C.V.; Fernandes, F. *Biological Wastewater Treatment in Warm Climate Regions*; IWA: London, UK, 2005; Volume 2.
19. Jimenez, B.; Maya-Rendon, C.; Salgado-Velzquez, G. The elimination of helminth ova, faecal coliforms, salmonella and protozoan cysts by various physicochemical processes in wastewater and sludge. *Water Sci. Technol.* **2001**, *43*, 179–182.

20. Koné, D.; Cofie, O.; Zurbrügg, C.; Gallizzi, K.; Moser, D.; Drescher, S.; Strauss, M. Helminth eggs inactivation efficiency by faecal sludge dewatering and co-composting in tropical climates. *Water Res.* **2007**, *41*, 4397–4402.
21. Cabaret, J.; Geerts, S.; Madeline, M.; Ballandonne, C.; Barbier, D. The use of urban sewage sludge on pastures: The cysticercosis threat. *Vet. Res.* **2002**, *33*, 575–597.
22. World Health Organization (WHO). *Health Guidelines for the Use of Wastewater in Agriculture and Aquaculture: Reports of a Who Scientific Group*; Technical Report Series no. 778; World Health Organization: Geneva, Switzerland, 1989.
23. Brownell, S.A.; Nelson, K.L. Inactivation of single-celled ascaris suum eggs by low-pressure uv radiation. *Appl. Environ. Microbiol.* **2006**, *72*, 2178–2184.
24. Von Sperling, M. Comparison among the most frequently used systems for wastewater treatment in developing countries. *Water Sci. Technol.* **1996**, *33*, 59–72.
25. Von Sperling, M.; Chernicharo, C.; Soares, A.; Zerbini, A. Coliform and helminth egg removal in a combined uasb reactor-baffled pond system in brazil: Performance evaluation and mathematical modelling. *Water Sci. Technol.* **2002**, *45*, 237.
26. Chernicharo, C. Post-treatment options for the anaerobic treatment of domestic wastewater. *Rev. Environ. Sci. Biotechnol.* **2006**, *5*, 73–92.
27. Von Sperling, M.; Chernicharo, C.; Soares, A.; Zerbini, A. Evaluation and modelling of helminth egg removal in baffled and unbaffled ponds treating anaerobic effluent. *Water Sci. Technol.* **2003**, *48*, 113–120.
28. Jorsaraei, A.; Gougol, M.; van Lier, J.B. A cost effective method for decentralized sewage treatment. *Process Saf. Environ. Prot.* **2013**, *92*, 815–821.
29. Uemura, S.; Harada, H. Treatment of sewage by a uasb reactor under moderate to low temperature conditions. *Bioresour. Technol.* **2000**, *72*, 275–282.
30. Van Lier, J.B.; Tilche, A.; Ahring, B.K.; Macarie, H.; Moletta, R.; Dohanyos, M.; Pol, L.W.; Lens, P.; Verstraete, W.; Management Committee of the IWA Anaerobic Digestion Specialised Group. New perspectives in anaerobic digestion. *Water Sci. Technol.* **2001**, *43*, 1–18.
31. Van Lier, J.B.; Vashi, A.; van der Lubbe, J.; Heffernan, B.; Fang, H. *Anaerobic Sewage Treatment Using Uasb Reactors: Engineering and Operational Aspects*; Imperial College Press: London, UK, 2010.
32. Mahmoud, N.; Zandvoort, M.; van Lier, J.; Zeeman, G. Development of sludge filterability test to assess the solids removal potential of a sludge bed. *Bioresour. Technol.* **2006**, *97*, 2383–2388.
33. Boes, J.; Helwigh, A. Animal models of intestinal nematode infections of humans. *Parasitology* **2000**, *121*, 97–111.
34. Yaya-Beas, R.E.; Zeeman, G.; van Lier, J.B. Helminth ova removal using uasb reactors at 4 °C. In Proceedings of the 3rd International Congress Smallwat 11, Seville, Spain, 26 April 2010.
35. Yaya-Beas, R.E.; D'engremont, M.; Kujawa, K.; Zeeman, G.; van Lier, J.B. Filtration capacity of an anaerobic sludge bed for the removal of helminth eggs. *Water Environ. J.* submitted for publication, 2015.
36. O'Lorcain, P.; Holland, C. The public health importance of ascaris lumbricoides. *Parasitology* **2000**, *121*, S51–S71.

37. Quilès, F.; Balandier, J.Y.; Capizzi-Banas, S. *In situ* characterisation of a microorganism surface by raman microspectroscopy: The shell of ascaris eggs. *Anal. Bioanal. Chem.* **2006**, *386*, 249–255.
38. Da Cruz, D.G.; Araújo, F.B.; Molento, M.B.; DaMatta, R.A.; de Paula Santos, C. Kinetics of capture and infection of infective larvae of trichostrongylides and free-living nematodes panagrellus sp. By duddingtonia flagrans. *Parasitol. Res.* **2011**, *109*, 1085–1091.
39. Arias, M.S.; Suárez, J.; Cazapal-Monteiro, C.F.; Francisco, I.; López-Arellano, M.E.; Piñeiro, P.; Suárez, J.L.; Sánchez-Andrade, R.; Mendoza de Gives, P.; Paz-Silva, A. Trematodes enhance the development of the nematode-trapping fungus *arthrobotrys* (*duddingtonia*) *flagran*. *Fungal Biol.* **2013**, *117*, 540–544.
40. Federica, S.M.; Alberto, F.L.; Emilia, I.L.; Carina, M.F.; Alfredo, S.C. Optimization of production of chlamydospores of the nematode-trapping fungus duddingtonia flagrans in solid culture media. *Parasitol. Res.* **2013**, *112*, 1047–1051.
41. Larsen, M. Prospects for controlling animal parasitic nematodes by predacious micro fungi. *Parasitology* **2000**, *120*, 121–131.
42. Manzanilla-López, R.H.; Esteves, I.; Finetti-Sialer, M.M.; Hirsch, P.R.; Ward, E.; Devonshire, J.; Hidalgo-Díaz, L. Pochonia chlamydosporia: Advances and challenges to improve its performance as a biological control agent of sedentary endo-parasitic nematodes. *J. Nematol.* **2013**, *45*, 1.
43. Seyssiecq, I.; Ferrasse, J.-H.; Roche, N. State-of-the-art: Rheological characterisation of wastewater treatment sludge. *Biochem. Eng. J.* **2003**, *16*, 41–56.
44. Pevere, A.; Guibaud, G.; van Hullebusch, E.; Lens, P.; Baudu, M. Viscosity evolution of anaerobic granular sludge. *Biochem. Eng. J.* **2006**, *27*, 315–322.
45. Mori, M.; Seyssiecq, I.; Roche, N. Rheological measurements of sewage sludge for various solids concentrations and geometry. *Process Biochem.* **2006**, *41*, 1656–1662.
46. Johansen, A.; Nielsen, H.B.; Hansen, C.M.; Andreasen, C.; Carlsgart, J.; Hauggard-Nielsen, H.; Roepstorff, A. Survival of weed seeds and animal parasites as affected by anaerobic digestion at meso-and thermophilic conditions. *Waste Manag.* **2013**, *33*, 807–812.
47. Seghezzo, L. Anaerobic Treatment of Domestic Wastewater in Subtropical Regions. Ph.D. Thesis, Wageningen Universiteit, Wageningen, The Netherlands, 2004.
48. De Graaff, M.S.; Temmink, H.; Zeeman, G.; Buisman, C.J. Anaerobic treatment of concentrated black water in a uasb reactor at a short hrt. *Water* **2010**, *2*, 101–119.
49. Mendez, J.; Jimenez, B.; Barrios, J. Improved alkaline stabilization of municipal wastewater sludge. *Water Sci. Technol.* **2002**, *46*, 139–146.
50. Jiménez, B. Treatment technology and standards for agricultural wastewater reuse: A case study in mexico. *Irrig. Drain.* **2005**, *54*, S23–S33.
51. Jiménez, B.; Drechsel, P.; Koné, D.; Bahri, A.; Raschid-Sally, L.; Qadir, M. Wastewater, sludge and excreta use in developing countries: An overview. In *Wastewater Irrigation and Health: Assessing and Mitigating Risk in Low-Income Countries*; Taylor & Francis: London, UK, 2010; pp. 3–29.
52. Jiménez, B.; Mara, D.; Carr, R.; Brissaud, F. Wastewater treatment for pathogen removal and nutrient conservation: Suitable systems for use in developing countries. In *Wastewater Irrigation and Health. Assessing and Mitigating Risk in Low-Income Countries*; Drechsel, P., Scott, C.A., Raschid-Sally, L., Redwood, M., Bahri, A., Eds.; International Water Management Institute and International Development Research Centre (IDRC): London, UK, 2010; pp. 149–169.

53. Chernicharo, C.; da Silva Cota, R.; Zerbini, A.; von Sperling, M. Post-treatment of anaerobic effluents in an overland flow system. *Water Sci. Technol.* **2001**, *44*, 229–236.
54. Diawara, A.; Drake, L.J.; Suswillo, R.R.; Kihara, J.; Bundy, D.A.; Scott, M.E.; Halpenny, C.; Stothard, J.R.; Prichard, R.K. Assays to detect β-tubulin codon 200 polymorphism in trichuris trichiura and ascaris lumbricoides. *PLoS Negl. Trop. Dis.* **2009**, *3*, e397.
55. Johnson, P.; Dixon, R.; Ross, A. An *in vitro* test for assessing the viability of ascaris suum eggs exposed to various sewage treatment processes. *Int. J. Parasitol.* **1998**, *28*, 627–633.
56. De Victorica, J.; Galván, M. Preliminary testing of a rapid coupled methodology for quantitation/viability determination of helminth eggs in raw and treated wastewater. *Water Res.* **2003**, *37*, 1278–1287.
57. Ayres, R.M.; Mara, D.D. *Analysis of Wastewater for Use in Agriculture. A Laboratory Manual of Parasitological and Bacteriological Techniques*; World Health Organization: Geneva, Switzerland, 1996; Volume 1.
58. Bailenger, J. Mechanisms of parasitological concentration in coprology and their practical consequences. *J. Am. Med. Technol.* **1979**, *41*, 65–71.
59. Bodík, I.; Herdová, B.; Drtil, M. The use of upflow anaerobic filter and ansbr for wastewater treatment at ambient temperature. *Water Res.* **2002**, *36*, 1084–1088.
60. American Public Health Association; American Water Works Association; Water Environment Federation. *Standard Methods for the Examination of Water and Wastewater*, 20th ed.; American Public Health Association: Washington, DC, USA, 1998; p. 1220.
61. Hach Company. *Hach Water Analysis Handbook*, 5th ed.; Hach Company: Loveland, CO, USA, 2008.
62. Lettinga, G.; Pol, L.H.; Koster, I.; Wiegant, W.; de Zeeuw, W.; Rinzema, A.; Grin, P.; Roersma, R.; Hobma, S. High-rate anaerobic waste-water treatment using the uasb reactor under a wide range of temperature conditions. *Biotechnol. Genet. Eng. Rev.* **1984**, *2*, 253–284.
63. Abdelgadir, A.; Chen, X.; Liu, J.; Xie, X.; Zhang, J.; Zhang, K.; Wang, H.; Liu, N. Characteristics, process parameters, and inner components of anaerobic bioreactors. *BioMed. Res. Int.* **2014**, *2014*, doi:10.1155/2014/841573.
64. Sengupta, M.E.; Thamsborg, S.M.; Andersen, T.J.; Olsen, A.; Dalsgaard, A. Sedimentation of helminth eggs in water. *Water Res.* **2011**, *45*, 4651–4660.
65. Elmitwalli, T.A.; Zandvoort, M.H.; Zeeman, G.; Bruning, H.; Lettinga, G. Low temperature treatment of domestic sewage in upflow anaerobic sludge blanket and anaerobic hybrid reactors. *Water Sci. Technol.* **1999**, *39*, 177–185.
66. Lew, B.; Tarre, S.; Belavski, M.; Green, M. UASB reactor for domestic wastewater treatment at low temperatures: A comparison between a classical uasb and hybrid uasb-filter reactor. *Water Sci. Technol.* **2004**, *49*, 295–301.
67. Mahmoud, N. *Anaerobic Pre-treatment of Sewage under Low Temperature (15 °C) Conditions in an Integrated UASB-Digester System*; Agricultural Wageningen University: Wageningen, The Netherlands, 2002.
68. Bolle, W.; van Breugel, J.; van Eybergen, G.; Kossen, N.; Zoetemeyer, R. Modeling the liquid flow in up-flow anaerobic sludge blanket reactors. *Biotechnol. Bioeng.* **1986**, *28*, 1615–1620.
69. Ojha, C.; Singh, R. Flow distribution parameters in relation to flow resistance in an upflow anaerobic sludge blanket reactor system. *J. Environ. Eng.* **2002**, *128*, 196–200.

70. Jeison, D.; Chamy, R. Comparison of the behaviour of expanded granular sludge bed (EGSB) and upflow anaerobic sludge blanket (UASB) reactors in dilute and concentrated wastewater treatment. *Water Sci. Technol.* **1999**, *40*, 91–97.
71. Mahmoud, N.; Zeeman, G.; Gijzen, H.; Lettinga, G. Solids removal in upflow anaerobic reactors, a review. *Bioresour. Technol.* **2003**, *90*, 1–9.
72. Dong, C.; Geng, Z.; Wang, Y. Expanded granular sludge bed (egsb) reactor treating actual domestic wastewater: Temperature influence. In Proceedings of the International Conference on Remote Sensing, Environment and Transportation Engineering, Nanjing, China, 26–28 July 2013.
73. Ozgun, H.; Dereli, R.K.; Ersahin, M.E.; Kinaci, C.; Spanjers, H.; van Lier, J.B. A review of anaerobic membrane bioreactors for municipal wastewater treatment: Integration options, limitations and expectations. *Sep. Purif. Technol.* **2013**, *118*, 89–104.
74. Rulkens, W. Increasing the environmental sustainability of sewage treatment by mitigating pollutant pathways. *Environ. Eng. Sci.* **2006**, *23*, 650–665.
75. Zeeman, G.; Kujawa-Roeleveld, K. Anaerobic treatment of source-separated domestic wastewater. In *Source Separation and Decentralization for Wastewater Treatment*; Larsen, T.A., Udert, K.M., Lienert, J., Eds.; IWA: London, UK, 2013; pp. 307–319.
76. Udert, K.M.; Lienert, J. *Source Separation and Decentralization for Wastewater Management*; IWA: London, UK, 2013.
77. Kujawa-Roeleveld, K.; Zeeman, G. Anaerobic treatment in decentralised and source-separation-based sanitation concepts. *Rev. Environ. Sci. Biotechnol.* **2006**, *5*, 115–139.

10

Impact of DEM Resolution on Puddle Characterization: Comparison of Different Surfaces and Methods

Jianli Zhang [1,2,†] and Xuefeng Chu [2,†,*]

[1] State Key Laboratory of Simulation and Regulation of Water Cycle in River Basin, China Institute of Water Resources and Hydropower Research, Beijing 100048, China; E-Mail: jl.z@163.com

[2] Department of Civil and Environmental Engineering (Dept 2470), North Dakota State University, PO Box 6050, Fargo, ND 58108-6050, USA

† These authors contributed equally to this work.

* Author to whom correspondence should be addressed; E-Mail: xuefeng.chu@ndsu.edu;

Academic Editor: Miklas Scholz

Abstract: DEM-based topographic characterization and quantification of surface depression storage are critical to hydrologic and environmental modeling. Mixed conclusions have been obtained from previous studies on the relationship between maximum depression storage (MDS) and DEM grid spacing, which is affected by different factors, such as topographic characteristics, surface delineation methods and DEM interpolation/aggregation methods. The objective of this study was to evaluate the effects of DEM resolution on topographic characterization with the consideration of these three factors. Twenty-three topographic surfaces (including ideal surfaces, laboratory-scale soil surfaces and watershed-scale land surfaces) were selected, and five software packages, ArcHydro, PCRaster, HEC-GeoHMS, TauDEM and PD (puddle delineation), were used for surface delineation. Our results indicated that MDS, maximum ponding area (MPA) and the number of puddles (NP) decreased with increasing grid spacing for most smoother surfaces due to the loss of topographic detail. For most rough surfaces (e.g., mountain-type surfaces with significant variations in surface elevations), however, the changing patterns of MDS and MPA varied with an increase in grid spacing mainly due to the unreal "artificial depressions/puddles" generated during the interpolation/aggregation process. This study emphasizes the importance of topographic characteristics, DEM resolution and surface delineation methods.

Keywords: digital elevation model (DEM); watershed delineation; grid spacing; depression storage; interpolation method

1. Introduction

Spatially-distributed GIS data, such as digital elevation models (DEMs), have been widely used for hydrotopographic analysis and modeling. DEM-based characterization of surface topography and quantification of surface depression storage are critical to watershed hydrologic and environmental modeling due to the important role of depressions in water retention on topographic surfaces [1–3]. The retention process of surface depressions can delay and reduce surface runoff and enhance infiltration [4–6]. Surface depression storage and surface roughness also affect soil erosion and crust formation [7]. In addition, surface depression storage is an important factor in preventing soil erosion in agricultural fields [8–10].

Since it is impractical to measure surface depression storage, it is often indirectly estimated [1,8]. Various DEM-based methods have been developed for watershed delineation and computation of maximum depression storage (MDS) [1,11–16]. Most of these methods implemented similar procedures by first identifying local minima and then filling the corresponding depressions/puddles up to their thresholds. The related watershed delineation software packages, such as ArcHydro [17], HEC-GeoHMS [18,19] and PCRaster [20], also have been developed.

Research efforts have been made to investigate the effects of DEM resolution on surface delineation and the computation of MDS. Huang and Bradford [7] examined depressional storage for Markov–Gaussian surfaces with three grid sizes (5, 10 and 20 mm) and concluded that MDS decreased as grid size increased due to the loss of surface details. Kamphorst *et al.* [8,9] calculated MDS for laboratory and field soil surfaces using PCRaster [20,21] and found that MDS did not structurally decrease with an increase in grid spacing. The results from their field plot study indicated that MDS stabilized in the range of sample spacings (10–30 mm) after an initial increase. Further linear regression analysis showed that MDS slightly increased with sample spacing for a rough surface, while it decreased slightly for a smooth surface [9]. Carvajal *et al.* [22] selected three soil surfaces (tilled, non-tilled and gullied soil surfaces). They used Jenson and Domingue's method [12] to calculate MDS and found that MDS decreased with an increase in grid size. The decrease in the calculated MDS can be attributed to an artificial smoothing effect related to larger grid sizes (or lower resolutions). Abedini *et al.* [3] selected three plots from Southern Ontario and modified a FORTRAN program from Huang and Bradford [7] to calculate MDS. Their results showed varied relationships between MDS and grid size. Maximum ponding area (MPA), however, increased with an increase in grid size, which was consistent with the finding from Ullah and Dickinson [23]. Yang and Chu [24] examined the relationships between two dimensionless parameters (DEM representation scale λ_L and surface roughness scale λ_R) and a series of puddle property parameters, including MDS, MPA, number of puddles (NP), number of puddle levels, mean of maximum puddle depths and mean of average puddle depths. They found that MDS and MPA followed a similar increasing/decreasing trend with an increase in λ_L, and the relationships of λ_L and MDS/MPA were relevant to surface topographic characteristics or roughness (λ_R).

DEMs and their accuracy depend on their resolutions or grid sizes and many other factors (e.g., acquisition technologies, sources of the original data and interpolation/aggregation methods). Advanced LiDAR (light detection and ranging) technology provides high-accuracy and high-resolution DEMs. Generally, lower resolution (or larger grid size) DEMs are generated from original higher resolution DEMs by using certain interpolation/aggregation methods. Use of such resampled lower-resolution DEMs can result in significant changes in surface topographic characteristics (e.g., slopes) and delineation of stream networks and watershed boundaries [25]. Although higher resolution LiDAR-derived DEMs provide more accurate representation of topographic features, Yang *et al.* [26] found that such improvements in hydrography did not necessarily result in improved watershed-scale hydrologic modeling. Thus, in addition to the complexity and variability in surface topography and the methods used for surface delineation, the varying results of surface delineation and the computed topographic parameters from previous studies also can be attributed to DEM resolution (or grid sizing) and the interpolation methods. However, few existing studies have systematically examined how these factors jointly affect the computation of MDS, an important topographic parameter. This study is aimed to evaluate the effects of DEM resolution on puddle characterization and to discuss, in detail, how these factors influence the computation of MDS.

2. Materials and Methods

2.1. Surface Delineation Methods

Various DEM-based surface delineation methods have been developed and widely used in hydrologic and water quality modeling [27]. These methods also have been used to calculate MDS for topographic surfaces (e.g., [7–9,22]). In this study, four watershed delineation software packages, including ArcHydro [17], PCRaster [20,21], HEC-GeoHMS [18,19] and TauDEM [28], were used for surface delineation and computation of MDS. As an essential preprocessing step, “filling sinks” or “pit removal” is implemented in these programs. That is, depressions or pits are filled to their overflow levels in order to create a depressionless DEM and to develop a uniform, well-connected drainage network for further watershed-scale hydrologic and environmental modeling. In these programs, flow direction of a DEM cell is assigned to its steepest downslope neighbor cell based on the D8 method [29]. The results from the four software packages were further compared with those from the puddle delineation (PD) program [30,31].

ArcHydro and ArcGIS: The ArcGIS software is widely used for watershed delineation in hydrologic modeling [17]. Watershed delineation can be conducted using the grid functions in ArcInfo, the spatial analyst extension in ArcGIS or ArcHydro, an extension of ArcGIS. The delineation function in ArcGIS is primarily based on an automated watershed delineation method [12,32], in which depression filling is implemented by the FILL algorithm in the Arc GRID module and the ArcView Spatial Analyst extension. As an ArcGIS tool, ArcHydro facilities terrain analysis and DEM-based watershed processing. In the surface delineation procedure, ArcHydro directly provides the MDS of a topographic surface.

Geospatial Hydrologic Modeling Extension (HEC-GeoHMS): HEC-GeoHMS is an extension of ArcGIS, which is specifically designed for surface delineation and preprocessing for HEC-HMS hydrologic modeling [18,19]. HEC-GeoHMS extracts the drainage paths and watershed boundaries from a surface DEM to represent the hydrologic structure that can be further used for simulating the watershed response to precipitation. The results delineated by HEC-GeoHMS then can be imported into HEC-HMS for watershed hydrologic modeling.

PCRaster: The PCRaster Environmental Modeling language is used for construction of iterative spatio-temporal environmental models [20,21]. The central concept of PCRaster is discretization of the landscape in space. PCRaster is based on an algorithm by Van Deursen [33]. In PCRaster and HEC-GeoHMS, the MDS of a topographic surface can be estimated indirectly as the volume difference between the original DEM and the filled, depressionless DEM.

Terrain Analysis Using Digital Elevation Models (TauDEM): TauDEM is designed as a toolbox in ArcGIS [28]. It consists of a set of tools for DEM-based hydrotopographic analysis. In addition to the D8 method, the D-infinity flow method is also incorporated into TauDEM, in which a flow direction is defined as an angle counterclockwise from east along the steepest downward slope of a triangular facet [34]. The major functions of the TauDEM toolbox include basic grid analysis (e.g., pit removal, D8/D-infinity flow directions and contributing area) and stream network analysis. In this study, both D8 and D-infinity methods were used for delineation, and the results were compared.

PD Software: An improved method has been developed for DEM-based puddle delineation [30,31]. This method is capable of: (1) characterizing puddles, including their cells, centers and thresholds; (2) computing the puddle property parameters, such as MDS for individual puddles and the entire surface; (3) determining the hierarchical relationships of puddles; and (4) handling special topographic features (e.g., flats). In this study, the delineation results from the PD software were compared with those from the four other methods.

2.2. Surfaces with Varying Scales and Topographic Features

Twenty-three surfaces with different scales and topographic characteristics were selected to evaluate the effects of DEM resolution on the quantification of surface topography (e.g., MDS and MPA). These surfaces included: (1) six ideal artificial surfaces (Surfaces I-1–I-6, Figure 1); (2) three laboratory-scale soil surfaces (Surfaces II-1–II-3, Figure 2); and (3) fourteen watershed-scale land surfaces (Surfaces III-1–III-14, Figure 3).

The six ideal surfaces (Figure 1), featuring varying numbers and sizes of depressions and peaks, were artificially created and specially used to better understand and demonstrate how various DEM resolutions alter topographic characteristics and further influence the computation of topographic parameters (e.g., MDS and MPA). Surfaces I-1–I-4 (Figure 1a–d) were characterized by evenly-distributed puddles. The number of puddles for these four surfaces increased from one to sixteen, as the radius of the puddles decreased from 10.0 to 2.5 cm. Surfaces I-5 and I-6 (Figure 1e,f) were characterized by a number of peaks on a flat surface. Surface I-5 had fewer peaks than surface I-6, and both surfaces had a zero MDS. These six surfaces, I-1–I-6, had an area of 25.6 × 25.6 cm^2, and the resolution of their original DEMs was 1.0 mm.

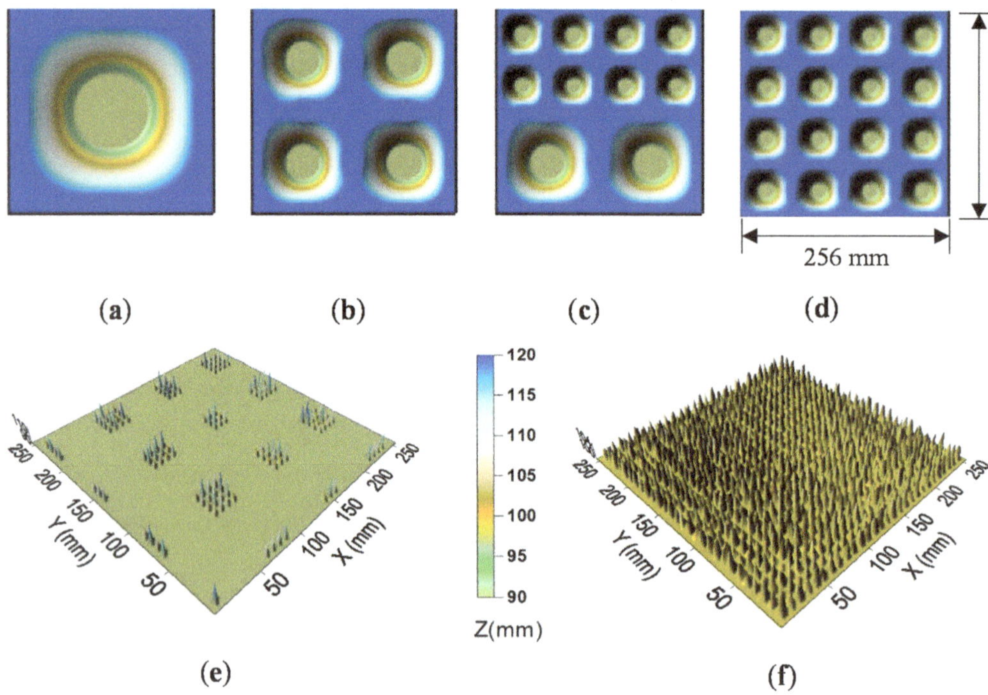

Figure 1. Ideal Surfaces I-1–I-6: (**a**) Surface I-1 (1 puddle); (**b**) Surface I-2 (4 puddles); (**c**) Surface I-3 (10 puddles); (**d**) Surface I-4 (16 puddles); (**e**) Surface I-5 (some peaks); (**f**) Surface I-6 (more peaks).

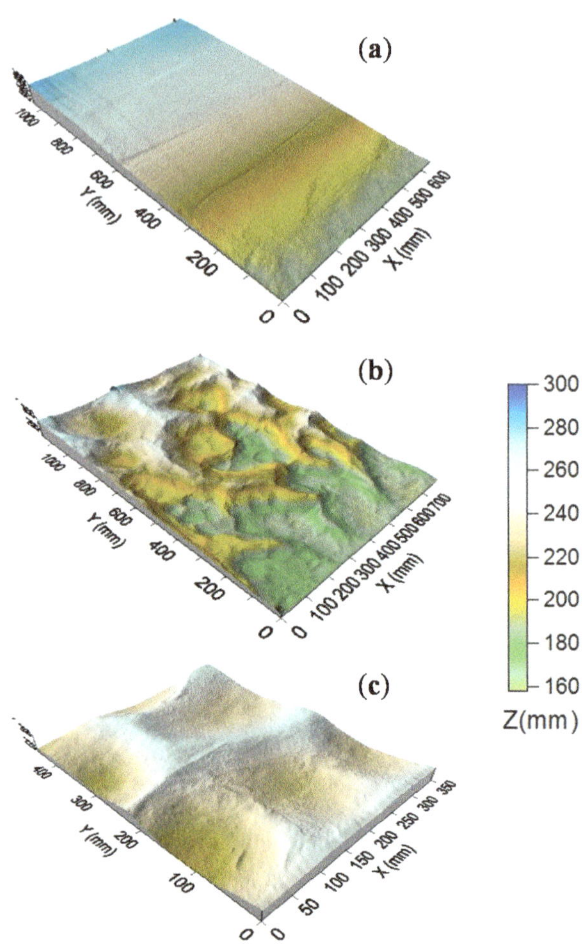

Figure 2. Laboratory-scale Surfaces II-1–II-3: (**a**) Surface II-1; (**b**) Surface II-2; (**c**) Surface II-3.

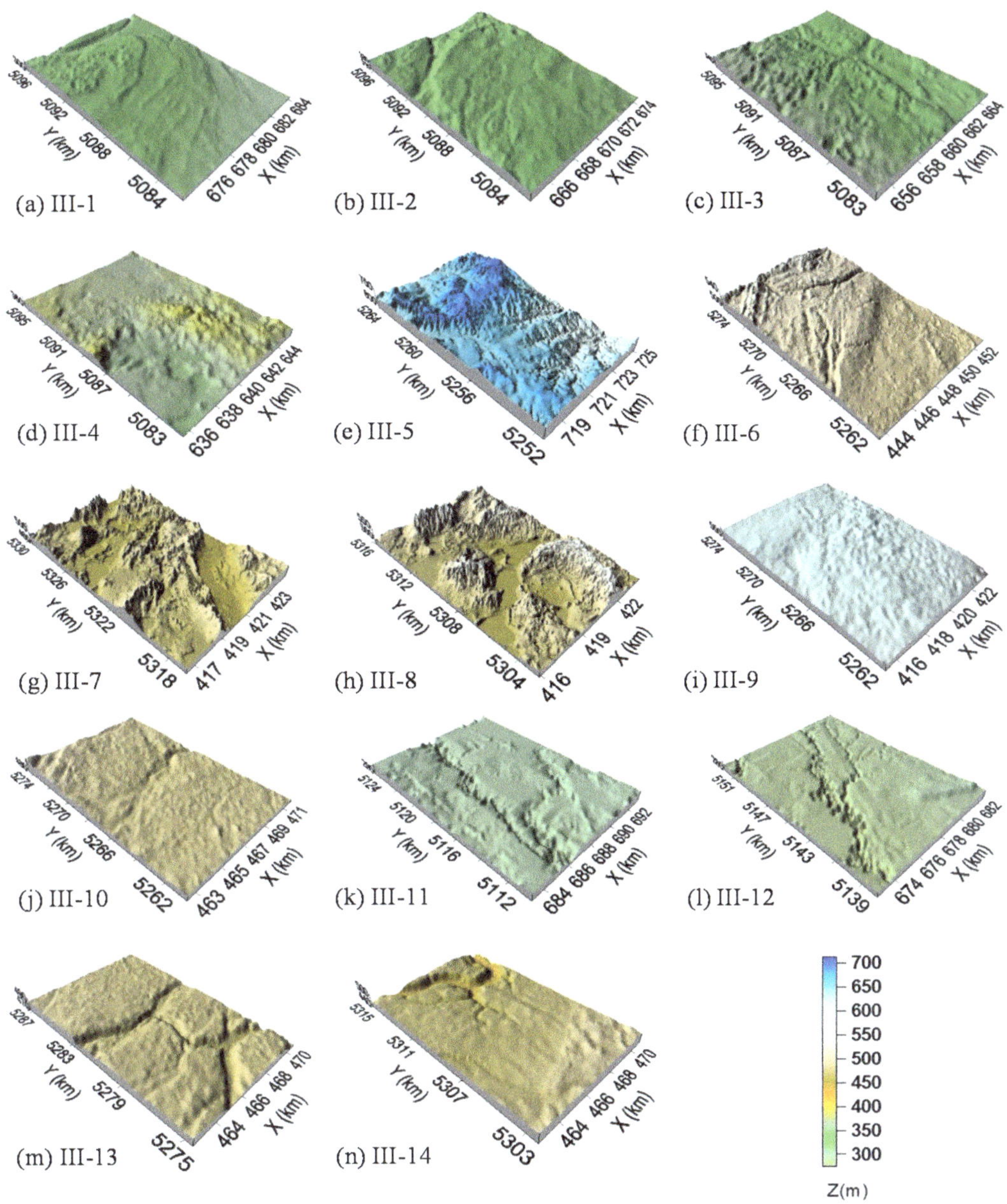

Figure 3. Watershed-scale Surfaces III-1–III-14.

The three laboratory-scale surfaces (Surfaces II-1–II-3, Figure 2) were created by using soil boxes. Surface II-1 (Figure 2a) was a smoother soil surface characterized by numerous small depressions/puddles, while Surface II-2 (Figure 2b) was a rougher soil surface featuring larger puddles and mounds. Surface II-3 (Figure 2c) was a special rough soil surface that had some depressions along the boundaries of the surface (*i.e.*, partial depressions). It was used to examine the boundary effect on surface delineation and computation of topographic parameters. An instantaneous-profile laser scanner was utilized to scan these three laboratory soil surfaces and acquire their 1-mm resolution DEMs.

In addition to the small-scale surfaces, fourteen watershed-scale land surfaces (III-1–III-14, Figure 3) located in North Dakota were selected to examine the relationships of DEM grid spacing and topographic parameters. The size of theses fourteen surfaces was 10 × 10 km^2. Their original 10-m DEMs were obtained from the USGS National Map Viewer. These fourteen land surfaces represented a variety of larger-scale topographic characteristics, including both plain and mountain land surfaces.

Surfaces III-1–III-4 (Figure 3a–d) had relatively flat surface topography, and their elevations were lower than 400 m. In contrast, Surfaces III-5–III-9 (Figure 3e–i) featured typical mountain-type topography with significant elevation variations (up to 141.5 m), and their elevations were mostly greater than 500 m. Surface III-10 (Figure 3j) also had mountain-type topography, but its elevations varied within a smaller range. Surfaces III-11–III-14 (Figure 3k–n) showed various river channels with lower elevations.

2.3. Interpolation/Aggregation Methods

Various interpolation/aggregation methods can be used to acquire DEMs with lower resolutions or larger grid sizes from an original higher resolution DEM. In this study, the kriging and averaging methods were selected to create DEMs with varying grid sizes to examine the relationship between DEM resolution or grid sizing and topographic characteristics (e.g., MDS). For the six ideal surfaces (I-1–I-6, Figure 1a–f), the averaging method was utilized to generate lower resolution DEMs with grid sizes of 2, 4, 6, 8, 16, 32, 64, 128 and 256 mm based on their original 1-mm resolution DEMs. It is expected that during this resampling process, some important topographic details may be lost as a result of aggregation, especially when the grid spacing is greater than the size of puddles. The MDS values were calculated for all surfaces. Note that Surfaces I-5 and I-6 had a zero MDS.

Similarly, based on the original 1-mm resolution DEMs, a series of lower-resolution DEMs were created by using the kriging method for the three laboratory soil surfaces, II-1–II-3 (Figure 2). For the fourteen watershed-scale Surfaces III-1–III-14 (Figure 3), the kriging method was also utilized to generate lower-resolution DEMs based on their original 10-m DEMs. The MDS, MPA and NP values of the laboratory and watershed-scale surfaces were calculated by the PD software for all original and derived/aggregated DEMs with varying grid sizes.

3. Results and Discussion

3.1. DEM Resolution Effects for Ideal Surfaces

The relationships between grid size and MDS for Surfaces I-1–I-4 are shown in Figure 4. The MDS values were constant for smaller grid sizes ($\Delta d < 4$ mm) for all four surfaces. That is, there was a range of grid sizes, within which MDS did not change or only had minor changes. Such a range of grid sizes could vary, depending on the surface topographic characteristics. This information can be used to determine the "best" representative DEM resolution for hydrologic modeling and analysis. As the grid size increased, different surfaces exhibited dissimilar MDS decreasing patterns due to the changes in DEM representations and the loss of surface topographic details (Figure 4). Eventually, the MDS of a surface decreased to zero at a certain grid size (Figure 4). The grid size corresponding to the zero MDS depended on the surface topographic characteristics (e.g., number, shape and size of puddles). As expected, the surfaces with smaller depressions (e.g., Surface I-4, Figure 1d) were more sensitive to the change in grid size, showing an earlier decrease in MDS and a smaller zero-MDS grid size. Surface I-1 with the single largest depression (Figure 1a) had the greatest zero-MDS grid size (Figure 4). The maximum grid sizes for Surfaces I-1–I-4 were 128, 64, 64 and 32 mm, respectively (Figure 4).

Figure 5 shows the changes in DEM representations and the loss of topographic detail for Surface I-4 (Figure 1d; 16 puddles) induced by the changes in DEM resolution (grid sizes Δd increased from 2 to 4, 8, 16 and 32 mm). The depression storage gradually decreased with an increase in grid spacing. As a result, Surface I-4 was smoothened out, eventually resulting in a zero MDS at the largest grid size $\Delta d = 32$ mm (Figure 5e). This example demonstrates the significant influence of DEM resolution on the topographic properties of a surface and related parameters (e.g., MDS) that are often used for hydrologic and environmental modeling.

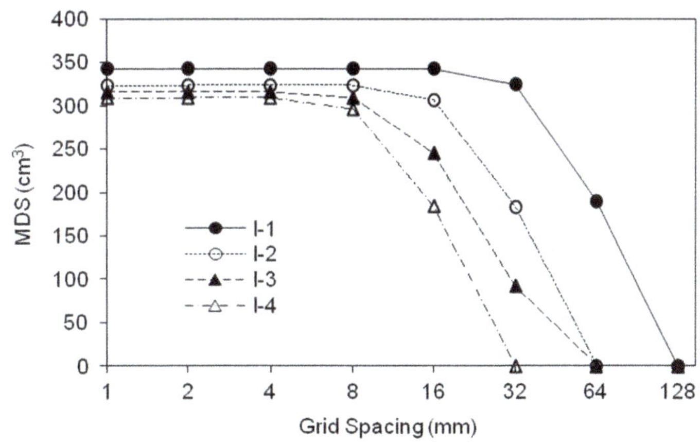

Figure 4. Relationships of grid spacing and maximum depression storage (MDS) determined by the averaging interpolation method for ideal Surfaces I-1–I-4.

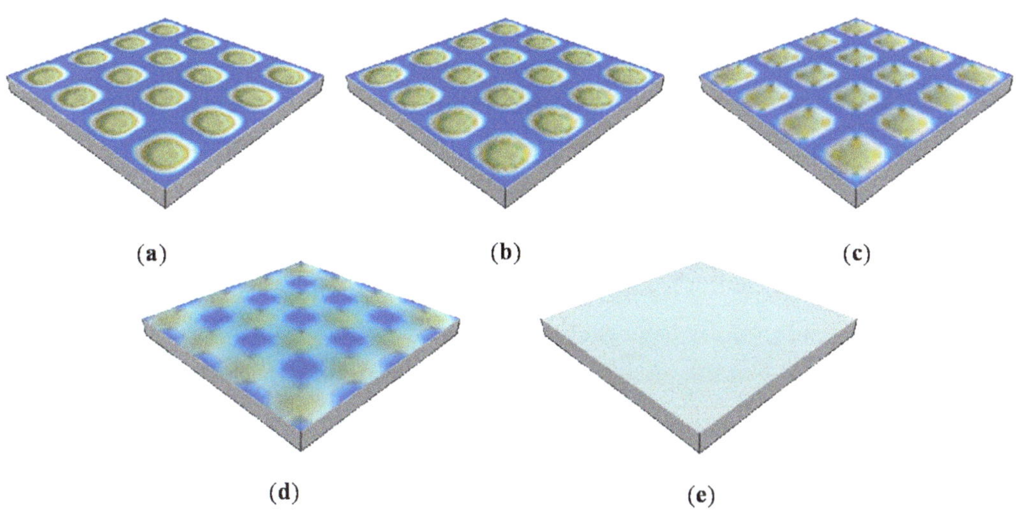

Figure 5. Demonstration of the loss of surface topographic details determined by the averaging interpolation method for ideal Surface I-4: **(a)** $\Delta d = 2$ mm; **(b)** $\Delta d = 4$ mm; **(c)** $\Delta d = 8$ mm; **(d)** $\Delta d = 16$ mm; **(e)** $\Delta d = 32$ mm.

The above analyses revealed that due to the loss of surface topographic detail, MDS decreased with an increase in grid size, and correspondingly, the number of puddles also decreased. This finding is similar to the one from Huang and Bradford [7], but different from several other researchers [3,8,9]. To verify the related conclusions and to better understand the underlying reasons, Surfaces I-5 and I-6 (Figure 1e,f) were used for further testing. As shown in Figure 6, MDS increased with an increase in the grid spacing for both surfaces, indicating that certain artificial depressions were created during the interpolation/aggregation process, which further affected the computed depression storage. However, the MDS of Surface I-5 started to increase beyond $\Delta d = 4$ mm. Then, MDS gradually decreased down to zero at Δd = 64 mm for both surfaces (Figure 6). Surface I-6 had much greater MDS values than Surface I-5 for smaller grid sizes (e.g., Δd = 2 and 4 mm) (Figure 6). Hence, it can be concluded that the extent of the influence of DEM resolution on the MDS computation also depends on the surface topographic characteristics.

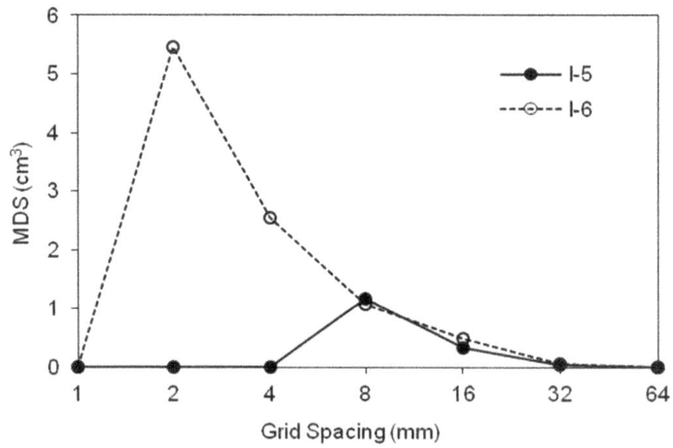

Figure 6. Relationships of grid spacing and maximum depression storage (MDS) determined by the averaging interpolation method for Surfaces I-5 and I-6.

To better demonstrate how the "artificial depressions" were generated during the interpolation/aggregation process, the derived DEMs for a small portion of Surface I-5 for Δd = 2, 4 and 8 mm were compared with the original DEM (Δd = 1 mm). As shown in Figure 7, for the grid sizes of 1–4 mm, the surfaces exhibited similar topographic features, such as peaks and a zero MDS, although the peaks were at different elevations for the three surfaces (Figure 7a–c). However, an artificial puddle was formed when the grid size was increased to 8 mm (Figure 7d), which resulted in a non-zero MDS. In summary, as DEM grid sizes change, "artificial puddles" can be created, and real small puddles may disappear, which consequently alters the surface topographic characteristics. This helps explain why different researchers (e.g., [3,8,9]) reached distinct conclusions concerning the effects of DEM resolution on MDS.

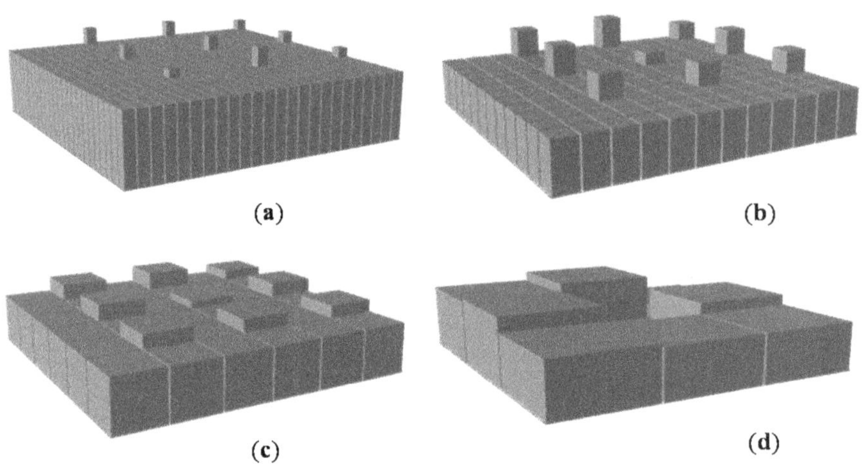

Figure 7. Demonstration of the generation of "artificial puddles" using the averaging interpolation method for a portion of Surface I-5: (**a**) Δd = 1 mm; (**b**) Δd = 2 mm; (**c**) Δd = 4 mm; (**d**) Δd = 8 mm.

3.2. DEM Resolution Effects for Laboratory-Scale Soil Surfaces

The puddle delineation results for Surfaces II-1 and II-2 (Figure 2a,b) are shown in Figure 8. The NP decreased quickly with an increase in grid spacing due to the loss of microrelief detail (Figure 8a). For example, as grid spacing increased from 2 to 3 mm, Surfaces II-1 and II-2 lost 2235 and 283 puddles, respectively. Both MDS and MPA of Surfaces II-1 and II-2 decreased with an increase in grid spacing and eventually approached zero (Figure 8b,c). The MDS and MPA of the smoother Surface II-1 decreased more quickly than the rougher Surface II-2 (Figure 8b,c). The variations in the MDS of Surface II-2 at $\Delta d = 80$ and 100 mm (Figure 8b) can be attributed to certain "artificial puddles" generated during the aggregation process. The MDS and MPA approached zero at Δd = 30 and 200 mm for Surfaces II-1 and II-2, respectively (Figure 8b,c).

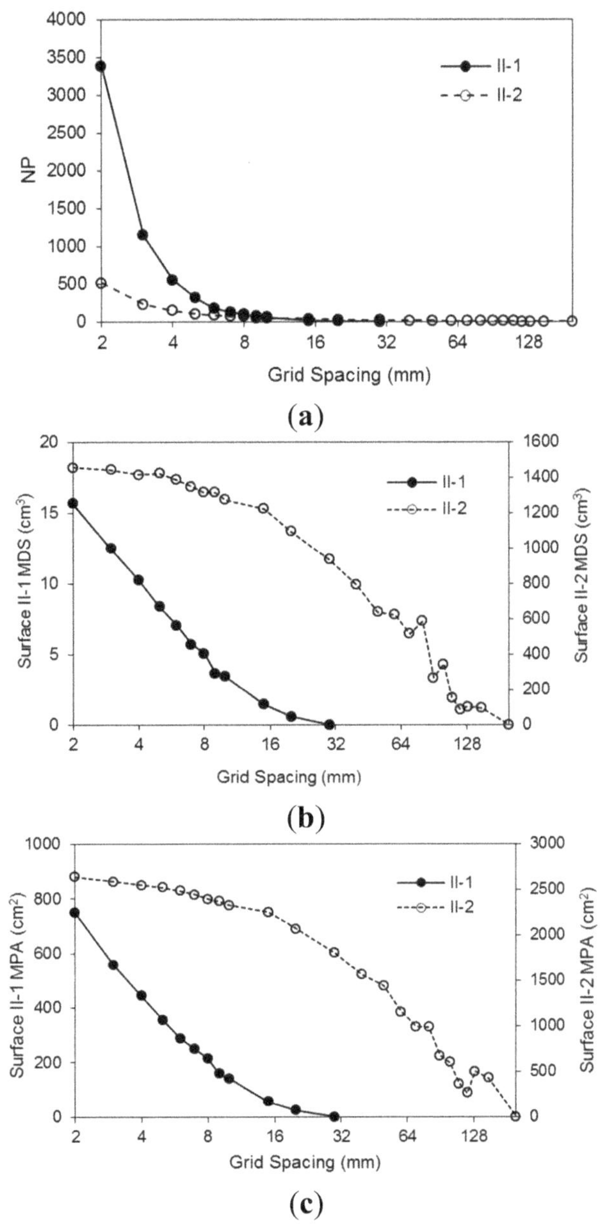

Figure 8. Relationships of grid spacing and (**a**) the number of puddles (NP); (**b**) the maximum depression storage (MDS) and (**c**) the maximum ponding area (MPA) determined by using the kriging method and the PD software for Surfaces II-1 and II-2.

3.3. DEM Resolution Effects for Watershed-Scale Surfaces

Figure 9 shows the comparison between grid spacing and MDS, MPA and NP for the watershed-scale Surfaces III-1–III-14 (Figure 3). Dissimilar changing patterns of MDS, MPA and NP can be observed for these fourteen surfaces (Figure 9). For Surfaces III-1–III-4, MDS and MPA decreased with increasing grid spacing and approached zero at Δd = 1280 m (Figure 9a,b). As expected, the rougher Surfaces III-3 and III-4 had greater MDS and MPA values. The NP values of Surfaces III-1–III-4 decreased rapidly with an increase in grid spacing (Figure 9c). These findings were similar to the results for the laboratory-scale soil surfaces shown in Figure 8.

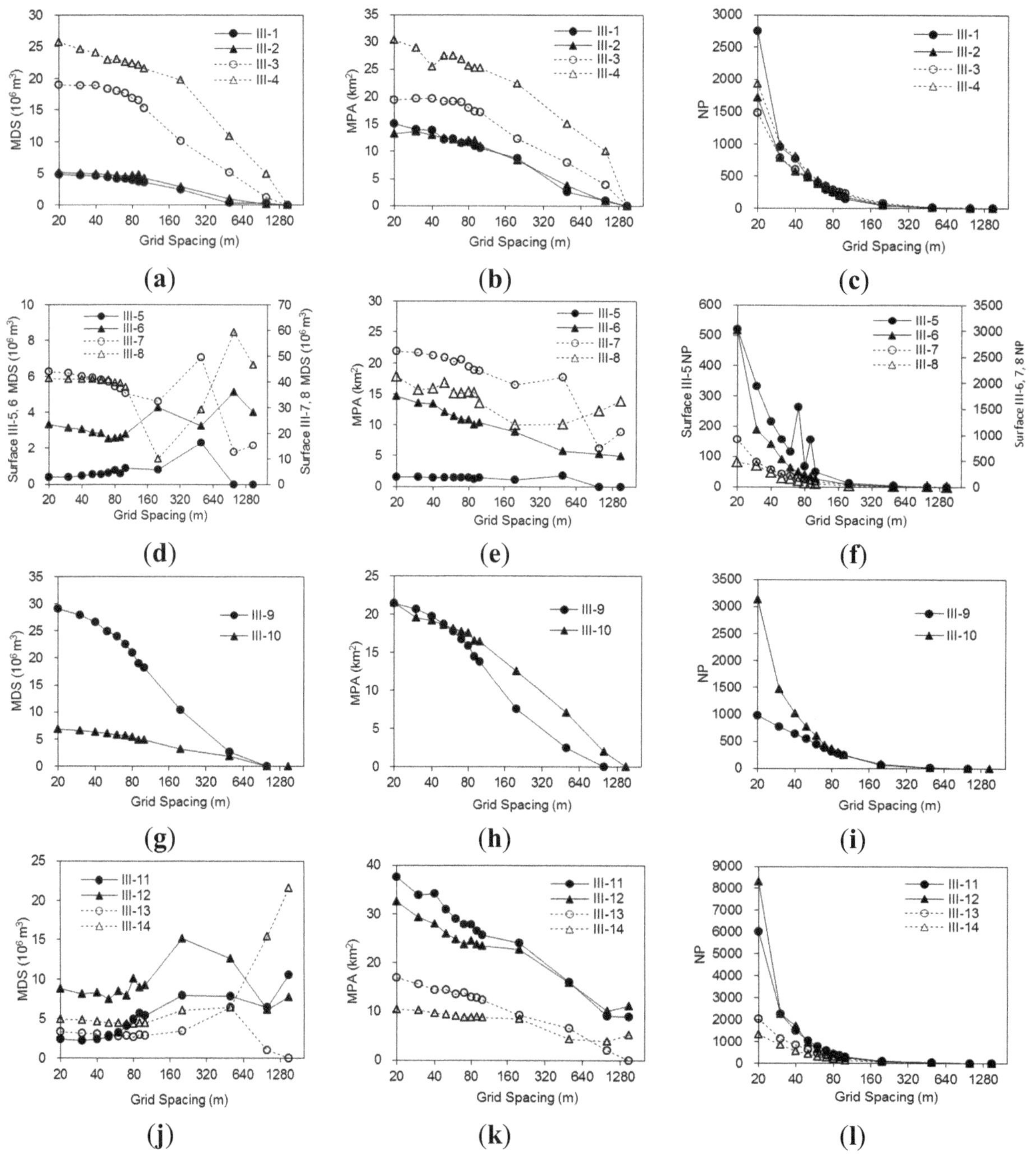

Figure 9. Relationships of grid spacing and maximum depression storage (MDS), maximum ponding area (MPA) and number of puddles (NP) determined by using the kriging method and the PD software for Surfaces III-1–III-14.

For Surfaces III-5–III-8, however, their MDS and MPA values varied with an increase in the grid size (Figure 9d,f). The MDS and MPA of Surface III-5 increased initially, reached the maximum values at $\Delta d = 500$ m and then decreased to zero at $\Delta d = 1000$ m (Figure 9d,e). The MDS of Surface III-6 varied with grid spacing (Figure 9d), while its MPA exhibited a decreasing pattern (Figure 9e). Surfaces III-7a and III-8 displayed significant variations in both MDS and MPA as the grid size changed. The NP curves of Surfaces III-5–III-8 showed an overall decreasing pattern as the grid size increased. However, the NP of Surface III-5 displayed some variations at $\Delta d = 70$–90 m (Figure 9f). This can be attributed to the "unreal" depressions induced during the interpolation/aggregation process. Comparisons of the MDS, MPA and NP between the four plain surfaces (III-1–III-4) (Figure 9a–c) and the four mountain-type surfaces (III-5–I III-8) (Figure 9d–f) suggested that the changing patterns of these topographic parameters with grid spacing were different, primarily depending on their topographic characteristics.

Comparison between Surfaces III-9 (Figure 3i) and III-10 (Figure 3j) revealed some interesting results. The elevations of Surface III-9 varied in a greater range from 511.4 to 620.3 m (*i.e.*, the difference was 108.9 m), while the elevations of Surface III-10 ranged from 475.3 to 508.5 m (the difference was only 33.2 m). As expected, the MDS of Surface III-9 was much greater than that of Surface III-10 (Figure 9g). However, both surfaces had very comparable MPA curves (both quantity and distribution) (Figure 9h). Like Surfaces III-1–III-4 (Figure 9c), the NP curves of Surfaces III-9 and III-10 also showed a decreasing trend with increasing grid spacing, even though Surface III-10 included more depressions (Figure 9i).

Similar to Surfaces III-7 and III-8 (Figure 9d), Surfaces III-11–III-12 also showed substantial variability in their MDS (Figure 9j). The changing pattern of MDS for Surface III-13 (Figure 9j) was similar to that of Surface III-5 (Figure 9d), and its MDS reached zero at $\Delta d = 1500$ m. In contrast, Surface III-14 showed a significant increase in MDS, especially for $\Delta d > 500$ m, and eventually reached the highest MDS at the coarsest resolution ($\Delta d = 1500$ m) (Figure 9j). The MPA values of Surfaces III-11–III-12 were greater than those of Surfaces III-13 and III-14, and the four MPA curves exhibited an overall decreasing trend with increasing grid spacing (Figure 9k). Similar to Surfaces III-1–III-4 (Figure 9c), substantial decreases can be observed in the NP curves of Surfaces III-11–III-14, as the grid size increased (Figure 9l).

To look into more detail at the alteration of DEM representations and the loss of topographic detail due to the changes in DEM resolution, a portion of the plain-type Surfaces III-1 and III-3 (zoomed-in areas) and four different grid sizes ($\Delta d = 20$, 80, 200 and 500 m) were selected, and the changes in their MDS and MPA values were analyzed. Figure 10a,b respectively show the changes in their MDS and MPA. Basically, both MDS and MPA decreased with increasing grid size. As a result, the puddles/depressions on Surfaces III-1 and III-3 were smoothened out, eventually resulting in a zero depression storage (Figure 10c,d). The MDS and MPA became zero (*i.e.*, no depression/puddle) at $\Delta d = 200$ m for Surface III-1 (Figure 10a–c) and at $\Delta d = 500$ m for Surface III-3 (Figure 10a,b,d). The above analyses revealed that due to the loss of surface topographic detail, MDS and MPA decreased with an increase in grid size, and correspondingly, the number of depressions/puddles also decreased. Similar to the small-scale surfaces (Figure 4), MDS and MPA appeared to be changed insignificantly for smaller grid sizes ($\Delta d < 80$ m). Again, this finding can be useful for multi-scale hydrologic modeling and analysis.

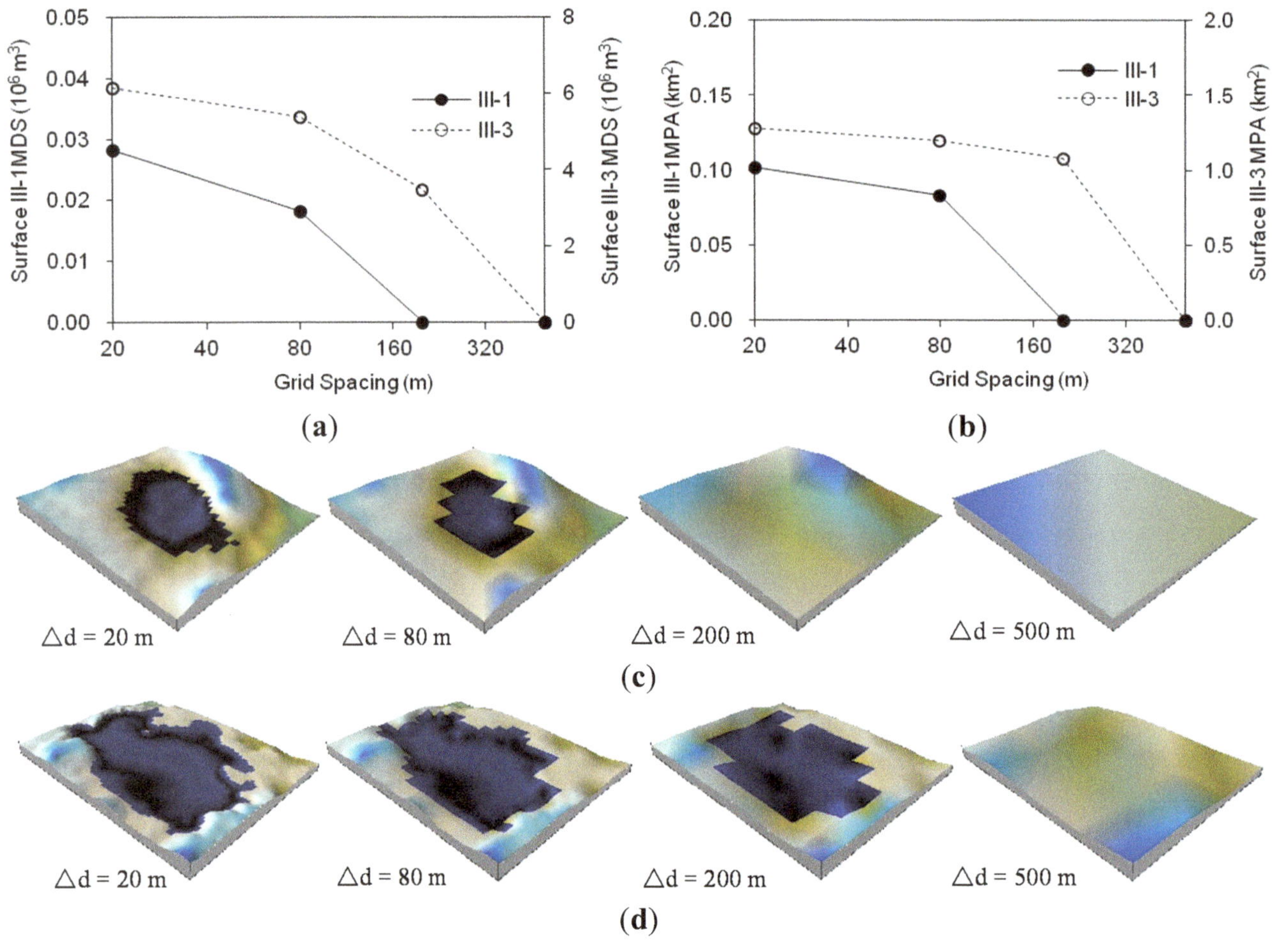

Figure 10. Demonstration of the loss of surface topographic details (maximum depression storage MDS and maximum ponding area MPA) determined by the kriging method and the PD software for Surfaces III-1 and III-3: (**a**) MDS; (**b**) MPA; (**c**) Surface III-1; (**d**) Surface III-3.

Figure 11a and 11b respectively show the changes in MDS and MPA for a selected portion of the mountain-type Surfaces III-5 and III-7, while Figure 11c and 11d display the delineated surfaces and puddle distributions for the four different resolutions (Δd = 20, 80, 200 and 500 m). Different from the selected plain Surfaces III-1 and III-3 (Figure 10a,b), as the grid size increased from 20 to 500 m, the MDS and MPA of the small piece of Surface III-5 increased from 0.06×10^6 to 2.11×10^6 m^3 and from 0.20 to 0.99 km^2, respectively (Figure 11a,b). Figure 11c also shows increases in the MDS and MPA. The MDS and MPA of the small piece of Surface III-7 displayed slight decreases for smaller grid sizes ($\Delta d \leq 200$ m) and then increased to 44.68×10^6 m^3 and 14.89 km^2, respectively (Figure 11a,b). Such changing trends of MDS and MPA also can be observed in Figure 11d. This result suggested that the computation of MDS was strongly affected by DEM resolution and the complexity of surface topography. The changes in the computed topographic parameters can be attributed to the introduced artificial depressions and the smoothing effects.

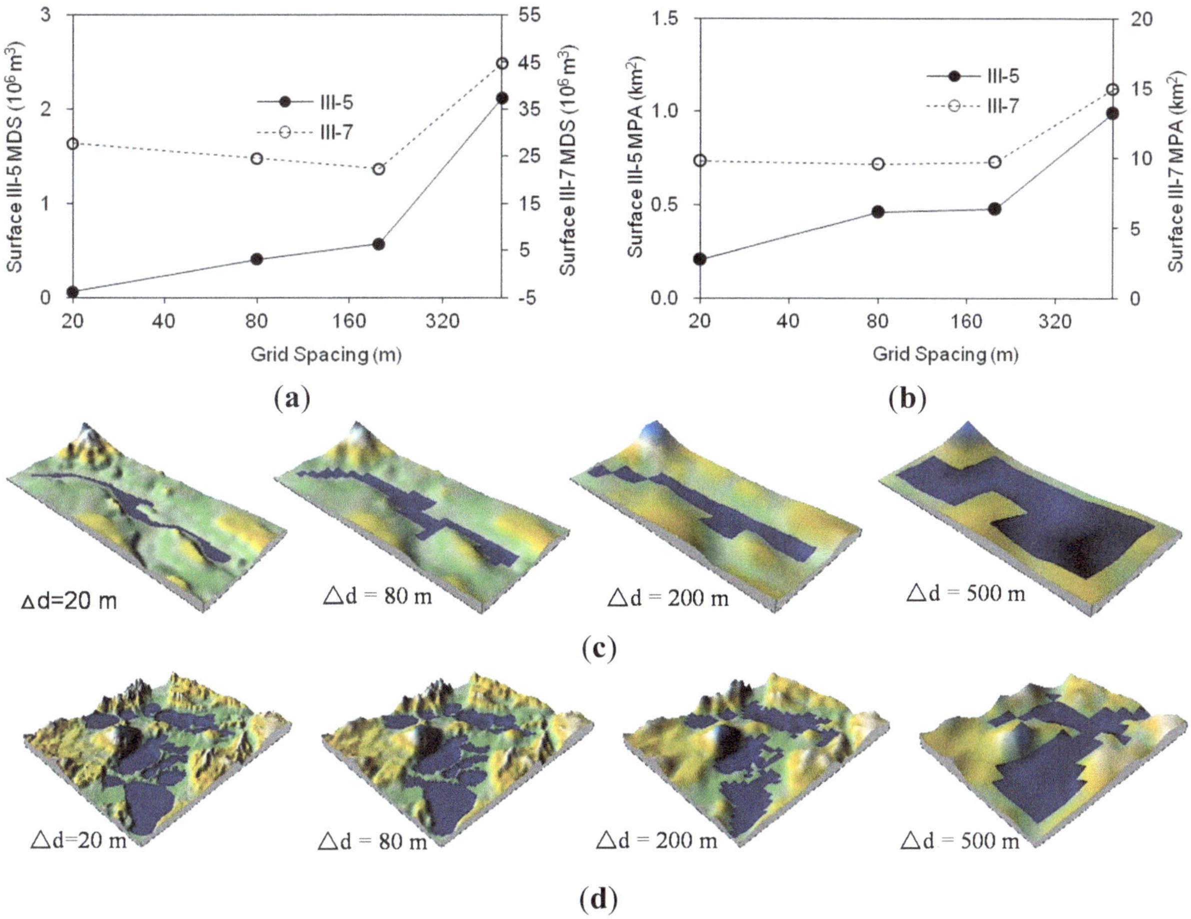

Figure 11. Demonstration of the changes of surface topographic characterization (maximum depression storage MDS and maximum ponding area MPA) determined by the kriging method and the PD software for Surfaces III-5 and III-7: (**a**) MDS; (**b**) MPA; (**c**) Surface III-5; (**d**) Surface III-7.

Figure 12a,b shows the changes in the MDS and MPA of the selected pieces of Surfaces III-13 and III-14 that were characterized by river channels. As the grid size increased from 20 to 500 m, the MDS and MPA of the small piece of Surface III-13 increased from 0.04×10^6 to 5.11×10^6 m^3 and from 0.51 to 3.46 km^2, respectively (Figure 12a,b). Such increases in the MDS and MPA also can be observed in Figure 12c. The MDS and MPA of the selected piece of Surface III-14 also increased from 0.07×10^6 to 5.13×10^6 m^3 and from 0.39 to 2.35 km^2, respectively (Figure 12a,b,d). This analysis, again, demonstrates the dominant role of surface topography and DEM resolution in the computation of topographic parameters.

3.4. Influences of Different Delineation Methods

Figure 13 shows the comparison of MDS values calculated by the PD, ArcHydro, HEC-GeoHMS, PCRaster and TauDEM software packages for varying resolution DEMs of the laboratory Surfaces II-1 and II-2 (Figure 2a,b), as well as the watershed-scale Surface III-5 (Figure 3e). For the laboratory-scale smooth Surface II-1 (Figure 2a), the five programs yielded almost identical MDS values for all grid sizes (Figure 13a), showing that the MDS values decreased with increasing DEM grid size. For the laboratory-scale rough Surface II-2, the MDS values from the five programs followed similar decreasing patterns with an increase in the grid size. However, the MDS curve from PCRaster exhibited minor differences at certain grid size points (Figure 13b). The MDS values were close to zero at $\Delta d = 200$ mm for all five methods (Figure 13b). For the watershed-scale Surface III-5, a significant difference in MDS can be observed at $\Delta d = 500$ m (Figure 13c). For Surface III-5, PCRaster yielded a peak MDS (7.23×10^6 m^3) at $\Delta d = 500$ m, while the four other programs reached a peak MDS (6.29×10^6 m^3) at $\Delta d = 700$ m (Figure 13c).

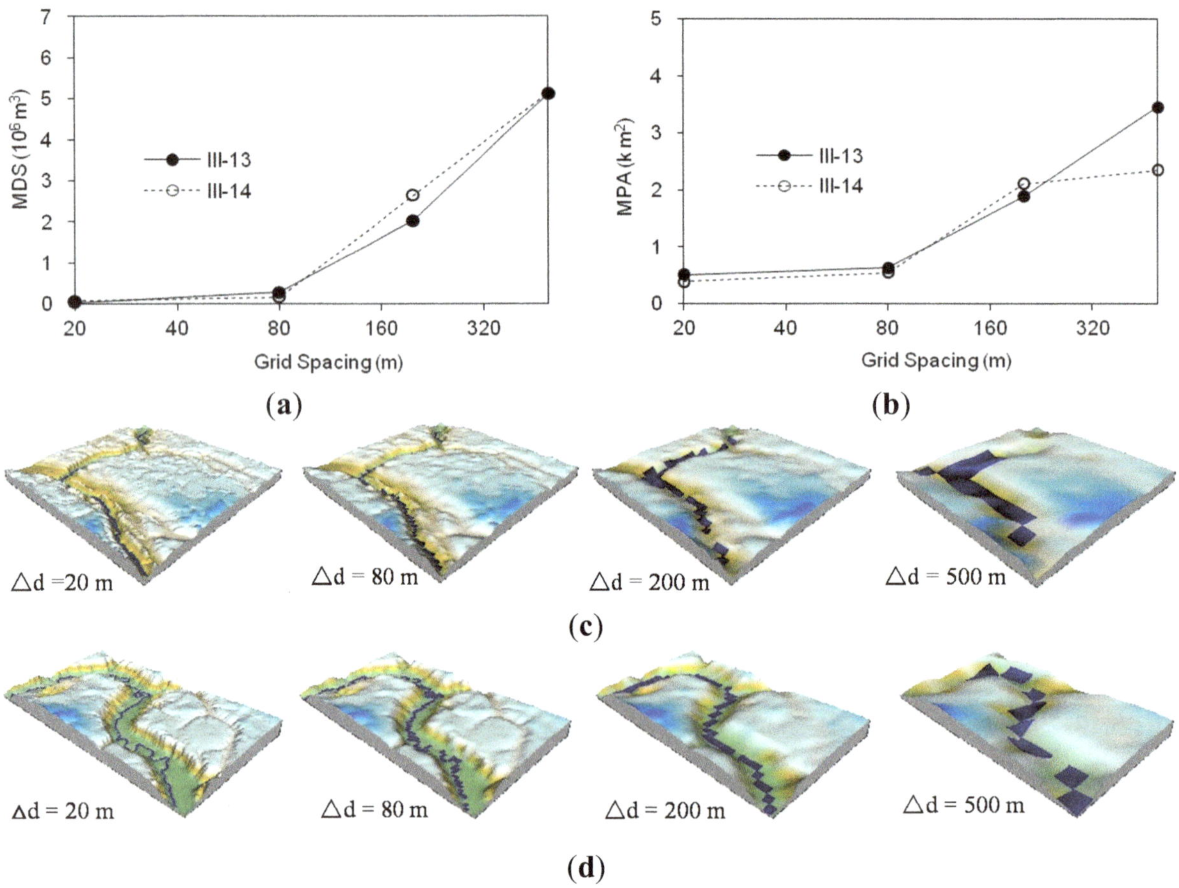

Figure 12. Demonstration of the changes of surface topographic characterization (maximum depression storage MDS and maximum ponding area MPA) determined by the kriging method and the PD software for Surfaces III-13 and III-14: (**a**) MDS; (**b**) MPA; (**c**) Surface III-13; (**d**) Surface III-14.

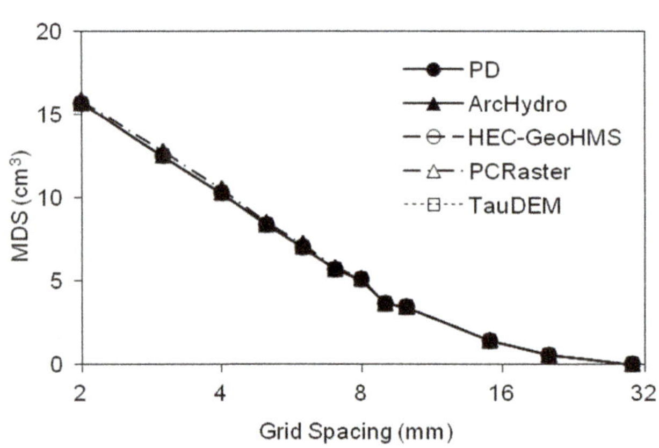

(a)

Figure 13. *Cont.*

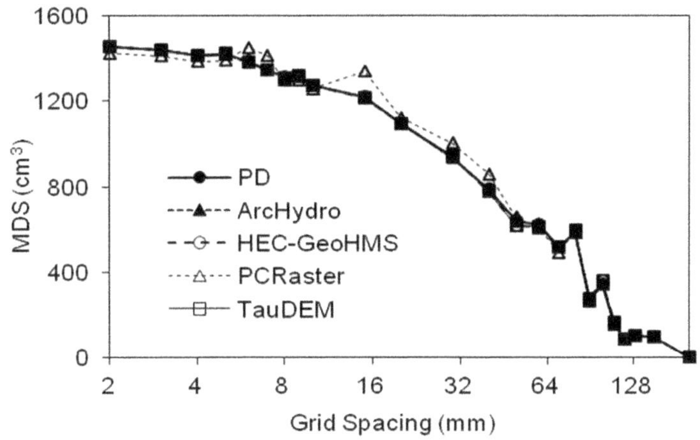

(b)

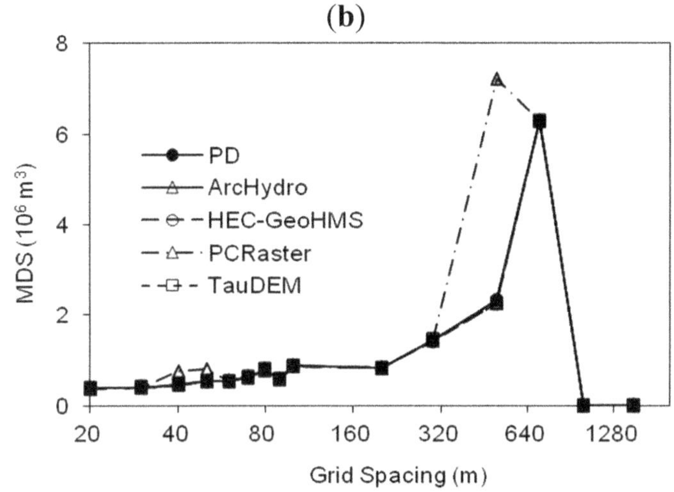

(c)

Figure 13. Relationships of grid spacing and maximum depression storage (MDS) determined by the PD, ArcHydro, HEC-GeoHMS, PCRaster and TauDEM software packages for: (**a**) Surface II-1; (**b**) Surface II-2; (**c**) Surface III-5.

Boundary conditions (e.g., closed or open boundaries) often affect the computation of MDS in a surface delineation program [8]. This boundary-related issue involves how to deal with the depressions adjacent to surface boundaries and computation of related depression storages. Surface II-3 (Figure 2c), featuring four major boundary depressions/puddles, was used to address this issue. Based on the original DEM of Surface II-3, eleven coarser resolution DEMs (Δd = 2–15 mm) were generated for this study. PD, ArcHydro, PCRaster, HEC-GeoHMS and TauDEM were applied to compute the MDS of this laboratory surface. As shown in Figure 14, PCRaster yielded much higher MDS values for all grid sizes, and the MDS remained nearly steady (Figure 14). This can be attributed to the fact that a closed boundary condition was assumed in PCRaster. The four other programs provided similar MDS values, which were nearly constant for all grid sizes (Figure 14). The aforementioned results suggest that different methods can be utilized in the surface delineation programs to deal with the boundary depressions or incomplete depressions/puddles, which could lead to different results. It is critical to fully understand the underlying assumptions in the programs.

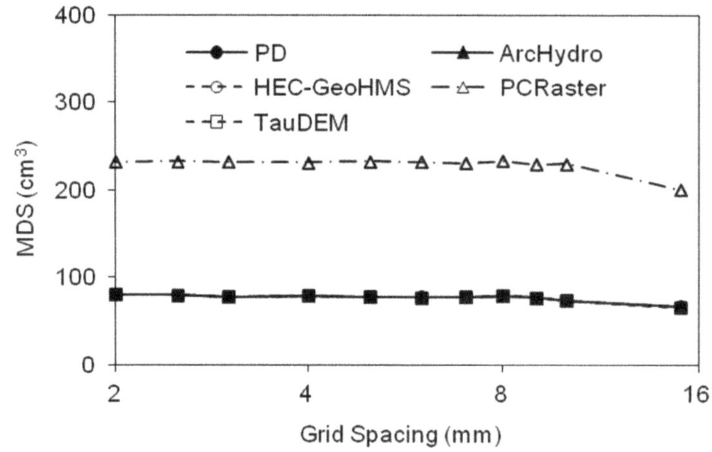

Figure 14. Relationships of grid spacing and maximum depression storage (MDS) determined by the PD, ArcHydro, HEC-GeoHMS, PCRaster and TauDEM software packages for Surface II-3 with boundary puddles.

For various DEM resolutions, TauDEM yielded MDS values similar to those from PD, ArcHydro and HEC-GeoHMS. However, the D8 and D-infinity methods resulted in dissimilar flow directions. For comparison purposes, the eight flow directions from the D8 method were converted to angles (east: 0°; northeast: 45°; north: 90°; northwest: 135°; west: 180°; southwest: 225°; south: 270°; southeast: 315°). Figure 15 shows the comparison of flow directions determined by the D-infinity and D8 methods for the laboratory Surface II-2 (Figure 2b) and the watershed Surface III-5 (Figure 3e). Although the overall distributions and changing patterns of flow directions from the two methods were similar for both surfaces (II-2 and III-5), the D-infinity method did provide more grey scales (*i.e.*, more flow directions). Additionally, certain discrepancies in flow directions at many local areas can be observed for both laboratory and watershed surfaces (Figure 15). These differences and variations in flow directions would potentially alter flow accumulations and hydrologic connectivity and further affect watershed modeling.

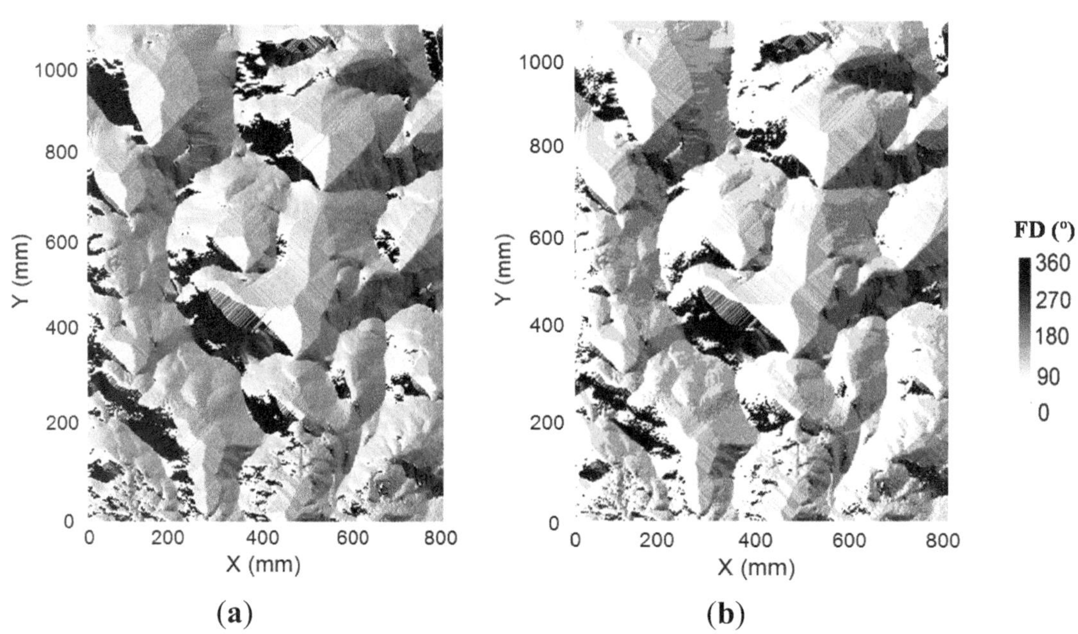

Figure 15. *Cont.*

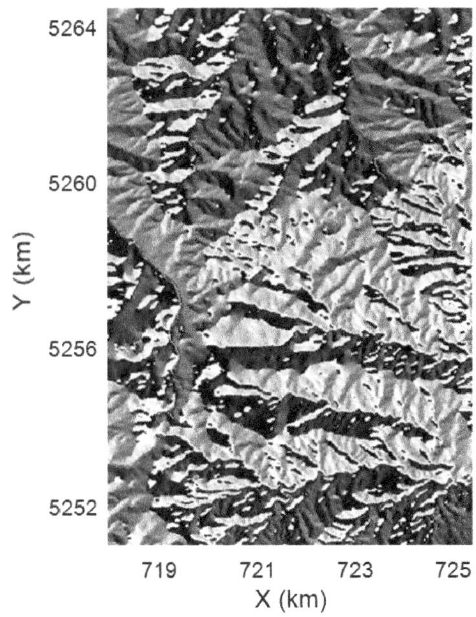

(c)

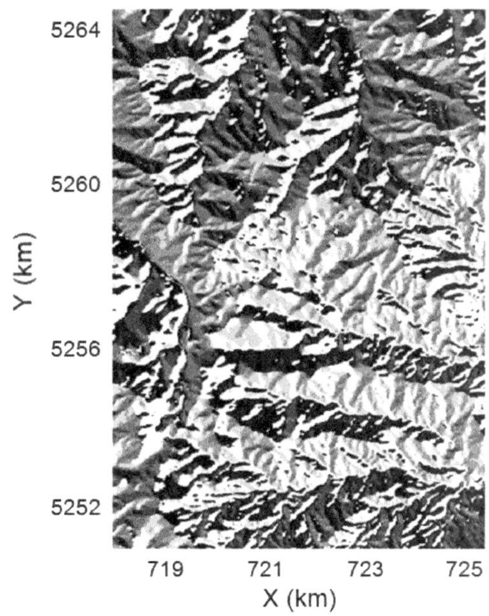

(d)

Figure 15. Comparison of flow directions computed by the D8 and D-infinity methods for laboratory Surface II-2 and watershed Surface III-5: (**a**) Surface II-2, D-infinity; (**b**) Surface II-2, D8; (**c**) Surface III-5, D-infinity; (**d**) Surface III-5, D8.

4. Conclusions

In this study, we systematically evaluated the impacts of DEM resolution on puddle characterization and computation of topographic parameters (e.g., MDS, MPA and NP) by analyzing the related influence factors, including surface topography and surface delineation methods. Twenty-three surfaces with varying spatial scales and topographic characteristics (including ideal artificial surfaces, laboratory-scale soil surfaces and real watershed-scale land surfaces) were selected and delineated by using five different methods/software packages: PD, ArcHydro, HEC-GeoHMS, PCRaster and TauDEM. Their performances in the computation of MDS were compared and evaluated.

For most small-scale surfaces (e.g., the ideal Surfaces I-1–I-4 and the laboratory soil Surfaces II-1–II-2) and watershed-scale plain-type surfaces (e.g., III-1–III-4), their computed MDS, MPA and NP values decreased with an increase in DEM grid spacing. This changing pattern primarily can be attributed to the smoothing effect for coarser resolution DEMs. For some special surfaces with topographic peaks, however, their MDS values increased initially, reached a maximum value and then decreased, which was mainly attributed to "unreal" depressions/puddles generated during the DEM interpolation/aggregation process. The MDS, MPA and NP of the watershed-scale mountain-type surfaces that had a wide range of elevations showed a distinct changing pattern, featuring significant variations with grid spacing, which can be attributed to both the smoothing effect and the artificial depressions from the interpolation/aggregation process. In addition, a certain threshold of DEM grid size existed for most surfaces, beyond which significant changes in topographic properties were observed. Such a DEM grid threshold varied depending on the real surface topographic characteristics (e.g., size and distribution of depressions/puddles). This information about the DEM scale can be useful for selecting the effective or "optimal" DEM resolution for hydrologic and environmental modeling.

Development of mathematical relationships between DEM resolution and topographic parameters, quantification of the DEM representation scale and surface roughness scale and identification of the optimal DEM resolution are of both theoretical and practical importance. Some relevant studies (e.g., [24]) have been performed. As an important future research direction, however, more efforts are needed to fully address this issue. In summary, DEM resolution had significant impacts on the characterization and quantification of surface topography. In the meantime, it should be emphasized that the extent and significance of the DEM resolution effect also depended on the complexity and variability of surface topography.

It was found that the PD, ArcHydro, HEC-GeoHMS and TauDEM programs provided similar delineation results and comparable topographic parameters for both large- and small-scale surfaces. However, due to the different methods used for boundary depressions/puddles (closed boundary approach), PCRaster yielded much higher MDS values for surfaces with boundary depressions/puddles. The D-infinity method in TauDEM provided improved the computation of flow directions, which had potential to further improve the quantification of flow accumulations and watershed modeling.

Acknowledgments

This material is based upon work supported by the National Science Foundation under Grant No. EAR-0907588 and NSF Experimental Program to Stimulate Competitive Research (EPSCoR) Award IIA-1355466. The authors would like to thank Jun Yang and Jessica Nelis for their contributions to the related work.

Author Contributions

Both authors contributed equally to this work, including the related research and writing.

Conflicts of Interest

The authors declare no conflict of interest.

References

1. Ullah, W.; Dickinson, W.T. Quantitative description of depression storage using a digital surface model. I determination of depression storage. *J. Hydrol.* **1979**, *42*, 63–75.
2. Darboux, F.; Davy, P.; Gascuel-Odoux, C.; Huang, C. Evolution of soil surface roughness and flowpath connectivity in overland flow experiments. *Catena* **2001**, *46*, 125–139.
3. Abedini, M.J.; Dickinson, W.T.; Rudra, R.P. On depressional storages: The effect of DEM spatial resolution. *J. Hydrol.* **2006**, *318*, 138–150.
4. Hansen, B. Estimation of surface runoff and water-covered area during filling of surface microrelief depressions. *Hydrol. Process.* **2000**, *14*, 1235–1243.
5. Govers, G.; Takken, I.; Helming, K. Soil roughness and overland flow. *Agronomie* **2000**, *20*, 131–146.
6. Darboux, F.; Huang, C. An instantaneous-profile laser scanner to measure soil surface microtopography. *Soil Sci. Soc. Am. J.* **2003**, *67*, 92–99.

7. Huang, C.; Bradford, J.M. Depressional storage for Markov-Gaussian surfaces. *Water Resour. Res.* **1990**, *26*, 2235–2242.
8. Kamphorst, E.C.; Jetten, V.; Guerif, J. Predicting depressional storage from soil surface roughness. *Soil Sci. Soc. Am. J.* **2000**, *64*, 1749–1758.
9. Kamphorst, E.C.; Duval, Y. Validation of a numerical method to quantify depression storage by direct measurements on moulded surfaces. *Catena* **2001**, *43*, 1–14.
10. Kamphorst, E.C.; ChadKuf, J.; Jetten, V.; Guerif, J. Generating 3D soil surfaces from 2D height measurements to determine depression storage. *Catena* **2005**, *62*, 189–205.
11. Moore, I.D.; Larson, C.L. Estimating micro-relief surface storage from point data. *Trans. ASAE* **1979**, *22*, 1073–1077.
12. Jenson, S.K.; Domingue, J.O. Extracting topographic structure from digital elevation data for geographic information systems analysis. *Photogramm. Eng. Remote Sens.* **1988**, *54*, 1593–1600.
13. Martz, L.M.; Garbrecht, J. Automated extraction of drainage network and watershed data from digital elevation models. *Water Resour. Bull.* **1993**, *29*, 901–908.
14. Garbrecht, J.; Martz, L.W. *TOPAZ: An Automated Digital Landscape Analysis Tool for Topographic Evaluation, Drainage Identification, Watershed Segmentation and Subcatchment Parameterization: Overview*; ARS-NAWQL 95–1; United States Department of Agriculture, Agricultural Research Service (USDA-ARS): Durant, OK, USA, 1997.
15. Martz, L.M.; Garbrecht, J. An outlet breaching algorithm for the treatment of closed depressions in a raster DEM. *Comput. Geosci.* **1999**, *25*, 835–844.
16. Planchon, O.; Darboux, F. A fast, simple and versatile algorithm to fill the depressions of digital elevation models. *Catena* **2001**, *46*, 159–176.
17. Maidment, D.R. *Arc Hydro: GIS for Water Resources*; ESRI Press: Redlands, CA, USA, 2002.
18. *User's Manual, HEC-GeoHMS, Geospatial Hydrologic Modeling Extension*; Version 4.2, CPD-77; U.S. Army Corps of Engineers, Hydrologic Engineering Center: Davis, CA, USA, 2009.
19. *User's Manual, HEC-GeoHMS, Geospatial Hydrologic Modeling Extension*; Version 10.1, CPD-77; U.S. Army Corps of Engineers, Hydrologic Engineering Center: Davis, CA, USA, 2013.
20. Van Deursen, W.P.A.; Wesseling, C.G. *The PC Raster package. Department of Physical Geography*; Utrecht University: Utrecht, The Netherlands, 1992.
21. Wesseling, C.G.; Karssenberg, D.; van Deursen, W.P.A.; Burrough, P.A. Integrating dynamic environmental models in GIS: The development of a dynamic modelling language. *Trans. GIS* **1996**, *1*, 40–48.
22. Carvajal, F.; Aguilar, M.A.; Aguera, F.; Aguilar, F.J.; Giraldez, J.V. Maximum depression storage and surface drainage network in uneven agricultural landforms. *Biosyst. Eng.* **2006**, *95*, 281–293.
23. Ullah, W.; Dickinson, W.T. Quantitative description of depression storage using a digital surface model. II Characteristics of surface depressions. *J. Hydrol.* **1979**, *42*, 77–90.
24. Yang, J.; Chu, X. Effects of DEM resolution on surface depression properties and hydrologic connectivity. *J. Hydrol. Eng.* **2013**, *18*, 1157–1169.
25. Vaze, J.; Teng, J.; Spencer, G. Impact of DEM accuracy and resolution on topographic indices. *Environ. Model. Softw.* **2010**, 25, 1086–1098.

26. Yang, P.; Ames, D.P.; Fonseca, A.; Anderson, D.; Shrestha, R.; Glenn, N.F.; Cao, Y. What is the effect of LiDAR-derived DEM resolution on large-scale watershed model results? *Environ. Model. Softw.* **2014**, *58*, 48–57.
27. *GIS Modules and Distributed Models of the Watershed*; ASCE Task Committee on GIS Modules and Distributed Models of the Watershed; American Society of Civil Engineers: Reston, VA, USA, 1999.
28. Tarboton, D.G.; Mohammed, I.N. TauDEM 5.1 Quick Start Guide to Using the TauDEM ArcGIS Toolbox. Available online: http://hydrology.usu.edu/taudem/taudem5/TauDEM51GettingStartedGuide.pdf (accessed on 27 April 2015).
29. O'Callaghan, J.F.; Mark, D.M. The extraction of drainage networks from digital 389 elevation data. *Comput. Vis. Graph. Image Process.* **1984**, *28*, 323–344.
30. Chu, X.; Zhang, J.; Chi, Y.; Yang, J. An improved method for watershed delineation and computation of surface depression storage. In *Watershed Management 2010: Innovations in Watershed Management under Land Use and Climate Change*; Potter, K.W., Frevert, D.K., Eds.; American Society of Civil Engineers (ASCE): Reston, VA, USA, 2010.
31. Chu, X.; Yang, J.; Chi, Y.; Zhang, J. Dynamic puddle delineation and modeling of puddle-to-puddle filling-spilling-merging-splitting overland flow processes. *Water Resour. Res.* **2013**, *49*, 3825–3829.
32. Jensen, S.K. Applications of hydrologic information automatically extracted from digital elevation models. *Hydrol. Process.* **1991**, *5*, 31–44.
33. Van Deursen, W.P.A. Geographical Information Systems and Dynamic Models: Development and application of a prototype spatial modelling language. Doctor's Dissertation, Utrecht University, Utrecht, The Netherlands, 1995.
34. Tarboton, D.G. A new method for the determination of flow directions and contributing areas in grid digital elevation models. *Water Resour. Res.* **1997**, *33*, 309–319.

11

Spatiotemporal Characteristics of Dry-Wet Abrupt Transition Based on Precipitation in Poyang Lake Basin, China

Xianghu Li [1,*] and Xuchun Ye [2]

[1] Key Laboratory of Watershed Geographic Sciences, Nanjing Institute of Geography and Limnology, Chinese Academy of Sciences, 73 East Beijing Road, Nanjing 210008, China

[2] School of Geographical Sciences, Southwest University, 2 Tiansheng Road, Chongqing 400715, China; E-Mail: yxch2000@swu.edu.cn

* Author to whom correspondence should be addressed; E-Mail: xhli@niglas.ac.cn;

Academic Editor: Athanasios Loukas

Abstract: The dry-wet abrupt transition (DWAT) is a typical anomaly of precipitation at the subseasonal scale and may result in a severer and greater hydro-meteorological hazard. This paper identified and examined the DWAT events in terms of precipitation in Poyang Lake basin, including both from dry to wet (DTW) and from wet to dry (WTD), during the period of 1960–2010 and analyzed its characteristics of temporal and spatial distribution and intra- and inter-annual tendencies based on the Mann–Kendall (M–K) test. The results revealed that the DTW was inclined to occur in March, but WTD in July and September; the inter-annual variation of the DWAT index showed a feeble long-term increasing trend in July and a decreasing trend in September with a *Z*-statistic of 1.81 and −1.44, respectively, although none of these trends achieved the statistical significance level ($\alpha = 0.05$); the north parts of Poyang Lake basin experienced more DWAT events during the last 50 years in which the occurrence of WTD is more frequent than that of DTW. The outcomes of the study will help mitigate and regulate the flood and drought in Poyang Lake basin, as well as in other regions.

Keywords: dry-wet abrupt transition; precipitation; spatiotemporal distribution; Poyang Lake basin

---

1. Introduction

Flood and drought are the most common natural disasters recorded in the world, which have caused severe economic losses (accounted for 40% and 15% of global losses caused by natural disasters, respectively), with huge damage to the environment and the agricultural economy, and have threatened the life of people [1–4]. During the last few decades, severe flood or/and drought disasters have occurred frequently the world over. As a consequence of global warming, the resulting hydro-meteorological hazards (*i.e.*, floods and droughts) have become increasingly serious and gained the widespread concern of the international community [5].

In China, floods and droughts are also the most frequently occurring natural disaster, because of the strong influence of the East Asian monsoon [6,7]. Two thirds of the Chinese territory and over half of the total population are affected by a variety of flood or/and drought events almost every year [1,8]. Moreover, global warming has significantly increased the variability of East Asian summer monsoon precipitation [9], which enhanced not only the risks of flood or drought disaster, but also the occurrence frequency of the drought-flood abrupt alternation (DFAA) event [10]. Practically, the DFAA, including both drought-to-flood and flood-to-drought, is a typical anomaly of precipitation at the subseasonal scale and can result in severer and greater damage than a separate flood or drought [10,11]. A DFAA disaster is usual in many regions the world over; especially in the Yangtze River basin and Huai River basin, China, the DFAA has been intensified in recent years due to climate change and speedy urbanization [12–15]. In particular, the middle and lower reaches of the Yangtze River experienced a rapid, severe DFAA disaster during spring-summer in 2011. Its serious situation of flood and drought disaster and broad stricken areas are rare in recent years and caused huge economic losses and serious damage to towns and farms.

Numerous researchers have so far investigated the changing characteristic of DFAA, as well as the factors affecting it. For example, Wu *et al.* [10,16] quantified the long period of abrupt transitions over the middle and lower reaches of the Yangtze River, China, during the summer through a DFAA index and investigated its correlation with large-scale atmospheric circulation characteristics. Tang *et al.* [17] and Zhang *et al.* [18] discussed the occurrence characteristics of DFAA in the north area of Huai River. Wang *et al.* [19] discussed the climatic characteristics of these transitions during the principal flood season and described a new criterion of DFAA based on the percentage of rainfall anomaly and further analyzed its spatio-temporal distribution characteristics. In addition, Cheng *et al.* [13,20] defined the intensity index of DFAA and explored the spatio-temporal distribution and intensity difference of DFAA in the Huai River basin. More recently, Zhang *et al.* [21] and Luo *et al.* [15] also emphasized the DFAA in terms of runoff when they dealt with the abnormal changes in water flows.

Usually, severe floods or/and droughts result from abnormal climate change [22], *i.e.*, extreme precipitation and temperature. To further monitor floods and drought occurrence, it is worthwhile to investigate the variation of local dry/wet conditions [23], which has become an important prerequisite of flood/drought disaster prevention and mitigation [4]. There are many researches that have dealt with the spatiotemporal variability and long-term trends of dry and wet condition. For instance, Bordi *et al.* [24] analyzed the spatiotemporal variability of dry and wet periods during the last 50 years in eastern China and found that the northern part of eastern China had been experiencing dry conditions more frequently since the 1970s. The research by Zhang *et al.* [25] indicated that the

Pearl River basin in South China has become drier during the rainy season and wetter in winter. Moreover, Zhai *et al.* [26] found that the occurrence frequencies of both dry and wet years for the period of 1961–2005 were lower for the southern region of China than for the northwest. Zhai *et al.* [26] further noted that an increasing frequency of wet years was detected in the upper and lower reaches of the Yangtze River, China.

On the other hand, the warming global climate has increased concurrent climatic extremes, such as floods, droughts and heat waves [27,28]. The increase in global temperatures substantially increases the probability that multiple extremes will occur simultaneously [29–31]. Recent studies have shown that worldwide, there are extensive increases in heavy precipitation and decreases in light and moderate rain due to global warming [32–37]. Increased temperature and decreased precipitation may contribute to significant changes in drought, and this has been confirmed by many researchers using modeling results and observations [36–38]. Trenberth *et al.* [39] believe that in a warming environment, both the increase in heavy precipitation events, as well as the decrease in light rain events can be expected. Based on global climate models, Gregory *et al.* [40] found that in a warmer climate, droughts may become longer lasting and more severe in drought-prone regions, because of enhanced evaporation. Beniston [41] assessed the changes in the exceedances of joint extremes of precipitation and temperature quantiles in Europe and found that the combination of cool/dry, cool/wet, warm/dry and warm/wet modes revealed a systematic change in the course of the 20th century, with significant declines in the frequency of occurrence of the cold modes and a sharp rise in that of the warm modes. Hao *et al.* [28] also investigated the joint occurrences of precipitation and temperature extremes in CRU (Climate Research Unit) and UD (University of Delaware) observations, as well as in CMIP5 (Coupled Model Intercomparison Project Phase 5) global climate simulations. He noted that the CRU and UD observations showed substantial increases in the occurrence of joint warm/dry and warm/wet combinations for the period 1978–2004 relative to 1951–1977. Diffenbaugh *et al.* [42] and AghaKouchak *et al.* [43] also argued that the global warming and the associated rise in extreme temperatures is increasing the probability of co-occurring warm-dry conditions in California.

However, Poyang Lake basin, an important national rice-producing base in the middle and lower reaches of the Yangtze River, China, has received far less attention with respect to DFAA disasters. The Poyang Lake area has suffered from frequent floods and droughts in the last few decades, which have caused huge damage to the environment and the agricultural economy [44,45]. Moreover, it has recently been shown that the frequency and severity of the floods and droughts in Poyang Lake basin have increased since 1990 [46], and this change could be attributable to the increased fluctuation of warm season rainfall [47]. Therefore, the objectives of the study are designed to: (1) identify and examine the dry-wet abrupt transitions (DWAT) in terms of precipitation in Poyang Lake basin, both from dry to wet (DTW) and from wet to dry (WTD), during the period of 1960–2010; and (2) explore the intra-annual distribution characteristics and tendencies of DWAT, as well as their spatial distribution pattern in Poyang Lake basin. By doing so, the index of DWAT for precipitation is calculated at the monthly scale, and the inter-annual variation trends are investigated based on the Mann–Kendall (M–K) test. The study is expected to serve as a useful reference and valuable information for flood and drought mitigation and regulation in Poyang Lake basin, as well as in other regions.

2. Study Area and Data

Poyang Lake is located in the middle and lower reaches of the Yangtze River, China (28°22'–29°45' N and 115°47'–116°45' E), which connects with the Yangtze River and receives water flows from the five rivers in Poyang Lake basin: Xiushui River, Ganjiang River, Fuhe River, Xinjiang River and Raohe River (Figure 1). Among them, the Ganjiang is the largest river in the region, extending 750 km, and contributes almost 55% of the total discharge into Poyang Lake. The headwaters of these rivers are located in the boundaries of the east, south and west of the Jiangxi Province that is surrounded by high mountains. The total drainage area of the water systems is 16.22×10^4 km^2, accounting for 9% of the drainage area of the Yangtze River basin. Land use in the catchment consists of forest (46%), shrub land (25%), crop land (25%) and small areas of pasture, urban centers and open water [48]. The topography in the catchment varies from highly mountainous and hilly areas (with a maximum elevation of 2,200 m above mean sea level) to alluvial plains in the lower reaches of the primary watercourses. The stream gradient decreases as these rivers flow onto the relatively flat region surrounding Poyang Lake [49].

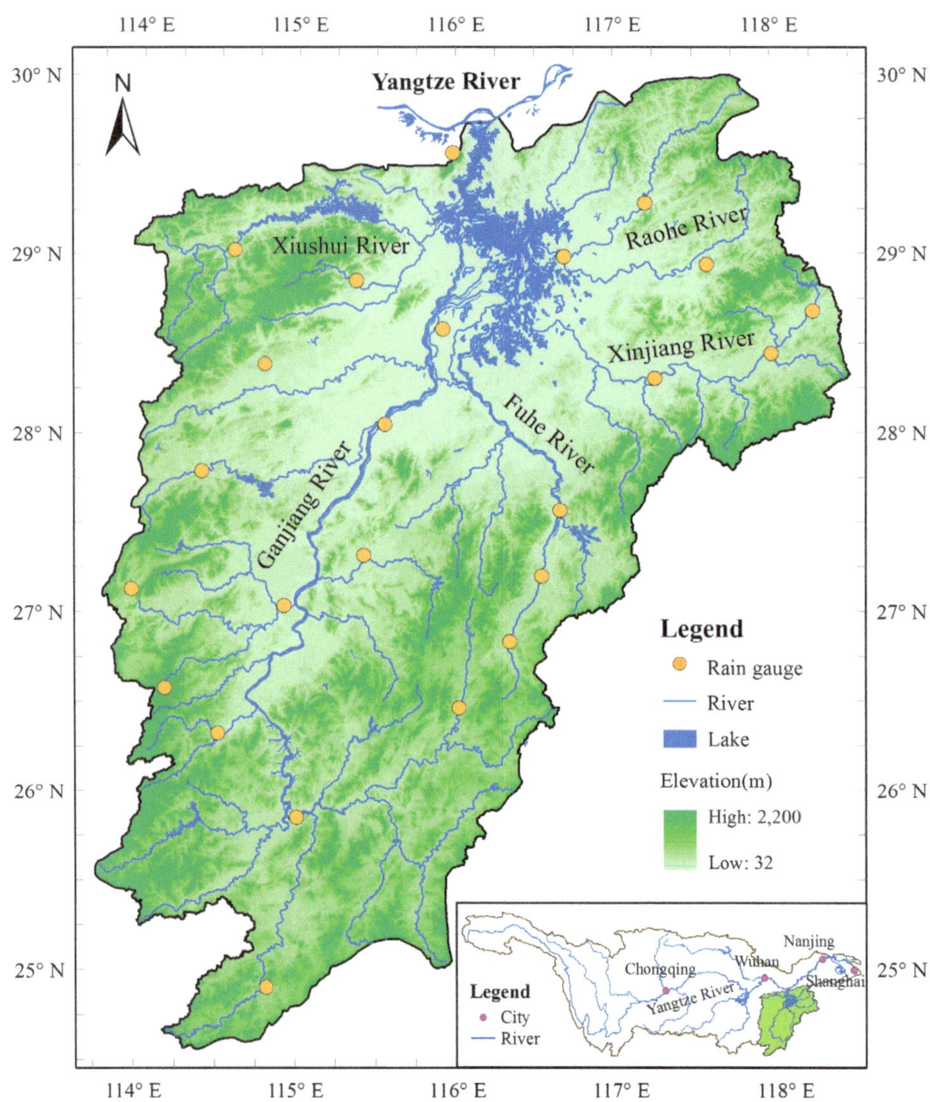

Figure 1. Location of the study area and the distribution of stations.

Poyang Lake basin has a subtropical wet climate dominated by the East Asia Monsoon. The mean annual precipitation is 1,630 mm for the period of 1960–2010, of which 55% occurs in March–June. Temperatures are highly seasonal, with June–August having an average of 27.3 °C and December–February 7.1 °C, with an annual average of 17.6 °C. Annual precipitation shows a wet season and a dry season and a short transition period in between. Precipitation increases quickly from January–June and decreases sharply in July, and after September, the dry season sets in and lasts through December, as Figure 2 shows.

Daily rain gauge data, during the period of 1960–2010 for 24 meteorological stations in the Poyang Lake basin are collected from the National Meteorological Information Center of China. The locations of these stations are shown in Figure 1. These data are used in the study to measure and examine the variation of precipitation and the characteristics of DWAT in the basin. Moreover, these data have been widely used for different studies previously [46–50], and the qualities have been proven to be reliable. In the study, the daily precipitation data from all of the stations are averaged to obtain the areal daily precipitation for the Poyang Lake basin, and the monthly rainfall is aggregated from the daily values.

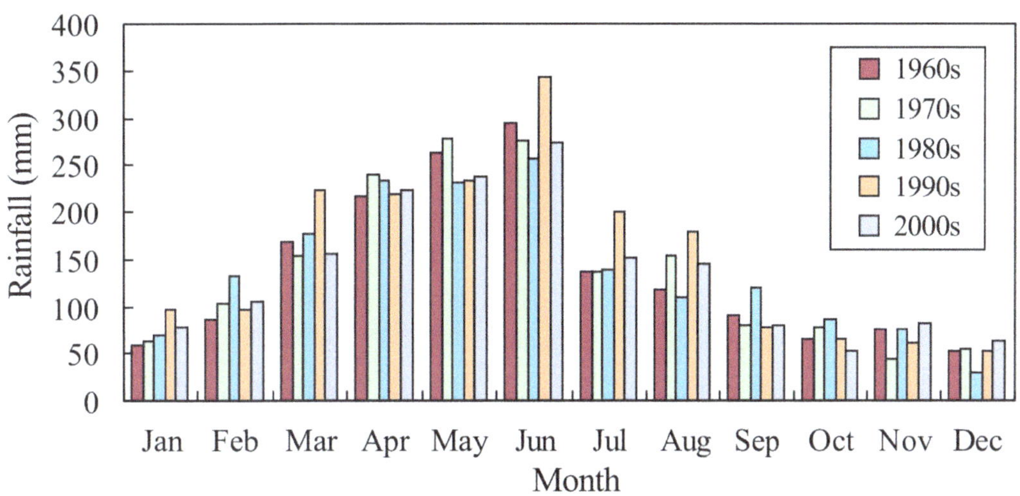

Figure 2. Variation of monthly precipitation in different decades.

3. Methods

To conveniently reflect the variation of rainfall and to explore the characteristics of the DWAT phenomenon in Poyang Lake basin, the normalization transition is needed for the original rain gauge data firstly. The normalization is done as follows:

$$NP_i = \frac{P_i - \overline{P}}{\sigma} \tag{1}$$

where NP_i are the normalized monthly precipitations; P_i are the monthly rain gauge data; $\overline{P}$ and σ are the mean and standard error of P_i, respectively; and i is the count of the months.

The DWAT phenomenon is described quantitatively in the study using a DWAT index, which is defined according to the studies by Wu *et al.* [11,17] and Zhang *et al.* [22] as follows:

$$DWATI = (NP_i - NP_{i-1}) \cdot (|NP_i| + |NP_{i-1}|) \cdot \alpha^{-|NP_i + NP_{i-1}|} \quad (i = 2, 3, 4, \ldots, n) \tag{2}$$

where *DWATI* is the index of the DWAT phenomenon; NP_i and NP_{i-1} are the normalized monthly precipitation in the i month and the $i-1$ month, respectively; α is a weight coefficient, and a value of 3.2 is suitable for the monthly scale according to the study by Zhang *et al.* [22]; and n is the total months.

Wu *et al.* [11,17] considered that the item ($NP_i - NP_{i-1}$) represents the intensity term of DWAT, the item ($|NP_i| + |NP_{i-1}|$) denotes the magnitude of the dry and wet and $\alpha^{-|NP_i+NP_{i-1}|}$ is the weight coefficient, which may decrease the weight of wet or dry in two consecutive months and increase the weight of DWAT events. Meanwhile, the absolute magnitude of those precipitation anomalies within 0.5 standard deviations is regarded as normal, while the anomalies over 0.5 standard deviations and under −0.5 standard deviations are defined as wet and dry, respectively. Furthermore, a positive *DWATI* value indicates that a DTW has occurred in that period, while a negative *DWATI* value shows the occurrence of WTD, and the larger absolute value means a stronger intensity of DTW or WTD. Numerous research works have shown that the *DWATI* is a vigorous index to describe the DWAT phenomenon quantitatively. Wu *et al.* [11,17] examined the validation of *DWATI* to represent the characteristics of the summer rainfall and found that the summers with high *DWATI* have more precipitation in July–August than in May–June, whereas the summers with low *DWATI* have more precipitation in May–June than in July–August. Wu *et al.* [11,17] and Zhang *et al.* [22] further noted that most anomalies of precipitation in May–June of high *DWATI* summers were under -1.0σ (anomalous droughts), but in July–August were over 1.0σ, whereas the situations for the low *DWATI* summers were just opposite. Moreover, Sun *et al.* [13] and Luo *et al.* [16] also used the *DWATI* with the different weight coefficient (α) value to investigate the DWAT in terms of runoff when they dealt with the abnormal changes in water flows. Similar results were also reached for which the DTW usually occurred in the high *DWATI* summers, and low *DWATI* summers were often related to the WTD. According to the above application, the *DWATI* might be regarded as a powerful quantitative index describing the summer DWAT events.

The M–K test [51,52] is also applied in this study to identify the inter-annual variation trends of DWAT events. The M–K test is a rank-based non-parametric method due to its robustness against the influence of abnormal data and especially its reliability for biased variables, and it has been widely applied for trend detecting in hydro-climatic time series (e.g., [49,53–56]).

To detect the existence of any step change points in the hydrological data $X_t = (x_1, x_2, x_3, \cdots, x_n)$, the accumulative number n_i of samples for which $x_i > x_j$ ($1 \le j \le i$) should be first calculated [48,49]. The normally distributed statistic d_k can be calculated as follows:

$$d_k = \sum_{i=1}^{k} n_i \quad (2 \le k \le n) \tag{3}$$

Under the null hypothesis of no trend, d_k is asymptotically normally distributed with expected mean value $E\,(d_k)$ and variance $Var\,(d_k)$ as follows:

$$E(d_k) = \frac{k(k-1)}{4} \tag{4}$$

$$Var(d_k) = \frac{k(k-1)(2k+5)}{72} \tag{5}$$

Under the above assumption, the normalized variable statistic $Uf\,(d_k)$ is calculated as:

$$Uf(d_k) = \frac{d_k - E(d_k)}{\sqrt{Var(d_k)}} \quad (k = 1, 2, 3, \cdots, n) \tag{6}$$

where $Uf(d_k)$ is the forward sequence, and the backward sequence $UB(d_k)$ is calculated using the same equation, but with a reversed series of data. When the null hypothesis is rejected (*i.e.*, if any of the points in the forward sequence are outside the confidence interval), the detection of an increasing ($Uf(d_k) > 0$) or a decreasing ($Uf(d_k) < 0$) trend is indicated.

In addition, to examine the long-term trend of the time series data, the standardized statistics (Z) for the one-tailed test is formulated as:

$$Z = \begin{cases} (S-1)/\sqrt{Var(S)} & (S > 0) \\ 0 & (S = 0) \\ (S+1)/\sqrt{Var(S)} & (S < 0) \end{cases} \tag{7}$$

where:

$$S = \sum_{i=1}^{n-1} \sum_{j=i+1}^{n} \operatorname{sgn}(x_j - x_i) \tag{8}$$

where x_i and x_j are the sequential data values and n is the length of the dataset; $\operatorname{sgn}(x_j - x_i)$ is a piecewise function, as:

$$\operatorname{sgn}(x_j - x_i) = \begin{cases} 1 & (x_j > x_i) \\ 0 & (x_j = x_i) \\ -1 & (x_j < x_i) \end{cases} \tag{9}$$

The null hypothesis of no trend is rejected if $|Z| > 1.96$ at the 0.05 significance level. A positive value of Z denotes an increasing trend, and a negative value corresponds to a decreasing trend. Because the presence of serial correlation can influence the identification of trends [57,58], it is necessary to do autocorrelation (serial correlation) analysis before the trend test.

4. Results and Discussion

4.1. Intra-Annual Distribution of DWAT in Poyang Lake Basin

The monthly *DWATI* based on the areal averaged precipitation during the period of 1960–2010 is calculated, and the results are shown in Figure 3. It can be seen that the *DWATI* oscillates around zero during the study period. In order to verify the validity of *DWATI* to describe the characteristics of the DWAT phenomenon in Poyang Lake basin, the status of precipitation in the corresponding month is also examined for the chosen five highest and lowest *DWATI*, as Table 1 shows. It is seen that the chosen high *DWATI* have more precipitation in the current month, with the normalized precipitation of over 0.7, even 1.0 in some months, which shows that they all appear as the anomalous wet event, while the low precipitation, observed in the previous month, ranging from −1.07 to −0.73, indicates that the anomalous dry event has occurred. Therefore, the large *DWATI* correctly reflect the occurrence of DTW. On the contrary, all chosen low *DWATI* reflect the opposite situation that the precipitations in the current month are under −1.0σ, but over 1.0σ in the previous month (Table 1). Therefore, the low

DWATI are often related to the WTD. The above analysis from Table 1 shows that the *DWATI* might be regarded as a vigorous index describing the DWAT phenomenon in Poyang Lake basin.

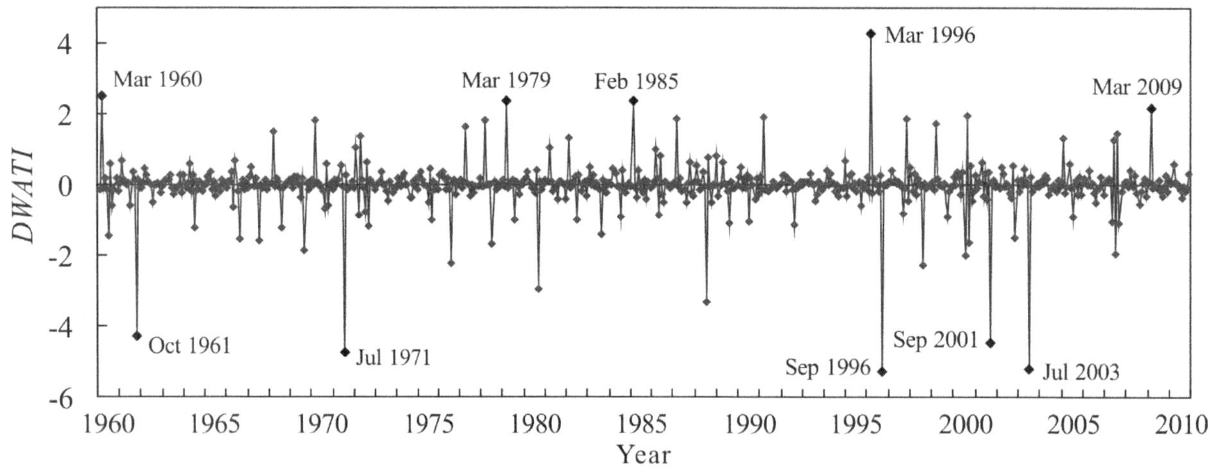

Figure 3. Variation of monthly *DWATI* (dry-wet abrupt transition index) during 1960–2010.

Table 1. Comparison of normalized precipitation for the chosen five highest and lowest *DWATI* in the current month and previous month.

Type	Date	*DWATI*	Precipitation in Current Month	Precipitation in Previous Month
	Mar. 1996	4.28	1.04	−1.04
	Mar. 1960	2.49	0.79	−1.07
High *DWATI*	Feb.1985	2.38	0.77	−1.05
	Mar. 1979	2.36	0.88	−0.76
	Mar. 2009	2.17	0.77	−0.73
	Sep. 1996	−5.25	−1.15	1.20
	Jul. 2003	−5.23	−1.15	1.19
Low *DWATI*	Jul. 1971	−4.71	−1.12	1.09
	Sep. 2001	−4.45	−1.28	1.08
	Oct. 1961	−4.29	−1.23	1.06

Subsequently, the *DWATI* is used for statistical analysis of the intra-annual distribution of the DWAT phenomenon. In descriptive statistics, a scatter plot is a convenient way of graphically depicting groups of numerical data through their distribution. Therefore, in this study, the scatter plot is used to present the distribution characteristics of *DWATI* in different months, and the results are shown in Figure 4. It is seen that the positive *DWATI* are mainly presented in the first half of the year, especially in March; both the frequency of high *DWATI* and the absolute values are larger than that in other months. In contrast, it is also clear that the negative *DWATI* are principally shown in the second half of the year, and especially, the lowest *DWATI* appear in July and September. Accordingly, Figure 4 demonstrates that the DTW is inclined to occur in March, while the WTD occurs in July and September in Poyang Lake basin.

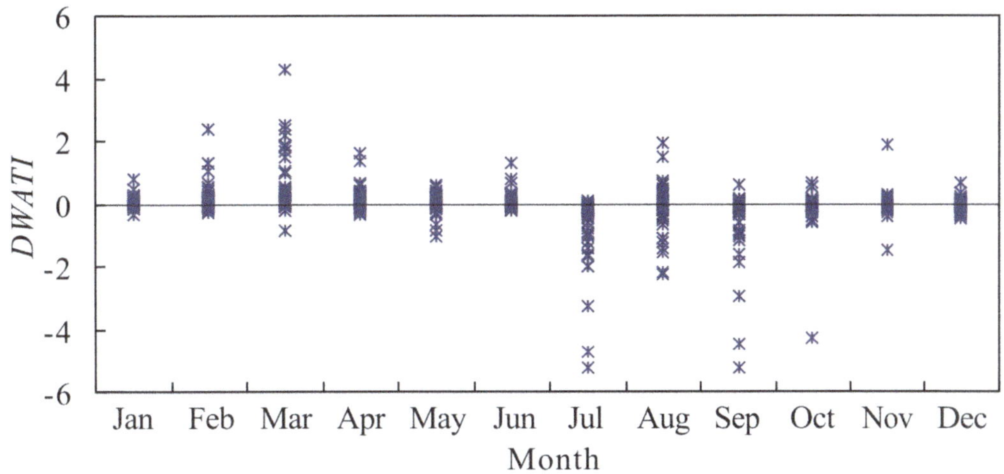

Figure 4. Scatter plot of *DWATI* in different months during 1960–2010.

4.2. Inter-Annual Variation Characteristics of DWAT in Poyang Lake Basin

As mentioned above, the DWAT events in Poyang Lake basin mainly occur in March, July and September; thereby, the *DWATI* in these months during 1960–2010 are selected for the long-term trends analysis using M–K test. Firstly, the independence of these selected time series data is tested, which is the underlying assumption of the M–K test. Figure 5 shows the variation of autocorrelation coefficient of *DWATI* with different lag intervals in March, July and September, respectively. It is obvious that the autocorrelation coefficients are very small and do not have statistically significance in March, July and September, regardless of the lag interval. Figure 5 indicates that the selected time series data are independent and appropriate for the M–K test.

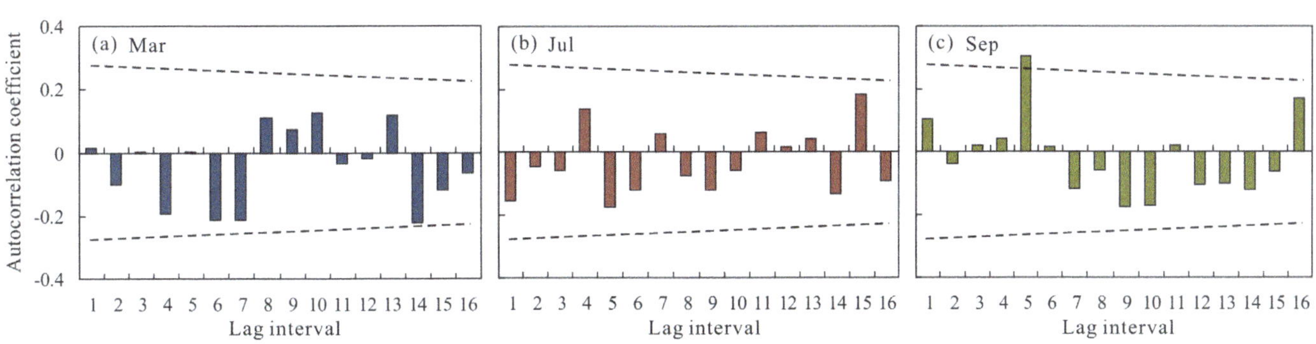

Figure 5. Autocorrelation coefficient of *DWATI* with different lag intervals in March (**a**); July (**b**) and September (**c**) (the dashed lines represent the confidence intervals at the 0.05 significance level).

In Figure 6, the inter-annual variation of *DWATI* in March, July and September, as well as their corresponding M-K sequential tests during 1960–2010 are shown. As explained in the Methods Section, the *Uf* curve shows the changing trend of *DWATI*. The time series has a downward trend if $Uf < 0$ and an increasing trend if $Uf > 0$. If the *Uf* values are greater than the critical values (the two dashed lines above and below zero), then this upward or downward trend is significant at the 95% significance level. It is seen from Figure 6 that the *DWATI* in March oscillates principally between 0 and 2.0 during the study period, and there is no clear trend with *Uf* fluctuating between the two critical value lines. Although the *DWATI* show a long-term upward trend with the *Z*-statistic of 1.81 for July and a downward trend with the *Z*-statistic of −1.44 for September, no statistically significant trend ($\alpha = 0.05$) was detected in July and September (Figure 6). Figure 6 indicates that both the DTW and WTD in Poyang Lake basin had no long-term increasing or decreasing trend at the 0.05 significance level.

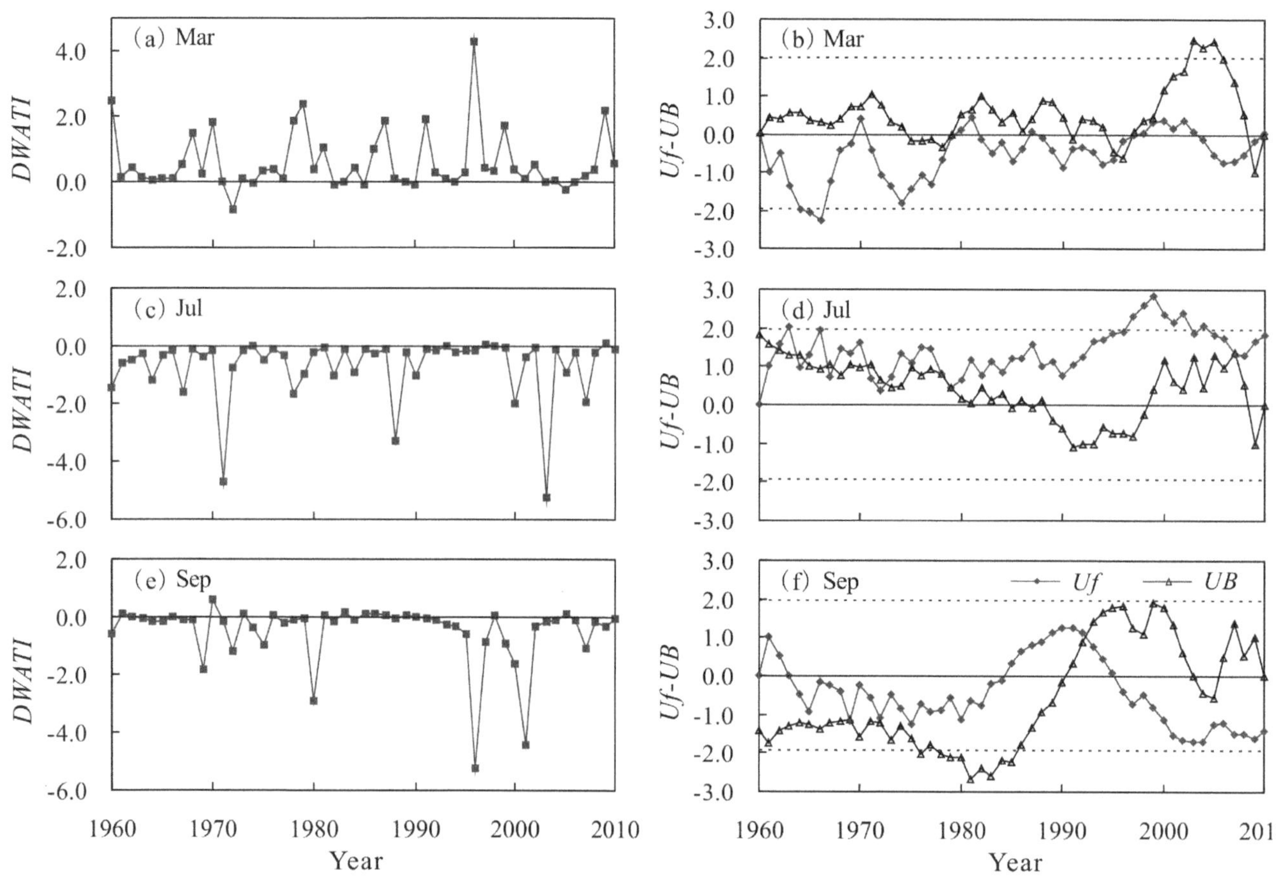

Figure 6. Inter-annual variation of *DWATI* in March (**a**), July (**c**) and September (**e**) and their corresponding M–K trend tests (**b**,**d**,**f**,) (the horizontal dashed lines in (**b**,**d**,**f**,) represent the critical value of the 0.05 significance level).

The variation characteristics of *DWATI* in Poyang Lake basin are also analyzed at the decadal scale to provide insight into the decadal variability of the DWAT phenomenon. Figure 7 shows the inter-decadal variation of the occurrence frequency of the severe DWAT events (*DWATI* > 1.0 or < −1.0) during the period of 1960–2010. It is seen that the DWAT events, both the DTW and WTD, are occurring with small differences in the different decades; even the number of DWATs remains changeless during long periods (*i.e.*, 1970s, 1980s and 1990s). In addition, the frequency of DTW per 10 years is not higher than that of WTD; especially in the 1960s and 2000s, the number of WTD is four-times and two-times more than the former, respectively. This conclusion indicates that the Poyang Lake basin is more likely to experience an anomalous dry following an anomalous wet event immediately.

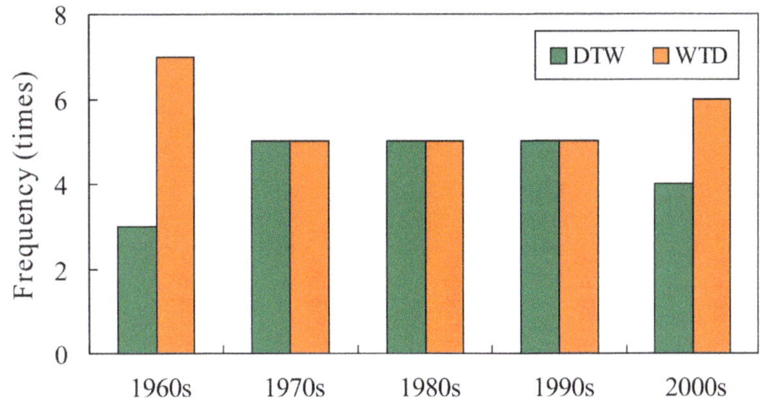

Figure 7. Inter-decadal variation of DWAT events in Poyang Lake basin.

4.3. Spatial Distribution Characteristics of DWAT in Poyang Lake Basin

Subsequently, the frequencies of the severe DWAT events, both the DTW ($DWATI > 1.0$) and the WTD ($DWATI < -1.0$), for each meteorological station in the basin are investigated during 1960–2010. The spatial distribution of their occurrence frequency is shown in Figure 8. A visual inspection found that the number of WTDs is observably more than that of the DTWs at nearly all stations, and the difference of the occurrence frequency between them is larger in the north parts than in the south parts of Poyang Lake basin. It is also seen from Figure 8 that the north parts experienced more DWAT events than the south parts, regardless of DTW or WTD.

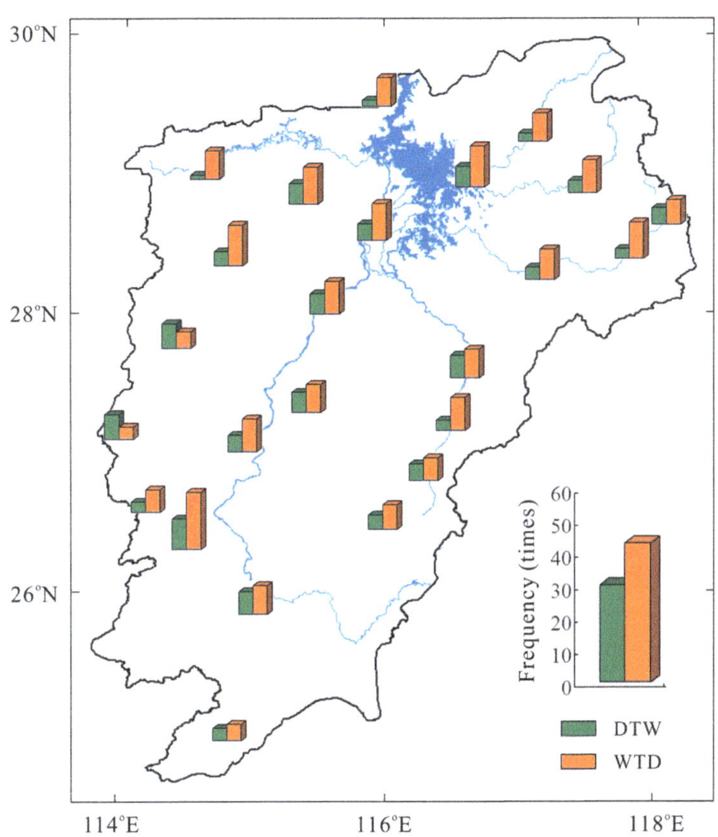

Figure 8. Spatial distribution of the occurrence frequency of DWAT events. DTW, dry to wet; WTD, wet to dry.

In addition, the spatial distribution of the DWAT tendency in Poyang Lake basin is also examined with the M–K test. Figure 9 shows the long-term trend of *DWATI* for each meteorological station in March, July and September, respectively. It is seen from Figure 9a that increasing and decreasing trends are coexistent in the basin, but none of them exceeds the significance level. Moreover, the stations with a decreasing trend are mainly located in the north and south of the basin, and others with increasing trends are principally located in the middle parts. In July, the *DWATI* at almost all stations present increasing trends, but only three of them have significant increasing trends ($\alpha = 0.1$) (Figure 9b). The opposite tendency with respect to Figure 9b is shown in Figure 9c. In September, almost all stations show a decreasing trend, and only three of them achieved the statistical significance level.

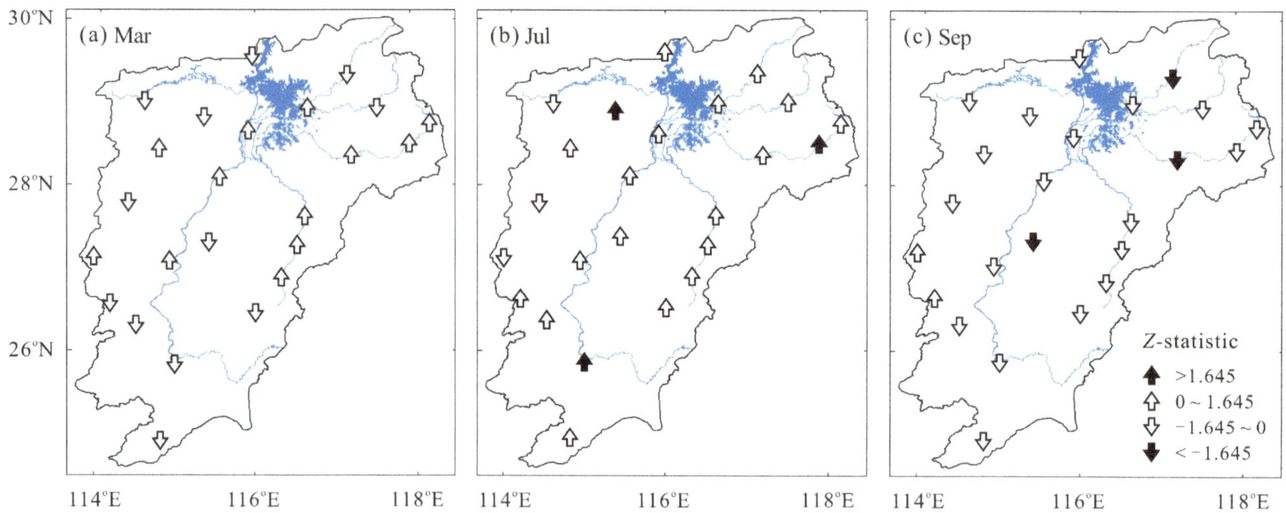

Figure 9. Spatial patterns of the long-term trends of *DWATI* in March (**a**); July (**b**) and September (**c**).

5. Conclusions

This paper identified and examined the dry-wet abrupt transitions in terms of precipitation in Poyang Lake basin during the period of 1960–2010 and analyzed its intra- and inter-annual distribution characteristics and tendencies, as well as their spatial distribution pattern. The results revealed that the *DWATI* might be regarded as a vigorous index describing the DWAT phenomenon in Poyang Lake basin, in which the high values indicated the occurrence of DTW, and the low values were often related to the WTD. Moreover, the DTW was inclined to occur in March, but July and September for intra-annual WTD. The inter-annual variation of *DWATI* showed no clear trend in March due to its fluctuation in the study period. While a feeble long-term increasing trend was detected in July and a decreasing trend in September with the *Z*-statistic of 1.81 and −1.44, respectively, none of these trends achieved the statistical significance level. As for the spatial distribution, it is found that the north parts of Poyang Lake basin experienced more DWAT events than the south parts; meanwhile, the number of WTDs is observably more than that of DTW at nearly all stations, and the difference in the occurrence frequency between them is larger in north parts than in south parts.

It is acknowledged that, as a consequence of the increase in greenhouse gas concentrations in the atmosphere, anthropogenic global warming has accelerated the regional water circulation and caused an asymmetrical distribution of precipitation, which has very likely increased the probability of floods, droughts and heat waves, especially in China, due to the strong influence of the East Asian monsoon. Under the background of such conditions, this paper is only a preliminary attempt to explore the DWAT event quantitatively, and further studies and research are still needed. It should also be pointed out that the methods used in the study and the results gained cannot fully demonstrate the spatial and temporal distribution characteristics of DWAT events in Poyang Lake basin. However, the *DWATI* has its inherent advantages in depicting the distribution characteristics and tendencies of the DWAT phenomenon in terms of precipitation, and the presented study provided a new index or approach to DWAT detecting for large-scale watersheds.

Acknowledgments

This work is jointly funded by the National Basic Research Program of China (973 Program) (2012CB417003), the National Natural Science Foundation of China (41201026) and the Science Foundation of Nanjing Institute of Geography and Limnology, Chinese Academy of Sciences (NIGLAS2012135001). The authors are grateful to the anonymous reviewers and the editor who helped us to improve the quality of the original manuscript.

Author Contributions

Xianghu Li conceived of and designed the experiments and wrote the manuscript. Xuchun Ye analyzed the data and improved the manuscript.

Conflicts of Interest

The authors declare no conflict of interest.

References

1. Nakayama, T.; Watanabe, M. Role of flood storage ability of lakes in the Changjiang River catchment. *Glob. Planet Change* **2008**, *63*, 9–22.
2. Garcia-Castellanos, D.; Estrada, F.; Jimenez-Munt, I.; Gorini, C.; Fernandez, M.; Verges, J.; de Vicente, R. Catastrophic flood of the Mediterranean after the Messinian salinity crisis. *Nature* **2009**, *462*, 778–781.
3. Zhang, J.; Li, N. *Quantitative Methods and Applications of Risk Assessment and Management on Main Meteorological Disasters*; Beijing Normal University Press: Beijing, China, 2007. (In Chinese)
4. Nie, C.J.; Li, H.R.; Yang, L.S.; Wu, S.H.; Liu, Y.; Liao, Y.F. Spatial and temporal changes in flooding and the affecting factors in China. *Nat. Hazards* **2012**, *61*, 425–439.
5. Li, L.P.; Zhang, K.M.; Luo, T.; Fu, X.H. An analysis of the drought and flood hazard characteristics and risks during the pre-rainy season in South China. *Nat. Hazards* **2014**, *71*, 1195–1213.

6. Liu, R.Y.; Liu, N. Flood area and damage estimation in Zhejiang, China. *J. Environ. Manag.* **2002**, *66*, 1–8.
7. Yu, F.; Chen, Z.; Ren, X.; Yang, G. Analysis of historical floods on the Yangtze River, China: Characteristics and explanations. *Geomorphology* **2009**, *113*, 210–216.
8. Wang, L.N.; Shao, Q.X.; Chen, X.H.; Li, Y.; Wang, D.G. Flood changes during the past 50 years in Wujiang River, South China. *Hydrol. Process.* **2012**, *26*, 3561–3569.
9. Feng, L.; Zhou, T.; Wu, B.; Li, T.; Luo, J.J. Projection of future precipitation changes over China with a high-resolution global atmospheric model. *Adv. Atmos. Sci.* **2010**, *28*, 464–476.
10. Wu, Z.W.; Li, J.P.; He, J.H.; Jiang, Z.H. Large-scale atmospheric singularities and summer long-cycle droughts-floods abrupt alternation in the middle and lower reaches of the Yangtze River. *Chin. Sci. Bull.* **2006**, *51*, 2027–2034.
11. Feng, G.L.; Yang, H.W.; Zhang, S.X.; Wang, K.; Shen, B.Z. A preliminary research on the reason of a sharp turn from drought to flood in the middle and lower reaches of the Yangtze River in late spring and early summer of 2011. *Chin. J. Atmos. Sci.* **2012**, *36*, 1009–1026. (In Chinese)
12. Sun, P.; Liu, C.L.; Zhang, Q. Spatio-temporal variations of drought-flood abrupt alternation during main flood season in East River Basin. *Pearl River* **2012**, *5*, 29–34. (In Chinese)
13. Cheng, Z.; Ding, X.J.; Xu, M.; Luo, L.S. Climate characters of typical droughts-floods abrupt alternation events in the middle-lower reaches of the Yangtze River. *Resour. Environ. Yangtze Basin* **2012**, *21*, 115–120. (In Chinese)
14. Wang, S.; Tian, H.; Ding, X. Climate characteristics of precipitation and phenomenon of drought-flood abrupt alternation during main flood season in Huaihe River basin. *Chin. J. Agrometeorol.* **2009**, *30*, 31–34. (In Chinese)
15. Luo, W.; Zhang, X.; Deng, Z.M.; Xiao, Y. Variation of the total runoff into Poyang Lake and drought-flood abrupt alternation during the past 50 years. *J. Basic Sci. Eng.* **2013**, *21*, 845–856. (In Chinese)
16. Wu, Z.W.; He, J.H.; Li, J.P.; Jiang, Z.H. The summer drought-flood coexistence in the middle and lower reaches of the Yangtze River and analysis of its air-sea background feathers in anomalous years. *Chin. J. Atmos. Sci.* **2006**, *30*, 570–577. (In Chinese)
17. Tang, M.; Shao, D.G.; Yao, C.L. Causes and countermeasures of sudden changing from drought to waterlogging in Huaibei Region. *J. China Inst. Water Resour. Hydropower Res.* **2007**, *5*, 26–32. (In Chinese)
18. Zhang, Q.; Chen, Y.D.; Chen, J.Q. Flood/drought variability in the Yangtze Delta and association with the climatic changes from the Guliya ice core: A wavelet approach. *Quat. Int.* **2008**, *189*, 163–172.
19. Wang, H.Q.; Zhao, G.N.; Peng, J.; Hu, J.F. Precipitation characteristics over five major river systems of Poyang drainage areas in recent 50 years. *Resour. Environ. Yangtze Basin* **2009**, *18*, 615–619. (In Chinese)
20. Cheng, Z.; Xu, M.; Luo, L.S.; Ding, X.J. Climate characteristics of drought-flood abrupt change events in Huaihe River basin. *J. China Hydrol.* **2012**, *31*, 73–79. (In Chinese)
21. Zhang, S.F.; Zhang, J.C.; Min, J.J.; Zhang, Z.X.; Zhuang, J.Y.; Lin, J. Drought-flood abrupt alternation based on runoff in the Huaihe River Basin during rainy season. *J. Lake Sci.* **2012**, *24*, 679–686. (In Chinese)

22. Nakayama, T.; Shankman, D. Impact of the Three-Gorges Dam and water transfer project on Changjiang floods. *Glob. Planet Change* **2013**, *100*, 38–50.
23. Du, J.; Fang, J.; Xu, W.; Shi, P.J. Analysis of dry/wet conditions using the standardized precipitation index and its potential usefulness for drought/flood monitoring in Hunan Province, China. *Stoch. Environ. Res. Risk Assess.* **2013**, *27*, 377–387.
24. Bordi, I.; Fraedrich, K.; Jiang, J.M.; Sutera, A. Spatio-temporal variability of dry and wet periods in eastern China. *Theor. Appl. Climatol.* **2004**, *79*, 81–91.
25. Zhang, Q.; Xu, C.Y.; Zhang, Z. Observed changes of drought/wetness episodes in the Pearl River basin, China, using the standardized precipitation index and aridity index. *Theor. Appl. Climatol.* **2008**, *98*, 89–99.
26. Zhai, J.; Su, B.; Gao, C.; Jiang, T. Spatial variation and trends in PDSI and SPI indices and their relation to streamflow in 10 large regions of China. *J. Clim.* **2010**, *23*, 649–663.
27. Leonard, M.; Westra, S.; Phatak, A.; Lambert, M.; van den Hurk, B.; McInnes, K.; Risbey, J.; Schuster, S.; Jakob, D.; Stafford-Smith, M. A compound event framework for understanding extreme impacts. *WIREs Clim. Change* **2014**, *5*, 113–128.
28. Hao, Z.; AghaKouchak, A.; Phillips, T.J. Changes in concurrent monthly precipitation and temperature extremes. *Environ. Res. Lett.* **2013**, *8*, 113–128.
29. Perkins, S.; Alexander, E.; Nairn, J.R. Increasing frequency, intensity and duration of observed global heatwaves and warm spells. *Geophys. Res. Lett.* **2012**, *39*, doi:10.1029/2012GL053361.
30. Easterling, D.R.; Meehl, G.A.; Parmesan, C.; Changnon, S.A.; Karl, T.R.; Mearns, L.O. Climate extremes: Observations, modeling, and impacts. *Science* **2000**, *289*, 2068–2074.
31. Sheffield, J.; Wood, E.F. Projected changes in drought occurrence under future global warming from multi-model, multi-scenario, IPCC AR4 simulations. *Clim. Dyn.* **2008**, *31*, 79–105.
32. Liu, B.; Xu, M.; Henderson, M.; Ye, Q. Observed trends of precipitation amount, frequency, and intensity in China, 1960–2000. *J. Geophys. Res.* **2005**, *110*, doi:10.1029/2004JD004864.
33. Goswami, B.N.; Venugopal, V.; Sengupta, D.; Madhusoodan, M.S.; Xavier, P.K. Increasing trends of extreme rain events over India in a warming environment. *Science* **2006**, *314*, 1442–1445.
34. Lau, K.M.; Wu, H.T. Climatology and changes in tropical oceanic rainfall characteristics inferred from Tropical Rainfall Measuring Mission (TRMM) data (1998–2009). *J. Geophys. Res.* **2011**, *116*, doi:10.1029/2011JD015827.
35. Villarini, G.; Smith, J.A.; Vecchi, G.A. Changing frequency of heavy rainfall over the Central United States. *J. Clim.* **2013**, *26*, 351–357.
36. Shiu, C.J.; Liu, S.C.; Fu, C.; Dai, A.; Sun; Y. How much do precipitation extremes change in a warming climate. *Geophys. Res. Lett.* **2012**, *39*, doi:10.1029/2012GL052762.
37. Liu, S.C.; Fu, C.; Shiu, C.J.; Chen, J.P.; Wu, F. Temperature dependence of global precipitation extremes. *Geophys. Res. Lett.* **2009**, *36*, doi:10.1029/2009GL040218.
38. Sun, Y.; Solomon, S.; Dai, A.; Portmann, R.W. How often will it rain? *J. Clim.* **2007**, *20*, 4801–4818.
39. Trenberth, K.E.; Dai, A.; Rasmussen, R.M.; Parsons, D.B. The changing character of precipitation. *Bull. Am. Meteorol. Soc.* **2003**, *84*, 1205–1217.
40. Gregory, J.M.; Mitchell, J.F.; Brady, A.J. Summer drought in northern midlatitudes in a time-dependent CO_2 climate experiment. *J. Clim.* **1997**, *10*, 662–686.
41. Beniston, M. Trends in joint quantiles of temperature and precipitation in Europe since 1901 and projected for 2100. *Geophys. Res. Lett.* **2009**, *36*, L07707.

42. Diffenbaugh, N.S.; Swain, D.L.; Touma, D. Anthropogenic warming has increased drought risk in California. *Proc. Natl. Acad. Sci. USA* **2015**, *112*, 3931–3936.
43. AghaKouchak, A.; Cheng, L.; Mazdiyasni, O.; Farahmand, A. Global warming and changes in risk of concurrent climate extremes: Insights from the 2014 California drought. *Geophys. Res. Lett.* **2014**, *41*, 8847–8852.
44. Li, X.H.; Zhang, Q.; Xu, C.Y.; Ye, X.C. The changing patterns of floods in Poyang Lake, China: Characteristics and explanations. *Nat. Hazards* **2015**, *76*, 651–666.
45. Li, X.H.; Zhang, Q. Variation of floods characteristics and their responses to climate and human activities in Poyang Lake, China. *Chin. Geogr. Sci.* **2015**, *25*, 13–25.
46. Guo, H.; Hu, Q.; Jiang, T. Annual and seasonal streamflow responses to climate and land-cover changes in the Poyang Lake basin, China. *J. Hydrol.* **2008**, *355*, 106–122.
47. Hu, Q.; Feng, S.; Guo, H.; Chen, G.; Jiang, T. Interactions of the Yangtze river flow and hydrologic processes of the Poyang Lake, China. *J. Hydrol.* **2007**, *347*, 90–100.
48. Ye, X.C.; Li, X.H.; Liu, J.; Xu, C.Y.; Zhang, Q. Variation of reference evapotranspiration and its contributing climatic factors in the Poyang Lake catchment, China. *Hydrol. Process.* **2014**, *28*, 6151–6162.
49. Ye, X.C.; Zhang, Q.; Liu, J.; Li, X.H.; Xu, C.Y. Distinguishing the relative impacts of climate change and human activities on variation of streamflow in the Poyang Lake catchment, China. *J. Hydrol.* **2013**, *494*, 83–95.
50. Li, X.H.; Zhang, Q.; Ye, X.C. Capabilities of satellite-based precipitation to estimate the spatiotemporal variation of flood/drought class in Poyang Lake basin. *Adv. Meteorol.* **2013**, *2013*, doi:10.1155/2013/901240.
51. Mann, H.B. Non-parametric test against trend. *Econometrica* **1945**, *13*, 245–259.
52. Kendall, M.G. *Rank Correlation Methods*; Charles Griffin: London, UK, 1975.
53. Zhao, G.J.; Hörmann, G.; Fohrer, N.; Zhang, Z.X.; Zhai, J.Q. Streamflow trends and climate variability impacts in Poyang Lake basin, China. *Water Resour. Manag.* **2010**, *24*, 689–706.
54. Novotny, E.V.; Stefan, H.G. Stream flow in Minnesota: Indicator of climate change. *J. Hydrol.* **2007**, *334*, 319–333.
55. Chen, H.; Guo, S.L.; Xu, C.Y.; Singh, V.P. Historical temporal trends of hydro-climatic variables and runoff response to climate variability and their relevance in water resource management in the Hanjiang basin. *J. Hydrol.* **2007**, *344*, 171–184.
56. Xu, Z.X.; Chen, Y.N.; Li, J.Y. Impact of climate change on water resources in the Tarim river basin. *Water Resour. Manag.* **2004**, *18*, 439–458.
57. Yue, S.; Pilon, P.; Phinney, B. Canadian streamflow trend detection: impacts of serial and cross-correlation. *Hydrol. Sci. J.* **2003**, *48*, 51–63.
58. Khaliq, M.N.; Ouarda, T.B.M.J.; Gachon, P.; Sushama, L.; St-Hilaire, A. Identification of hydrological trends in the presence of serial and cross correlations: a review of selected methods and their application to annual flow regimes of Canadian rivers. *J. Hydrol.* **2009**, *368*, 117–130.

12

Effects of *Misgurnus anguillicaudatus* and *Cipangopaludina cathayensis* on Pollutant Removal and Microbial Community in Constructed Wetlands

Pengfei Li [1], Jian Zhang [2,*], Huijun Xie [1], Zhen Hu [2], Haiyan He [2] and Wenxing Wang [1]

[1] Environment Research Institute, Shandong University, Jinan 250100, China;
E-Mails: lpflyss@163.com (P.L.); xiehuij@sdu.edu.cn (H.X.); wxwang@sdu.edu.cn (W.W.)

[2] School of Environmental Science and Engineering, Shandong University, Jinan 250100, China;
E-Mails: huzhen885@sdu.edu.cn (Z.H.); hhy636@126.com (H.H.)

* Author to whom correspondence should be addressed; E-Mail: zhangjian00@sdu.edu.cn;

Academic Editors: Say-Leong Ong and Jiangyong Hu

Abstract: Aquatic animals play an important role in the energy flow and matter cycling in the wetland ecosystem. However, little is known about their effects on pollutant removal performance and microbial community in constructed wetlands. This work presents an initial attempt to investigate the effects of *Misgurnus anguillicaudatus* (loach) and *Cipangopaludina cathayensis* (snail) on nutrient removal performance and microbial community of constructed wetlands (CWs). Compared with a control group, CW microcosms with aquatic animals exhibited better pollutant removal performance. The removal efficiencies of total phosphorus (TP) in the loach group were 13.1% higher than in the control group, and snails increased the ammonium removal most effectively. Moreover, the concentration of total organic carbon (TOC) and TP in sediment significantly reduced with the addition of loaches and snails ($p < 0.05$), whereas the concentration of total nitrogen (TN) showed an obvious increase with the addition of loaches. High-throughput sequencing showed a microbial community structure change. Loaches and snails in wetlands changed the microbial diversity, especially in the *Proteobacteria* and denitrifying community. Results suggested that benthic aquatic animals might play an important role in CW ecosystems.

Keywords: treatment wetlands; high-throughput sequencing; sediment; benthic animals

1. Introduction

Wetlands are one of the most productive ecosystems, providing unique habitats for birds, plants, fish, and benthic animals. Constructed wetlands (CWs) are artificial wetlands designed to intercept wastewater and remove a wide range of pollutants before being discharged into natural water bodies. Most studies focused on the design of CWs with high pollutant removal efficiency by cultivating different kinds of plants (e.g., *Typha orientalis*, *Phragmites australis*), changing the operating methods of CWs (e.g., subsurface low and free water surface flow systems), and using different kinds of sediment media (e.g., sand, coal gangue) [1–3]. Furthermore, an increasing number of studies have explored benthic animal communities and factors influencing them in CWs [4,5]. Nevertheless, little is known about the effects of aquatic benthic animals on pollutant removal in CWs.

Benthic animals can directly absorb pollutions and transfer organic detritus from sedimentary storage into dissolved nutrients, which are removed by plants and algae absorption or microbial transformation [6]. Through their interaction, such as uptake of food from sediment, burrowing, and defecation, benthic species link energy flow and matter cycling in wetland ecosystems [7]. Furthermore, benthic species increase the sediment–water interface, which facilitates particle exchange between sediment and the water column. As a result, benthic animals can directly and indirectly stimulate pollution removal in sediment by influencing microbial population and activity [8]. However, research about the effects of aquatic benthic animals on microbial community structure and diversity in CWs is scarce.

Information about microbial community structure and diversity is important for understanding the relationship between environmental and ecosystem functions [9]. Microbial communities are also known as major drivers of wetland functions, supporting elemental cycling and biodegradation of pollutants [6,10]. An understanding of the bacterial community composition and structure in wetlands can increase the likelihood of successfully constructing a treatment wetland. The currently available high-throughput sequencing (using 16S rRNA genes) of environmental DNA allows for the rapid analysis of diversity and functionality of microbial communities [11].

As an initial attempt, the effects of loaches and snails on pollutant removal in both water and sediment are evaluated in this study. Microbial community and diversity are evaluated during the process, based on Illumina technology. This work will contribute to a better understanding of pollutant removal and microbial community variation and the involvement of benthic animals in CWs.

2. Materials and Methods

2.1. Benthic Animal Selection

Misgurnus anguillicaudatus (loach) and *Cipangopaludina cathayensis* (snail) are the most common and typical benthic species in wetland systems in China. Those two species can survive on high nutrient levels and low dissolved oxygen conditions [12] and are able to influence the water or sediment condition directly or indirectly [13].

2.2. CWs Microcosm Set-Up

Wetland microcosms were designed in a surface flow constructed wetland (SFCWs) and established under a transparent rain shelter. The experiments lasted from July to November. Twelve experimental wetland units consisting of polyethylene tubs were operated side by side. The treatment units were 65 cm deep and 50 cm in diameter with an outlet at the bottom, filled with washed river sand (particle size <2 mm, 25 kg) as the sediment to a depth of 25 cm. A total of 20 L of water was kept approximately 10 cm above the sand surface. The schematic of wetland microcosms is shown in Figure 1. Samples of *Phragmites australis*, which is common in China and efficient in wastewater treatment, were collected from the natural wetlands of Baiyun Lake in Zhangqiu, Shandong province. *P. australis* stalks approximately 1 m in height were carefully and gently washed with tap water to remove adhering soil and dead plant tissue, then cultivated with 10% Hoagland solution for two weeks. Well-grown plants with similar sizes were chosen and moved into the laboratory-scale CWs systems in May. The planting density of *P. australis* was 65 ± 2.3 plants/m^2.

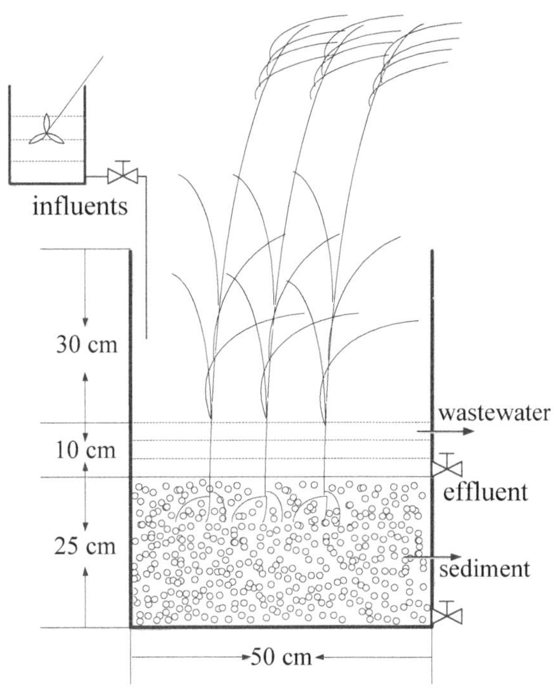

Figure 1. Schematic of the wetland microcosms.

Loaches and snails were bought from the aquarium market. We chose ones with similar size and weight (7.5 ± 0.7 g for loaches and 5.3 ± 0.5 g for snails). All animals were kept in tap water for one week before being put into the CWs systems. Then, animals were put into CWs systems for 30 days to adapt the environment. During this period, dead or unhealthy animals would be excluded. During the whole experimental period of nearly 4 months, CWs were well operated and both plants and animals in CWs were flourishing and healthy.

To minimize experimental variability, a simulation of polluted river water based on Grade II treatment standard of municipal sewage treatment plants in China was used in the experiment. The simulated polluted river water was composed of glucose, $(NH_4)_2SO_4$, KH_2PO_4, and KNO_3. The primary pollutants of the simulated polluted river water were organic matter, ammonia nitrogen, and phosphorus. The concentration of chemical oxygen demand (COD), ammonium (NH_4-N), total nitrogen (TN), total phosphorus (TP), and other water parameters are shown in Table 1. Sequencing batch-fed mode was applied in the study. Each wetland unit was drained through the effluent valve of each unit after a cycle. These units were immediately re-filled manually with the simulated polluted river water. Each unit was operated at hydraulic retention time (HRT) of 7 d, with theoretical hydraulic loading rate of 0.015 m^3/m^2d.

Table 1. Characteristics of influents from the wetland microcosm units (n = 30).

Parameter	Mean Concentration	Standard Deviation
COD (mg/L)	61.21	3.24
TN (mg/L)	21.53	1.12
NH_4-N (mg/L)	7.94	0.32
TP (mg/L)	1.22	0.14
DO (mg/L)	7.15	1.35
pH	7.56	0.31

2.3. Experimental Design and Sampling

The experiment was divided into four groups; each group had three parallel samples. Every group had a *P. australis* wetland unit with different benthic animals, as described in Table 2.

Table 2. Benthic animal addition in different groups.

Group	Benthic Animals	Density (g/m^2)
Control	-	-
Snail	*M. anguillicaudatus*	149.43 ± 3.34
Loach	*C. cathayensis*	150.62 ± 5.42
Mixture	*M. anguillicaudatus*	75.21 ± 2.31
	C. cathayensis	76.13 ± 3.53

Note: - no animals added.

Effluents at SFCWs were sampled to evaluate the transformations of organics, nitrogen, and phosphorus. Water samples were collected from 8:00 a.m. to 10:00 a.m., using a syringe tube at 5 cm depth below the water surface of each unit every 2 d. Each group comprised three parallel samples. Laboratory analysis was performed on the water samples for COD, NH_4-N, nitrate (NO_3-N), nitrite (NO_2-N), TN, and TP. All the parameters mentioned above were determined based on standard methods [14]. COD was measured by a HACH DR 2800TM Spectrophotometer, USA.

To evaluate TN, TP, and total organic carbon (TOC) storage in the sediment media during the experiment, for each barrel, sediments were collected at five individual locations (four corners and the center of the microcosms) in each parallel and then mixed into a composite sample. Sediment samples from the three parallel samples were subsequently combined with a composite sample before DNA extraction and then air-dried and ground before the TN, TP, and TOC measurement [15,16]. Analyses in media samples were conducted with a 3-AA3 Auto Analyzer (Bran-Luebbe, Hamburg, Germany).

2.4. DNA Extraction and Pyrosequencing

DNA was extracted using the Power Soil DNA isolation kit (MoBio Laboratories, Inc., Carlsbad, CA, USA), following the manufacturer protocols. The DNA concentrations were determined using a Nanodrops ND-1000 UV-vis spectrophotometer (NanoDrop Technologies, Wilmington, DE, USA). 16S rDNA gene amplification was generated using the universal bacterial primers 11f and 1428r. The Miseq pyrosequencing was performed at the Major Biocompany (Shanghai, China).

Sequences that were shorter than 250 bp in length and of a lower quality score than 30 were removed from the pyrosequencing-derived data sets. Clustering of the sequencing into operational taxonomic units (OTUs) was performed upon UCLUST software at 97% sequence identity [17]. Rarefaction analysis, Simpson diversity, Chao1 richness estimations, and Good's coverage were calculated by Mothur analysis (http://www.mothur.org) at 3% distance level.

2.5. Statistical Analysis

All statistical analyses were performed with the statistical program SPSS 17.0 (SPSS Inc., Chicago, IL, USA), including analysis of variance (ANOVA). In all tests, differences were considered statistically significant when $p < 0.05$.

3. Results and Discussion

3.1. Pollutant Removal in Water

The variation of pollutants removal efficiency is shown in Figure 2. The removal efficiencies of TP and NH_4-N were relatively high compared to those of COD and TN. COD and TN removal efficiency in all samples represented similar trends, with gradual decrease (Figure 2a,c). In general, TN removal increased early and then gradually decreased over time from the beginning to the end, with the maximum removal efficiency (80%) occurring in July. No significant difference between COD and TN removal efficency among treated and control samples ($p > 0.05$) was observed.

NH_4-N removal efficiencies of treated units indicated obvious differences from the control group ($p < 0.05$), highest at 86.85% in snail groups, followed by loach and mixture groups, and lowest at 70.09% in control samples. In the entire experimental time, the removal efficiency in the snail group showed significant difference from the control group ($p < 0.05$). In traditional aquaculture, aquaculture wastewater always has high concentrations of ammonia [18,19] because of high stocking density and ammonia excretion of aquatic animals. In the present research, the density of animals was lower than that in aquaculture, and the addition of animals into constructed wetland did not increase the concentions of ammonia. The density of animals is one of the important influences on water quality. In the wetland microcosms of the present study, the fauna population only slightly increased in the entire experimental time because of the limited food supplement. Removal efficiency of NH_4-N and TN indicated certain differences, whereas TN removal indicated no obvious difference from other groups. This result indicated a possibility of differences in NO_3-N or NO_2-N removal.

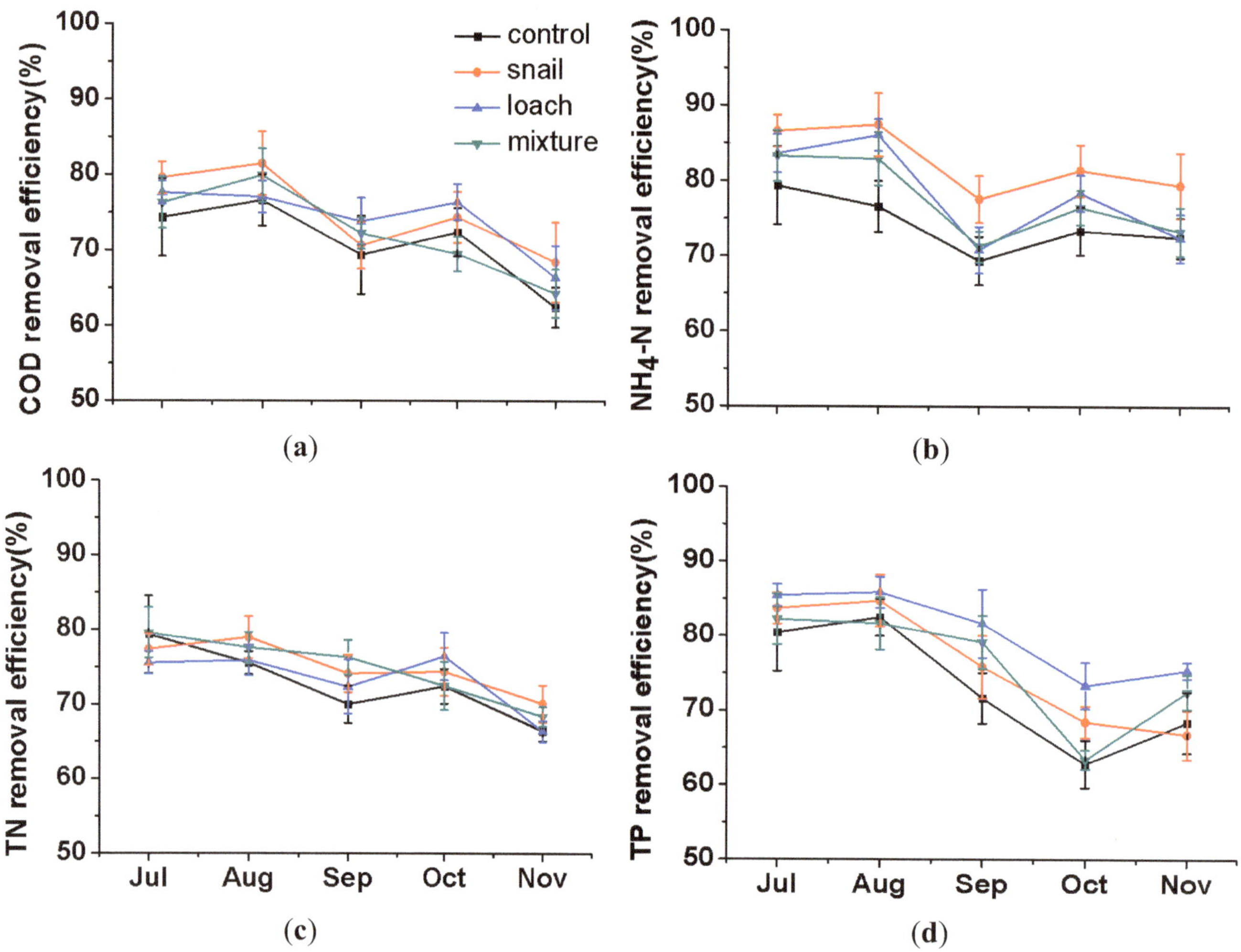

Figure 2. Removal efficiency of COD (**a**); NH_4-N (**b**); TN (**c**); TP (**d**).

Loach units showed highest removal performance of TP. The average TP removal efficiency ranged from 76.32% to 86.42%, which was much higher than the lowest level of 62% in the control unit ($p < 0.05$). The removal performance of TP in the loach group was also higher than that in the snail group. However, in July and August the difference was not significant ($p > 0.05$). Phosphorus was removed in CWs through substrate adsorption, chemical precipitation, bacterial action, plant and algal uptake, and incorporation into organic matter. Benthic animals in CWs may increase P uptake owing to their ingestion, bacterial action, and plant and algal uptake. Sediment adsorption also plays an important role [20].

The nutrient removal of all units in summer was higher than that in the winter. Consistent with other reports [21,22], we think temperature was one of the most important effects on the removal of nutrients. Temperatured in summer are much higher than in winter, and in general, high temperature usually increase the activity of the microorganisms, oxygen availability, and plant uptake.

The difference between NH_4-N and TN removal in each group showed certain differences in NO_3-N or NO_2-N removal. Thus, to further understand the effects of benthic animals on N removal, concentrations of NO_3-N and NO_2-N in effluent were analyzed (Figure 3). Different wetland animals play different roles in NO_3-N and NO_2-N concentration. The concentration of NO_2-N was highest in the snail group, followed by the mixture group and the control group. The loach group had the highest concentration of NO_3-N and the lowest concentration in NO_3-N among the four goups. During the study period, the average NO_3-N concentrations in the loach group were reduced from 3.21 ± 1.04 mg/L to a minimum of 2.64 ± 0.81 mg/L. The average NO_2-N concentrations in the loach group were reduced from 0.27 ± 0.07 mg/L to a minimum of 0.18 ± 0.03 mg/L. Blair *et al.* [23] found that the addition of earthworms increased the NO_3-N concentration in agroecosystems over a two-year period, corroborating the results of the present study.

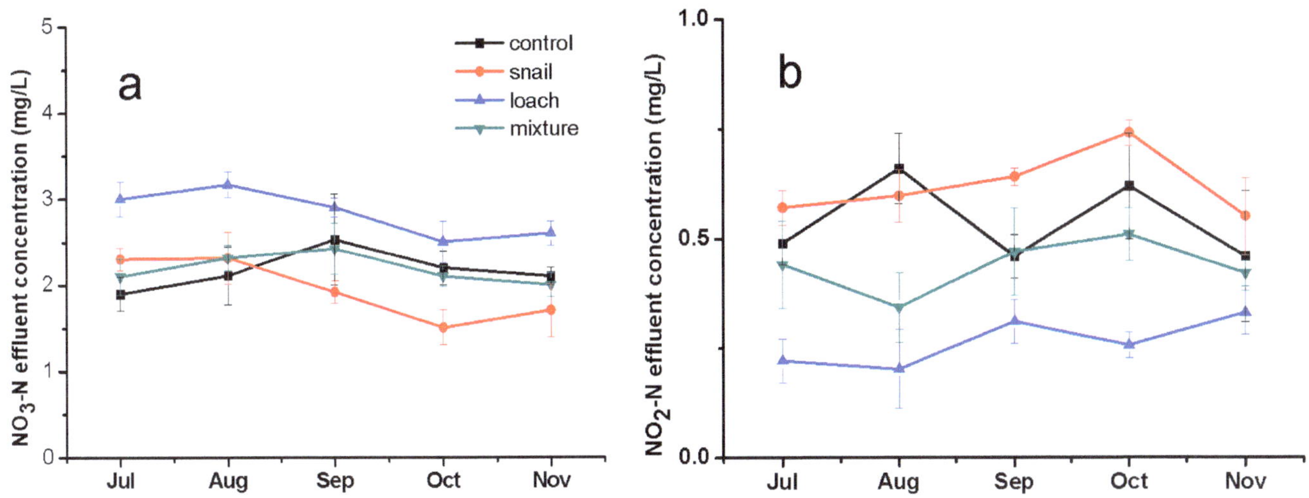

Figure 3. NO_3-N (**a**) and NO_2-N (**b**) concentrations of effluent.

3.2. Nutrient Changes in the Sediment

Notably, bioturbation can significantly change sediment characteristics. Results of nutrient changes in the sediment are shown in Figure 4.

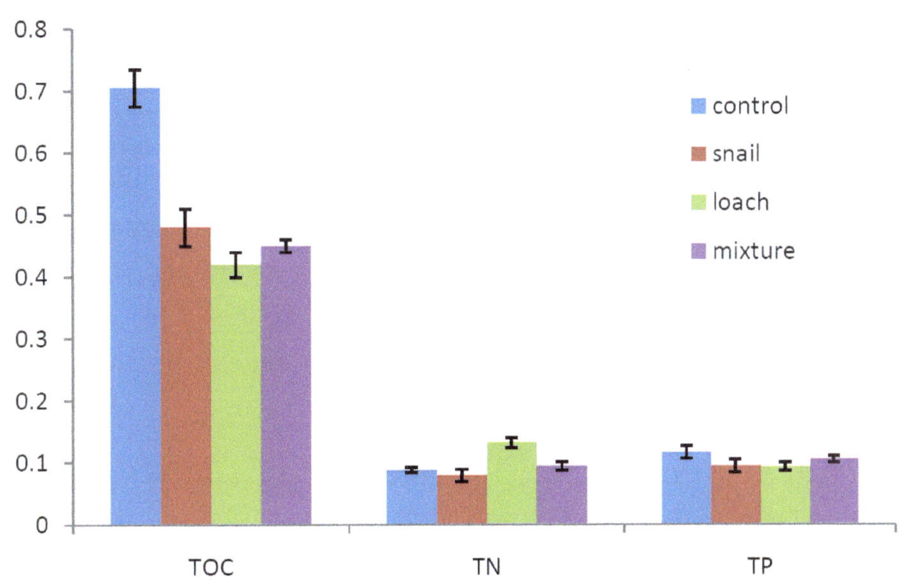

Figure 4. Nutrient and organic matter content of the sediment (average ± standard error) in the treatment and the control at the end of the experiment.

Our results showed that the four units have similar TP concentration. The average TP concentrations in all units throughout the experimental period ranged from 0.11 ± 0.09 mg/g to 0.16 ± 0.05 mg/g. The concentration of TP in sediment in treated groups was not significantly different from the control group ($p > 0.05$), but the removal efficiency of TP in water (Figure 2d) in the loach group was much higher than in the control group ($p < 0.05$). Therefore, the addition of loach in CWs may increase the removal efficiency of TP in water, which could contribute to higher P uptake owing to bacterial action and plant and algal uptake.

The concentration of TN in the loach group was much higher than in the three other groups, up to 0.149 mg/g ($p < 0.05$), followed by the mixture group, the control group, and the snail group. The high concentration of TN in the loach group may have been caused by the excretion of the loaches and the significant high concentration of NO_3-N in the water, which may cause adsorption in sediment.

The TOC concentration in the control group was much lower than in the other three groups ($p < 0.05$). The concentration of TOC was 0.71 ± 0.02, 0.48 ± 0.03, 0.42 ± 0.02, and 0.45 ± 0.01 mg/g in the control, snail, loach, and mixed animal units, respectively. TOC plays an important role in many biogeochemical processes in wetland systems. TOC can strongly affect the transport of contaminants from soils to surface water because of its complexation capacity [24]. The decrease of TOC in the experimental group may have been caused by the ingestion of benthic animals; different animals possess different ingestion abilities. The presence of benthic animals in CWs decreases the concentration of TOC in sediment, which could also decrease the release of organic carbon from sediments at the same time.

3.3. Microbial Communities

The key processes for nutrient removal in CWs include microbial conversion, decom position, plant uptake, sedimentation, and adsorption-fixation reactions [25]. Microbe activity is the main factor influencing the ammonia removal. Thus, more research must be conducted about the microorganism.

A total of 146,859 reads were obtained from the four samples through pyrosequencing analysis. The average read length is 401–418 base pairs (bp). Each library contained 32,378–40,422 reads. The good coverage index revealed that 93.60%–95.02% of the animals were obtained, with OTUs ranging from 3501 to 5191 (Table 3).

Chao1 and ACE were employed as community richness estimators at 3% dissimilarity. The Chao1 richness estimator showed that the richness ranged from 6852.16 to 9955.64. Compared with the control group, a decreasing trend was observed in the loach group. The mixture and snail groups, however, showed crosscurrent tendency. Therefore, the benthic species in the present study changed the community richness of this CWs. CWs with snails had the highest amount of bacteria among the four groups, which probably occurred because of the ingestion and movement of the snails [26].

Table 3. Comparison of phylotype coverage, diversity, and richness estimators at a phylogenetic distance of 3%.

Group	Cutoff	OTUs	ACE	Chao 1	Shannon	Simpson ($\times 10^{-3}$)	Coverage (%)
control	0.03	4,216	9,417.33	7,367.66	6.03	41.62	94.73
Snail	0.03	5,191	13,097.03	9,955.64	6.43	18.50	93.43
Loach	0.03	3,501	9,009.30	6,852.16	5.66	31.29	95.03
mixture	0.03	4,233	10,029.95	7,917.18	6.85	3.645	93.62

The community diversity was analyzed by calculating the Shannon and Simpson diversity index. For the four groups, an order of mixture > snail > control > loach was obtained in terms of community diversity. Based on these results, the snails had positive effects on the bacterial community. The loach group had less density of microbes, as it ingests microbes from sediment as food.

Figure 5 shows the relative abundance of the bacterial phylogenic groups for each wetland unit. Proteobacteria represented the majority of the community composition of all sediments. Nevertheless, the relative abundance of this phylum was particularly high in the environments of the CWs, ranging from 30% to 42% in four groups. The Proteobacteria phylum includes a very high level of bacterial metabolic diversity related to global carbon, nitrogen, and sulfur cycling [27]. The wide distribution of this phylum in nature [28,29] and in CWs [30,31] has been documented.

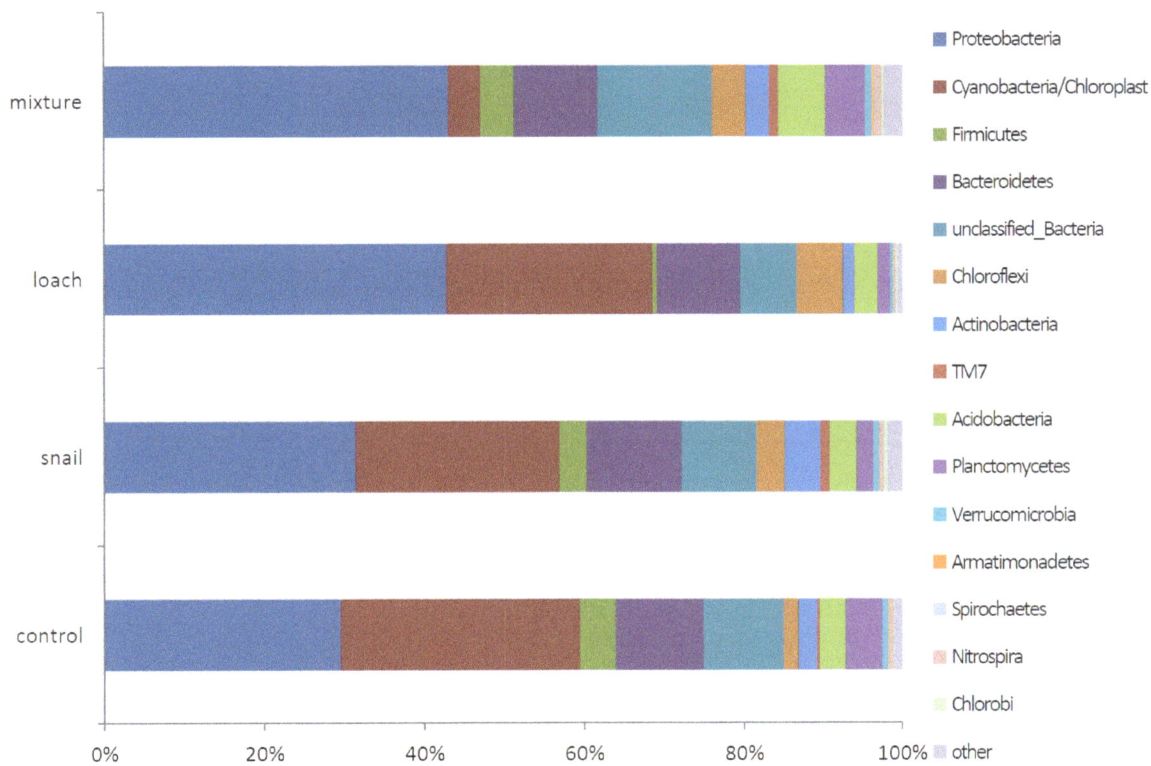

Figure 5. Abundances of the bacterial phylogenic groups.

Within the phylum Deltaproteobacteria, particularly the Syntrophorhabdus, Desulfonema, Smithellagenera, and Gammaproteobacteria, mainly Pseudomonas were enriched in the soil environments of the CWs (Figure 6). Bacteria belonging to the phylum Verrucomicrobia are nearly ubiquitous in soil [32]. The evidence available indicates that members of this phylum are globally distributed, abundant, and active groups of soil bacteria [33]. Apparently, the activities of loach lead to the increase of Proteobacteria phylum. The cyanobacteria in snail and loach groups had a significant decrease compared with the three other groups. The snails thrived and had much larger numbers than did the other groups. Snails ingested the cyanobacteria as food.

As known, the process of nitrate reduction is dependent on the denitrifier communities. *Bacillus*, *Micrococcus*, and *Pseudomona* are the main denitrifier communities in wetland sediment [34]. The consortiums of these three microbes were found to be efficient in nitrate reduction. Figure 7 shows the relative abundance of the denitrifier communities. *Bacillus*, *Micrococcus*, and *Pseudomona* in the loach group were lower than those in the three other groups. *Bacillus* in the control group was much higher than in the three other groups. The snail group had the most *Micrococcus* species, and *Pseudomona* was the highest in the mixture group.

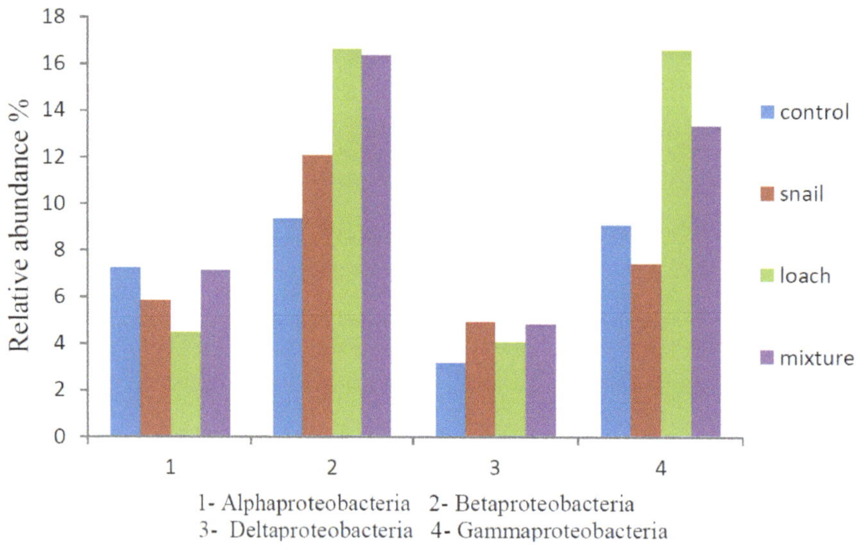

Figure 6. Relative bundances of the Proteobacteria groups.

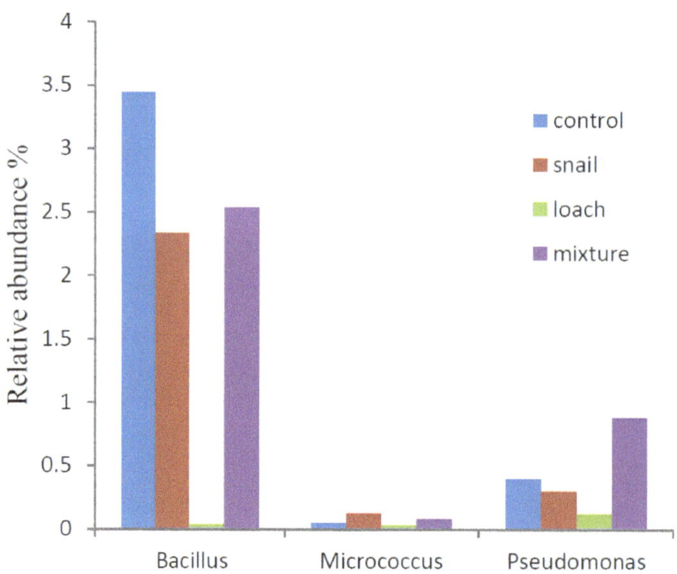

Figure 7. Relative abundances of the denitrifier communities.

Figure 2 shows that the loach group has high NO_3-N and low NO_2-N content, and has significant difference from the three other groups ($p < 0.05$). Investigation of the aerobic metabolism of *Bacillus*, *Micrococcus,* and *Pseudomona* revealed many interesting features of assimilatory nitrate reduction to ammonia. *Bacillus* species could carry out anaerobic dissimilatory reduction of nitrate to ammonia via nitrite [35]. *Pseudomonas* in the wetland system attained assimilatory denitrification [36]. The genera of *Bacillus*, *Micrococcus*, and *Pseudomonas* were found to be nitrate reducers. Among the three, *Pseudomonas* and *Bacillus* were found to be the most efficient in terms of nitrate reduction [37]. These two kinds of genera were just numerically superior. When the denitrifier communities were low, nitrate was reduced, and the concentration of NO_3-N was high.

4. Conclusions

Benthic animals in wetland systems can improve the removal efficiencies of TP, TN, and NH_4-N in effluent. Loach units showed the highest removal performance of TP and reduction of the concentration of TOC and TP in sediment. The loach activities led to the increase of the Proteobacteria phylum. The mixture group increased the community diversity of microorganisms. The presence of benthic animals in CWs changed the microbial structure and increased the biodiversity of the microcosms. The results suggest that benthic aquatic animals can reduce the concentration of pollutants and improve the quality of CWs ecosystems. The high diversity of the communities sampled in the natural ecosystem could be due to a higher number of ecosystem functions.

Acknowledgments

This work was supported by the Independent Innovation Foundation of Shandong University (2014JC023), the National Natural Science Foundation of China (No. 21307076), the National Water Special Project (2012ZX07203-004), and the Fundamental Research Funds of Shandong University (No. 2014TB003).

Author Contributions

The experimental work was conducted by Jian Zhang, Wenxing Wang, Huijun Xie and Pengfei Li. The manuscript was written by Pengfei Li, Huijun Xie and Zhen Hu. Data analysis was performed by Pengfei Li and Haiyan He. Microbial analysis was performed by Huijun Xie and Pengfei Li. All authors approved the final version of the article, including the authorship list.

Conflicts of Interest

The authors declare no conflict of interest.

References

1. Thullen, J.S.; Sartoris, J.J.; Nelson, S.M. Managing vegetation in surface-flow wastewater-treatment wetlands for optimal treatment performance. *Ecol. Eng.* **2005**, *25*, 583–593.
2. Nahlik, A.M.; Mitsch, W.J. Tropical treatment wetlands dominated by free-floating macrophytes for water quality improvement in Costa Rica. *Ecol. Eng.* **2006**, *28*, 246–257.
3. Ávila, C.; Reyes, C.; Bayona, J.M.; García, J. Emerging organic contaminant removal depending on primary treatment and operational strategy in horizontal subsurface flow constructed wetlands: Influence of redox. *Water Res.* **2013**, *47*, 315–325.
4. Spieles, D.J.; Mitsch, W.J. Macroinvertebrate community structure in high-and low-nutrient constructed wetlands. *Wetlands* **2000**, *20*, 716–729.
5. Jurado, G.B.; Johnson, J.; Feeley, H.; Harrington, R.; Kelly-Quinn, M. The potential of integrated constructed wetlands (ICWs) to enhance macroinvertebrate diversity in agricultural landscapes. *Wetlands* **2010**, *30*, 393–404.
6. Zak, D.R.; Holmes, W.E.; White, D.C.; Peacock, A.D.; Tilman, D. Plant diversity, soil microbial communities, and ecosystem function: Are there any links? *Ecology* **2003**, *84*, 2042–2050.

7. Fanjul, E.; Escapa, M.; Montemayor, D.; Addino, M.; Alvarez, M.F.; Grela, M.A.; Iribarne, O. Effect of crab bioturbation on organic matter processing in South West Atlantic intertidal sediments. *J. Sea Res.* **2014**, *95*, 206–216.
8. Covich, A.P.; Palmer, M.A.; Crowl, T.A. The role of benthic invertebrate species in freshwater ecosystems: Zoobenthic species influence energy flows and nutrient cycling. *Bioscience* **1999**, *49*, 119–127.
9. Sims, A.; Zhang, Y.; Gajaraj, S.; Brown, P.B.; Hu, Z. Toward the development of microbial indicators for wetland assessment. *Water Res.* **2013**, *47*, 1711–1725.
10. Adrados, B.; Sánchez, O.; Arias, C.A.; Becares, E.; Garrido, L.; Mas, J.; Brix, H.; Morató, J. Microbial communities from different types of natural wastewater treatment systems: Vertical and horizontal flow constructed wetlands and biofilters. *Water Res.* **2014**, *55*, 304–312.
11. İnceoğlu, Ö.; Al-Soud, W.A.; Salles, J.F.; Semenov, A.V.; van Elsas, J.D. Comparative analysis of bacterial communities in a potato field as determined by pyrosequencing. *PLoS ONE* **2011**, *6*, doi:10.1371/journal.pone.0023321.
12. Alonso, Á.; Camargo, J.A. Nitrate causes deleterious effects on the behaviour and reproduction of the aquatic snail Potamopyrgus antipodarum (Hydrobiidae, Mollusca). *Environ. Sci. Pollut. Res.* **2013**, *20*, 5388–5396.
13. Gonçalves, A.F.; Castro, L.F.C.; Pereira-Wilson, C.; Coimbra, J.; Wilson, J.M. Is there a compromise between nutrient uptake and gas exchange in the gut of *Misgurnus anguillicaudatus*, an intestinal air-breathing fish? *Comp. Biochem. Physiol. Part D Genomics Proteomics* **2007**, *2*, 345–355.
14. American Public Health Association (APHA). *Standard Methods for the Examination of Water and Wastewater*, 20th ed.; American Public Health Association/American Water Works Association/Water Environment Federation: Washington, DC, USA, 1998.
15. Wu, H.; Zhang, J.; Li, P.; Zhang, J.; Xie, H.; Zhang, B. Nutrient removal in constructed microcosm wetlands for treating polluted river water in northern China. *Ecol. Eng.* **2011**, *37*, 560–568.
16. Wu, S.Q.; Chang, J.J.; Dai, Y.; Wu, Z.B.; Liang, W. Treatment performance and microorganism community structure of integrated vertical-flow constructed wetland plots for domestic wastewater. *Environ. Sci. Pollut. Res.* **2013**, *206*, 3789–3798.
17. Edgar, R.C. Search and clustering orders of magnitude faster than BLAST. *Bioinformatics* **2010**, *26*, 2460–2461.
18. Lin, Y.; Jing, S.; Lee, D.; Wang, T. Nutrient removal from aquaculture wastewater using a constructed wetlands system. *Aquaculture* **2002**, *209*, 169–184.
19. Gong, L.J.; Yang, X.F.; Xiong, B.X.; Li, G.P.; Chen, X.L. Study on Nitrogen, Phosphor and Chemical Oxygen Demand of Differnt Categories of Aquaculture Lakes by Means of Principal Component Analysis, Factor Analysis and Cluster Analysis. *Adv. Mater. Res.* **2012**, *340*, 369–377.
20. Xu, D.; Xu, J.; Wu, J.; Muhammad, A. Studies on the phosphorus sorption capacity of substrates used in constructed wetland systems. *Chemosphere* **2006**, *63*, 344–352.
21. Bachand, P.A.; Horne, A.J. Denitrification in constructed free-water surface wetlands: II. Effects of vegetation and temperature. *Ecol. Eng.* **1999**, *14*, 17–32.
22. Poach, M.E.; Hunt, P.G.; Reddy, G.B.; Stone, K.C.; Johnson, M.H.; Grubbs, A. Swine wastewater treatment by marsh–pond–marsh constructed wetlands under varying nitrogen loads. *Ecol. Eng.* **2004**, *23*, 165–175.

23. Blair, J.M.; Parmelee, R.W.; Allen, M.F.; McCartney, D.A.; Stinner, B.R. Changes in soil N pools in response to earthworm population manipulations in agroecosystems with different N sources. *Soil Biol. Biochem.* **1997**, *29*, 361–367.
24. Tipping, E. Modeling the competition between alkaline earth cations and trace metal species for binding by humic substances. *Environ. Sci. Technol.* **1993**, *27*, 520–529.
25. Tchobanoglous, G. Constructed wetlands and aquatic plant systems: Research, design, operational, and monitoring issues. In *Constructed Wetlands for Water Quality Improvement*; Lewis Publishers: Boca Raton, FL, USA, 1993; pp. 23–34.
26. Xu, D.; Li, Y.; Howard, A. Influence of earthworm Eiseniafetida on removal efficiency of N and P in vertical flow constructed wetland. *Environ. Sci. Pollut. Res.* **2013**, *20*, 5922–5929.
27. Kersters, K.; de Vos, P.; Gillis, M.; Swings, J.; Vandamme, P.; Stackebrandt, E. Introduction to the Proteobacteria. In *The Prokaryotes: A Handbook on the Biology of Bacteria*; Springer: New York, NY, USA, 2006; Volume 5, pp. 3–37.
28. Wang, Y.; Sheng, H.; He, Y.; Wu, J.; Jiang, Y.; Tam, N.F.; Zhou, H. Comparison of the levels of bacterial diversity in freshwater, intertidal wetland, and marine sediments by using millions of illumina tags. *Appl. Environ. Microbiol.* **2012**, *78*, 8264–8271.
29. Peralta, R.M.; Ahn, C.; Gillevet, P.M. Characterization of soil bacterial community structure and physicochemical properties in created and natural wetlands. *Sci. Total Environ.* **2013**, *443*, 725–732.
30. Gorra, R.; Coci, M.; Ambrosoli, R.; Laanbroek, H.J. Effects of substratum on the diversity and stability of ammonia-oxidizing communities in a constructed wetland used for wastewater treatment. *J. Appl. Microbiol.* **2007**, *103*, 1442–1452.
31. Ahn, C.; Gillevet, P.M.; Sikaroodi, M. Molecular characterization of microbial communities in treatment microcosm wetlands as influenced by macrophytes and phosphorus loading. *Ecol. Indic.* **2007**, *7*, 852–863.
32. Zhang, L.; Xu, Z. Assessing bacterial diversity in soil. *J. Soils Sediments* **2008**, *8*, 379–388.
33. Sangwan, P.; Kovac, S.; Davis, K.E.; Sait, M.; Janssen, P.H. Detection and cultivation of soil Verrucomicrobia. *Appl. Environ. Microbiol.* **2005**, *71*, 8402–8410.
34. Chitra, A.V.; Lakshmanaperumalsamy, P. Biodegradation of nitrate in wastestreams from explosives manufacturing plants. *Res. J. Microbiol.* **2006**, *1*, 142–151.
35. Tiedje, J.M. Ecology of denitrification and dissimilatory nitrate reduction to ammonium. *Biol. Anaerob. Microorg.* **1988**, *717*, 179–244.
36. Ayyasamy, P.M.; Shanthi, K.; Lakshmanaperumalsamy, P.; Lee, S.; Choi, N.; Kim, D. Two-stage removal of nitrate from groundwater using biological and chemical treatments. *J. Biosci. Bioeng.* **2007**, *104*, 129–134.
37. Rajakumar, S.; Ayyasamy, P.M.; Shanthi, K.; Thavamani, P.; Velmurugan, P.; Song, Y.C.; Lakshmanaperumalsamy, P. Nitrate removal efficiency of bacterial consortium *(Pseudomonas* sp. KW1 and *Bacillus* sp. YW4) in synthetic nitrate-rich water. *J. Hazard. Mater.* **2008**, *157*, 553–563.

13

Stochastic Flocculation Model for Cohesive Sediment Suspended in Water

Hyun Jung Shin [1], Minwoo Son [2,*] and Guan-hong Lee [3]

[1] Department of Oceanography, Inha University, Incheon 402-751, Korea; E-Mail: hyunjung_@inha.edu

[2] Department of Civil Engineering, Chungnam National University, Daejeon 305-764, Korea

[3] Department of Oceanography, Inha University, Incheon 402-751, Korea; E-Mail: ghlee@inha.ac.kr

* Author to whom correspondence should be addressed; E-Mail: mson@cnu.ac.kr;

Academic Editor: Miklas Scholz

Abstract: Existing flocculation models for cohesive sediments are classified into two groups: population balance equation models (PBE) and floc growth models. An FGM ensures mass conservation in a closed system. However, an FGM determines only the average size of flocs, whereas a PBE has the capability to calculate a size distribution of flocs. A new stochastic approach to model the flocculation process is theoretically developed and incorporated into a deterministic FGM in this study in order to calculate a size distribution of flocs as well as the average size. A log-normal distribution is used to generate random numbers based on previous laboratory experiments. The new stochastic flocculation model is tested with three laboratory experiment results. It was found and validated with measured data that the new stochastic flocculation model has the capability to replicate a size distribution of flocs reasonably well under different sediment and carrier flow conditions. Three more distributions (normal; Pearson type 3; and generalized extreme value distributions) were also tested. From the comparison with results of different distribution functions, it is shown that a stochastic FGM using a log-normal distribution has a comparative advantage in terms of simplicity and accuracy.

Keywords: cohesive sediment; flocculation process; size distribution; stochastic approach; distribution function

---

1. Introduction

The flocculation process, a series of aggregation and disaggregation (so called breakup) processes, is one of the most important characteristics of suspended cohesive sediment such as mud and organic matter. Through the flocculation process, the size and density of cohesive sediment continuously vary, resulting in variation of settling velocity and particle response time. Because the settling velocity and particle response time significantly affect the advection and diffusion process of suspended sediment, an improved understanding of the flocculation process is paramount for cohesive sediment transport. Therefore, the flocculation process has been intensively studied. The flocculation process is mainly due to eletro- or biological-chemical attraction of fine sediment (e.g., [1]). When suspended particles collide with each other, there is a chance of aggregation by attraction. Collisions of suspended sediment are caused by Brownian motion, differential settling, and turbulent shear [2–4]. It has been concluded in many studies that turbulent shear is the dominant mechanism of suspended sediment collisions (e.g., [5,6]). Moreover, turbulent shear is considered as the main mechanism of floc breakup (e.g., [1,7,8]). Many laboratory studies on the effects of turbulent shear have been conducted utilizing a mixing tank and Couette viscometer in order to control the intensity of turbulent shear (e.g., [9–12]). From these studies, it is apparent that the flocculation process of aggregation (mainly due to particle collision) and breakup is significantly governed by turbulent shear.

With the aim of quantitatively modeling the flocculation process, many flocculation models have been proposed. Generally, flocculation models can be classified into two groups at present; a population balance equations model (PBE) and a floc growth model (FGM). A PBE, often called a size class flocculation model, is based on discrete size classes, and typically calculates the change of floc numbers in each size class (e.g., [13–15]). A PBE has the advantages of considering many mechanisms of flocculation and calculating a size distribution of flocs. However, a PBE is too numerically expensive to be combined with numerical sediment transport models [16] and demands many empirical parameters, increasing uncertainties. Moreover, a discrete PBE approach is limited in terms of ensuring the mass conservation of flocs in a closed system, although many schemes have been proposed to overcome this weakness (e.g., [17,18]). A FGM, often called a Lagrangian flocculation model, is a differential equation usually composed of two terms, an aggregation term and a breakup term, representing the increase in floc size due to aggregation and the decrease due to breakup, respectively. A FGM is simple enough to be combined with numerical sediment transport models. Furthermore, a FGM ensures the mass conservation of cohesive sediment in a closed system because it calculates the average size of flocs. Based on the collisional frequency equation of Levich [19] and dimensional analysis, Winterwerp [8] proposes a semi-empirical FGM under the assumption that the fractal dimension of flocs is constant. However, many studies show that the fractal dimension of natural flocs is variable (e.g., [20,21]). Based on Levich [19] and Khelifa and Hill [20], Son and Hsu [7] improved Winterwerp's FGM [8] by adopting an empirical equation with a variable fractal dimension. However, neither Son and Hsu [7] nor Winterwerp [8] appropriately address the temporal evolution of floc size, although both calculate the equilibrium floc size under constant turbulence intensity conditions. Son and Hsu [1] further theoretically and mathematically derive an equation for the yield strength of flocs, which depends on the floc size and fractal dimension. The equation is similar to an empirical equation from Sonntag and Russel [22] when one empirical parameter, r, is set to be about

0.7 (see [1] for more details). By incorporating this equation with a FGM from Son and Hsu [7], Son and Hsu [1] propose a new FGM using a variable fractal dimension and yield strength. The model from Son and Hsu [1] has been validated with laboratory experiments by Spice *et al.* [23] and Burban *et al.* [24]. From the simulation results, it is clearly established that a new FGM from Son and Hsu [1] has the capability to replicate the temporal evolution of floc size as well as the equilibrium floc size. In addition, from comparison with results calculated by models developed by Son and Hsu [7] and Winterwerp [8], it is also concluded that both the variable fractal dimension and the variable yield strength of flocs are critical to modeling the flocculation process. Son and Hsu [25] develop a sediment transport model for cohesive sediment and incorporate it into FGMs from Winterwerp [8], Son and Hsu [7], and Son and Hsu [1]. Through validation of models with field data measured in the Ems/Dollard estuary [26], it is shown that a sediment transport model combined with Son and Hsu's model [1] calculates the characteristics observed in the Ems/Dollard estuary. Son and Hsu's model [1] is used in Son and Lee [27] again in order to understand the role of flocculation under oscillatory flow representing short waves. It is found in Son and Lee [27] that the flux of sediment under fixed short wave conditions has opposite directions depending on the flocculation process. To the best knowledge of the authors, the model from Son and Hsu [1] is one of few flocculation models with results that are in good agreement with field data when combined with a high-resolution sediment transport model which quantitatively calculates interactions between turbulence and floc dynamics.

As mentioned above, a FGM is numerically efficient and ensures mass conservation. However, compared to PBE, a FGM is disadvantaged in the sense that it does not have the capability to calculate size distribution of flocs. This is because a FGM calculates the average size of flocs using a deterministic equation. In order to simulate sediment transport, Man and Tsai [28] propose the stochastic advection-diffusion equation based on the Langevin equation, solving the equation using numerical schemes for stochastic partial differential equations [29–31]. In simulation results, a stochastic model showed the capability to calculate the range and variation of sediment suspension [28]. Maggi [32] proposes a FGM incorporated with stochastic approach based on Winterwerp's FGM [8]. As mentioned in the previous paragraph, Winterwerp's FGM [8] assumes that the fractal dimension is constant. To overcome this weak point, Maggi [32] adopts the concept of a variable fractal dimension. However, when the key equation for the temporal change rate of primary particles within a floc is derived, the fractal dimension is assumed to be constant again (see Equations (3) and (4) in [32]). In addition, the FGM of Maggi [33] also assumes the yield strength of flocs to be constant, whereas Son and Hsu [1] showed that the temporal evolution of floc size is reasonably replicated when a variable yield strength of flocs is assumed.

This study aims to improve the existing FGM developed by Son and Hsu [1] to predict both temporal evolution of average size and size distribution of flocs by adopting a stochastic approach to flocculation modeling. A brief description of Son and Hsu's FGM [1] is given in Section 2.1. Based on this FGM, a new stochastic FGM is proposed in Section 2.2. The developed model is validated with data obtained from laboratory experiments in Section 3. Finally, the conclusions of this study are drawn in Section 4.

2. Materials and Methods

2.1. Previous FGM of Son Hsu [1]

The Son and Hsu FGM [1] calculates the temporal evolution of average floc size:

$$\frac{\mathrm{d}D}{\mathrm{d}t}=\frac{Gd^{\beta}}{\beta\ln\frac{D}{d}+1}\left[\frac{k_A^{'}}{3}\frac{c}{\rho_s}d^{F-3}D^{-F+4-\beta}-\frac{k_B^{'}}{3}\left(\frac{\mu G}{B_1}\right)^{q}d^{-p+\frac{2q}{3}F}D^{1-\beta+\frac{2q}{3}(3-F)}\left(D-d\right)^{p}\right] \tag{1}$$

where D is the average size of flocs; d is the size of primary particle; G is the dissipation parameter (shear rate); μ is the dynamic viscosity of the fluid; $k_A^{'}$ and $k_B^{'}$ are empirical coefficients for efficiency of aggregation and breakup; B_1 is the coefficient for cohesive force of one primary particle; c is the mass concentration of floc; ρ_s is the density of primary particle; and p and q are empirical coefficients set assumed to be p = 1 and q = 0.5 in this study based on the previous study by Winterwerp [8]. The fractal dimension of flocs, F, is calculated by a power law proposed by Khelifa and Hill [20]:

$$F=\alpha\left(\frac{D}{d}\right)^{\beta} \tag{2}$$

The empirical coefficients, α and β, are specified as $\alpha=3.0$ and $\beta=\frac{\log(F_c/3)}{\log(D_{fc}/d)}$. F_c is a characteristic fractal dimension and D_{fc} is a characteristic size of floc. The typical vales of F_c and D_{fc} are suggested to be F_c = 2.0 and D_{fc} = 2000 μm by Khelifa and Hill [20].

The first and second terms on the right hand side of Equation (1) calculate the increase and decrease of floc size due to aggregation and breakup, respectively. The Son and Hsu FGM [1] is an ordinary differential equation calculated by marching numerical methods such as Euler's method and Runge-Kutta method. In Son and Hsu's study [1], the FGM is solved by the fourth-order Runge-Kutta method. The fourth-order Runge-Kutta method is used again in this study in order to solve the stochastic FGM described in Section 2.2.

2.2. Application of Stochastic Approach to FGM

In many studies (e.g., [10,15,34–39]), it has been shown that floc size distribution is similar to a log-normal distribution containing a single peak. Contradictorily, it has also been shown from previous studies (e.g., [40–43]) that floc size distribution may be bimodal. Results from Gratiot and Manning [44] suggest that the irregular distribution of floc size result from organic matter associated with mud. When the organic content of mud is removed, the size distribution of floc is close to a log-normal distribution. Laboratory experiments using Kaolinite also show a log-normal distribution of floc size (e.g., [9,11,23,24]). In many field experiments (e.g., [15,35–37]), a log-normal distribution of floc size is found when the organic content of mud is less than 5%. Thus, the low organic content of mud and the log-normal distribution of floc size are assumed in this study.

In order to apply the stochastic approach to FGM, it is necessary to consider properties of the flocculation process and physical meanings of Equation (1). One of the main assumptions of the Son and Hsu FGM [1] is that a floc has the fractal structure quantified by the fractal dimension, F (see Equation (2)). That is, floc aggregation is assumed to maintain self-similarity. k'_A is a coefficient accounting for the efficiency of collision and diffusion (see [8] for more details). Gibbs [45] insists that the value of k'_A depends on the properties of sediment. Tsai and Hwang [11] insist that k'_A represents the property of the water type (sea water and fresh water) based on the previous studies by Lick and Lick [46] and Lick *et al.* [16]. From studies by Winterwerp [8] and Mietta [47], it is also known that the aggregation process is significantly affected by physical-chemical properties such as pH and salinity and the organic compounds in the sediment. Therefore, the aggregation process is not considered to be random. The mass concentration (c), the dissipation parameter (G), and properties of primary particle such as density (ρ_s) and size (d) are also deterministic because they are given conditions of the flocculation process. To the best of our knowledge, there is not a general consensus about breakup processes at present [32]. Deterministic PBEs include a breakup distribution function, although the function is used to simply calculate the type of breakup such as binary and ternary fragmentations (e.g., [48,49]). In the case of FGM, k'_B representing the efficiency of floc breakup plays a similar role. k'_B determines the proportion of floc breakup due to turbulent shear. This coefficient has a close relationship with the floc yield strength (F_y), the minimum stress for floc breakup (see [1,8] for more details). However, turbulent shear exceeding F_y does not necessarily cause the breakup of flocs (e.g., [20]). Therefore, k'_B is assumed to be a random number in this study. As the size of flocs increases, the density of flocs decreases, resulting in higher probability of breakup. Thus, the distribution of floc size should be skewed positively (skewed to the right; skewness ≥ 0). Based on this idea and previous studies mentioned in the first paragraph of this section, k'_B is empirically assumed to be a random number having a log-normal distribution (see Equation (3)).

$$k'_B = \exp\left\{X\sqrt{\log\left(\frac{\sigma_l^2}{\mu_l^2}\right) + 1} + \log\left(\mu_l^2 \Big/ \sqrt{\sigma_l^2 + \mu_l^2}\right)\right\} \tag{3}$$

To determine k'_B at each time step, a Monte Carlo simulation method is used in this study. A Monte Carlo method (or Monte Carlo experiment) is a computational algorithm using repeated random sampling in order to obtain numerical results. μ_l and σ_l are the mean and the standard deviation of log-normal distribution, respectively. In Equation (3), X is a pseudorandom number. Incorporating Equation (3) into Equation (1), the stochastic FGM of this study is developed. μ_l is equal to k'_B of the deterministic FGM. σ_l is calibrated with the distribution shape of experimental results. Compared to the previous deterministic Son and Hsu FGM [1], the new stochastic FGM has one more parameter (σ_l).However, the new FGM has a capability to calculate the size distribution of flocs ensuring the mass conservation. μ_l and σ_l are calibrated with three experimental results in Section 3.

3. Results and Discussion

During each case of the experiment, 30 numerical simulations were conducted. This study focuses on the distribution of floc size in the equilibrium state. Therefore, the values of floc size at the last 1000 time steps (2061 total) are analyzed under the assumption that the flocculation process approaches a dynamic equilibrium state prior to 1061 time steps (see Figure 1). The stochastic FGM of this study is validated with 3 laboratory experiments by [9,23,24]. Following Spicer *et al.* [23], polystyrene particles were mixed in a stirred tank with an impeller in order to measure the size distribution of aggregates with volumetric and mass concentrations (ϕ and c) of $1.4 \times 10^{-5.0}$ and 0.0147 kg/m^3, respectively. The average value of G produced by the impeller is 50 s^{-1}. The size and density of polystyrene particles (c and ρ_S) of the experiment are 0.87 μm and 1050 kg/m^3, respectively. In Biggs and Lant [9], the size distribution of activated sludge is reported under the experimental condition of $G = 19.4$ s^{-1} and $\phi = 0.05$. Under the assumption that ρ_S is 2650 kg/m^3 and the density of sludge is 1300 kg/m^3, the mass concentration is calculated to be 24.19 kg/m^3. The activated sludge is mixed in a baffled batch vessel using a flat six-blade impeller. Burban *et al.* [24] also report the results of laboratory experiments with Detroit River sediment. c and G of the experiment are 0.8 kg/m^3 and 200 s^{-1}, respectively. d and ρ_S of Detroit River sediment are assumed to be 4 μm and 2650 kg/m^3 due to absence of further information. The detailed conditions of experiments are summarized in Table 1. The calibrated values of empirical coefficients are shown in Table 2.

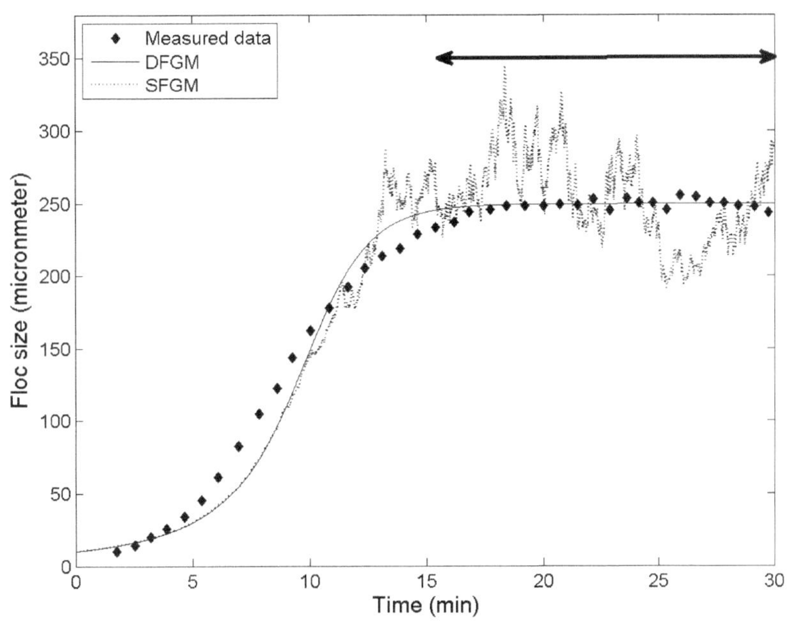

Figure 1. Example of numerical simulation and analysis data. Dots represent the experimental result of Spicer *et al.* [23]. Solid and dotted lines are the stochastic FGM (SFGM) of this study and the deterministic FGM (DFGM), respectively.

Table 1. Experimental conditions.

Experiment	ϕ	c (kg/m^3)	ρ_S (kg/m^3)	d (μm)	G (s^{-1})
Spicer *et al.*, (1998)	1.4×10^{-5}	0.0147	1050	0.87	50.0
Biggs and Lant (2000)	0.05	24.19 (Assumed)	2650 (Assumed)	4.00 (Assumed)	19.4
Burban *et al.*, (1989)	N.A.	0.05	2650 (Assumed)	4.00 (Assumed)	200.0

Note: N.A.—Not available.

Table 2. Calibrated coefficients of stochastic FGM.

Experiment	k'_A	μ_l	σ_l^2	B_1
Spicer *et al.*, (1998)	6.74	4.39×10^{-6}	4.47×10^{-6}	2.63×10^{-14}
Biggs and Lant (2000)	0.02	2.61×10^{-5}	5.62×10^{-5}	4.20×10^{-13}
Burban *et al.*, (1989)	0.30	3.95×10^{-5}	3.16×10^{-3}	4.10×10^{-12}

Figure 2 shows the simulation results of the stochastic FGM. The experiment from Spicer *et al.* [23] was carried out under the dilute condition (c = 0.0147 kg/m^3). The intensity of turbulent shear is not low (G = 50 s^{-1}). However, the mean size of flocs is about 200 μm. Therefore, it is clear that the polystyrene particles are easily aggregated. Under this assumption, the coefficient determining the aggregation efficiency has a relatively high value ($k'_A = 6.74$). In terms of the mean size of flocs, the simulation results are in satisfactory agreement with experimental results. The mean size is determined by calibrating k'_A and μ_l. That is, the model has the capability to calculate mean size of flocs reasonably well. It is shown in Figure 2 that the calculated size distribution is positively skewed slightly more than the measured result. The log-normal distribution function has only 2 parameters, the mean and variance. Therefore, the skewness and kurtosis of log-normal distribution is indirectly determined by the values of mean and variance. Three- or four-parameter distribution functions such as generalized extreme value distribution and Wakeby distribution have the capability to determine the skewness or kurtosis by calibrating additional parameters. However, more empirical and site-specific parameters need to be calibrated when a 3- or 4-parameter distribution function is incorporated into FGM.

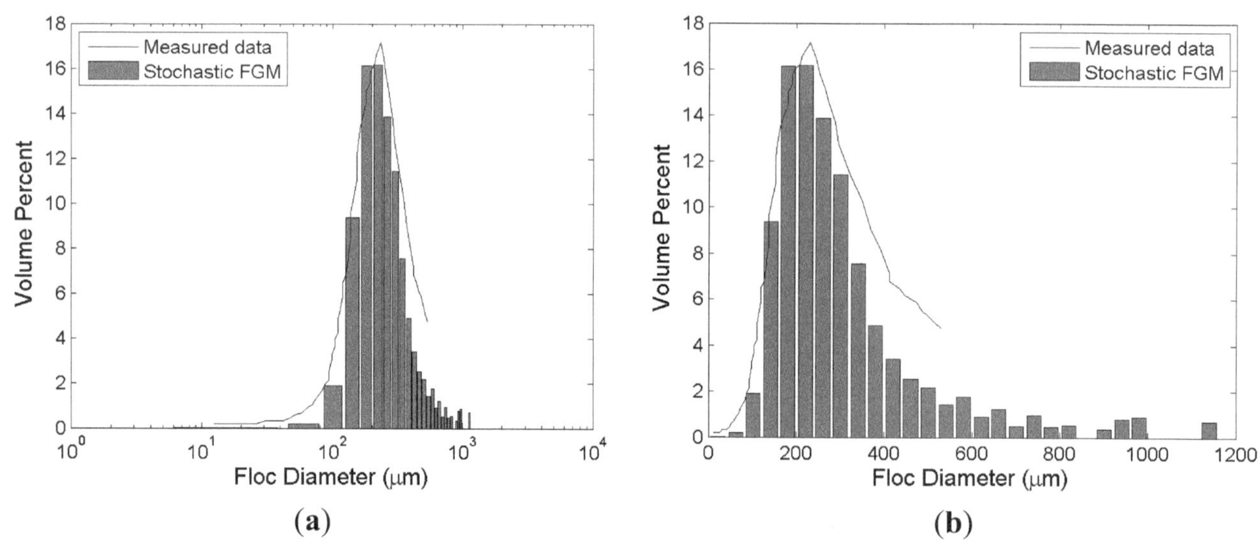

Figure 2. Simulation results of stochastic FGM and experimental results of Spicer *et al.* [23]. The bars and the solid lines represent the experimental and simulation results, respectively. (**a**) semi-log axis; (**b**) normal (Cartesian) axis.

Figure 3 represents the results of stochastic simulation and measured data from Biggs and Lant [9]. Their experiment was conducted under conditions with very high concentration (c = 24.19 kg/m^3). The intensity of turbulence is relatively low (G = 19.4 s^{-1}). Under these conditions, the mean size of flocs is not large (about 120 μm). Therefore, it is deduced that the aggregation process of activated sludge used for the experiment has a low efficiency resulting in a low value for the aggregation efficiency parameter

($k'_A = 0.02$). The simulation result is in a good agreement with laboratory measurements. Both the mean size and the distribution are reasonably replicated by the proposed stochastic FGM. The sudden increase of volume fraction around floc size 700 μm is due to the random property of stochastic simulation. Figure 4a shows the mean, maximum, and minimum floc sizes of 30 simulation cases. The dotted line is the equilibrium floc size calculated by the deterministic Son and Hsu FGM [1]. Figure 4b shows the temporal evolution of mean floc size measured by Biggs and Lant [9], a simulation result of deterministic FGM ([1]), and one case of simulation by stochastic FGM. As mentioned in the first paragraph of this section, the last 1000 values of floc size for each stochastic simulation are analyzed in this study. Therefore, the size distribution of this study is determined using 30,000 values for floc size. Figure 4 shows that a stochastic FGM calculates the floc size very randomly after the temporal evolution of floc size approaches the equilibrium state (around 15 min in Figure 4b). The maximum floc size among the values in the last 1000 time steps has a wide range (200 μm to 700 μm). Furthermore, the minimum floc size is also in the range of 10 μm to 50 μm whereas the mean values are between 70 and 120 μm in most cases.

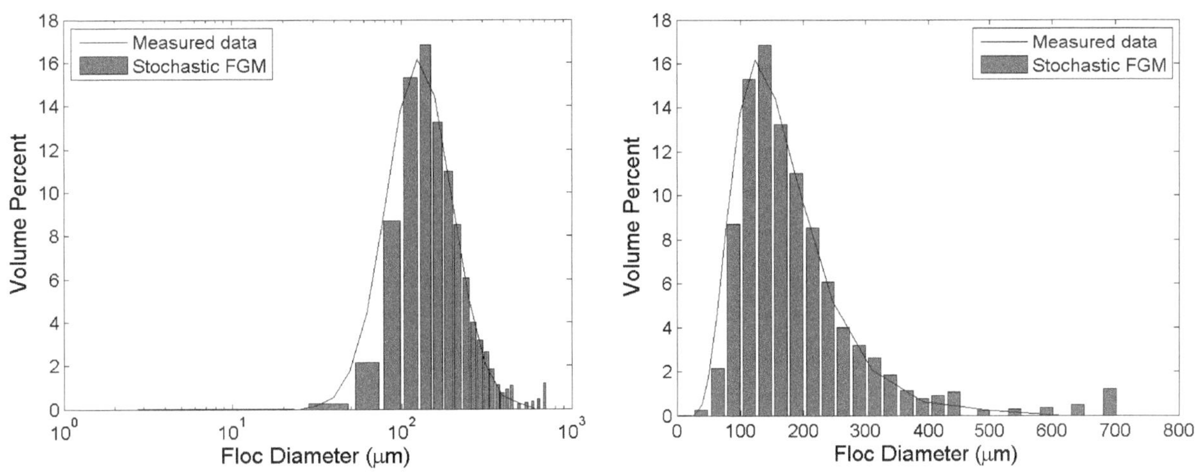

(a) **(b)**

Figure 3. Simulation results of stochastic FGM and experimental results of Biggs and Lant [9]. The bars and the solid lines represent the experimental and simulation results, respectively. (**a**) semi-log axis; (**b**) normal (Cartesian) axis.

Burban *et al.* [24] measured the size distribution of Detroit River sediments with a high dissipation parameter ($G = 200\ s^{-1}$) and a moderate mass concentration condition ($c = 0.8\ kg/m^3$). Due to the high turbulence intensity, a high breakup efficiency is expected. Thus, the mean of k'_B is calibrated to be relatively high ($\mu_l = 3.95 \times 10^{-5}$) compared to the two previous cases ([9,23]). Turbulent shear plays two conflicting and simultaneous roles in the flocculation process. The first one is to increase the chance of particle collisions resulting in floc aggregation. Turbulence is the most important mechanism of aggregation ([6,33,50,51]). The second role is disaggregation of flocs. Based on laboratory experiments, Tsai and Hwang [38] insist that a floc usually disaggregates into two roughly equal-sized daughter flocs. From this finding, it is deduced that the main mechanism of breakup is turbulent shear rather than particle collision. Figure 5 shows the effects of concentration and turbulent shear on mean floc size. Figure 5 shows that the high turbulent shear enhances the breakup process compared to aggregation process causing reduction of floc size ([2]). Therefore, the mean floc size of

Burban *et al.* [24] is relatively small (about 20 μm) resulting in a high value for μ_l. Figure 6 shows the results of numerical simulation by stochastic FGM and measurements from Burban *et al.* [24]. The volume fraction near the mean value is slightly overestimated by the stochastic FGM. However, the overall shape of the distribution is in good agreement with measurement results. The volume fraction slightly increases around floc size 80 μm due to the random nature of the stochastic model. A stochastic model is usually based on a random number generation technique (pseudorandom number generator in this study). Therefore, perturbations due to randomness are inevitable.

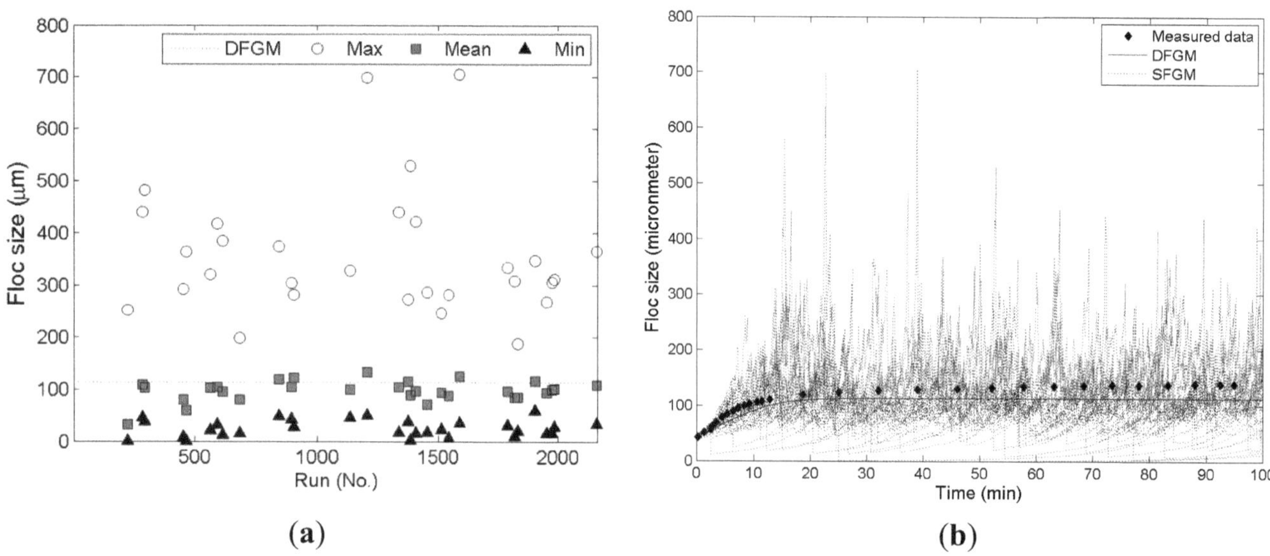

Figure 4. The random calculation of stochastic FGM. The dotted line of (**a**) shows the simulation results of deterministic FGM (DFGM). The circle, square, and triangle symbols represent the maximum, mean, and minimum values of each simulation, respectively; the solid and dotted lines of (**b**) are the results of deterministic FGM (DFGM) and stochastic FGM (SFGM), respectively; (a) mean, maximum, and minimum floc size (b) temporal evolution of floc size.

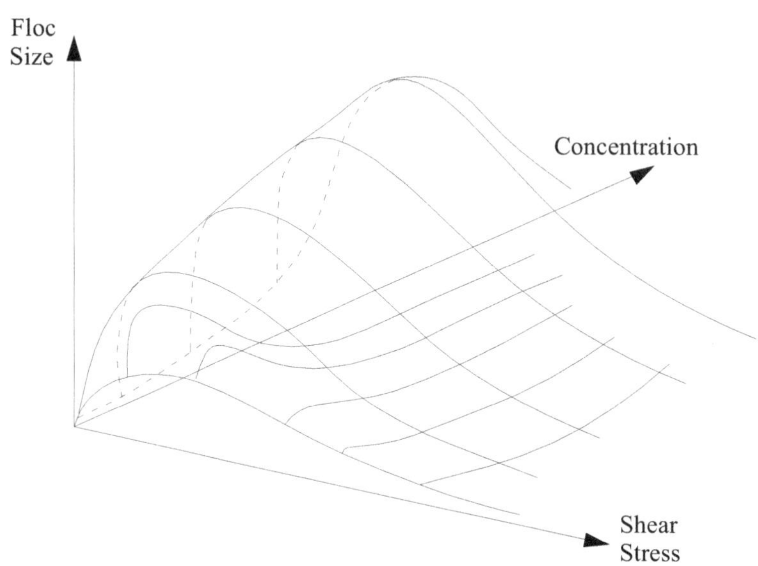

Figure 5. Conceptual diagram representing effect of turbulent shear on mean floc size.

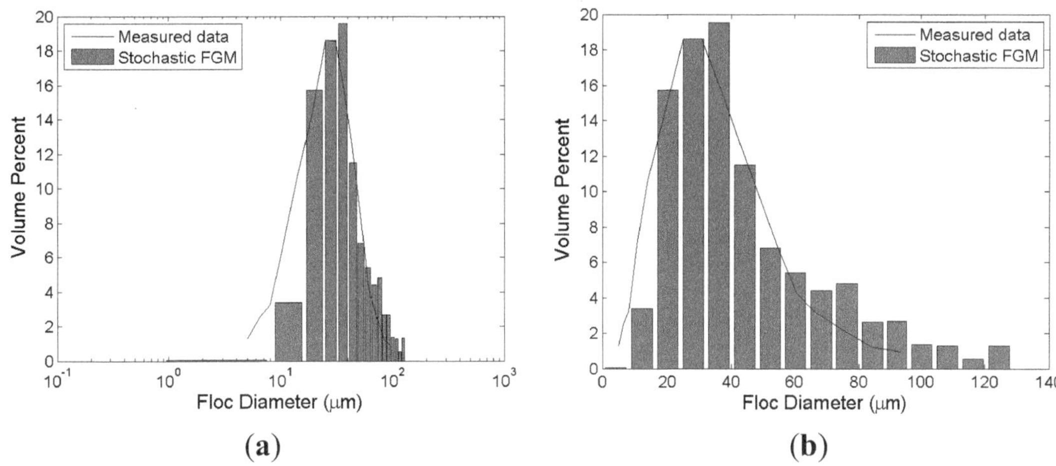

Figure 6. Simulation results of stochastic FGM and experimental results of Burban *et al.* [24]. The bars and the solid lines represent the experimental and simulation results, respectively. (**a**) semi-log axis; (**b**) normal (Cartesian) axis.

Three more probability distributions have been tested. The results of stochastic FGM using a normal distribution are drawn in Figure 7. The results are compared with Biggs and Lant [9]. A normal distribution has two parameters: mean and variance. The values of mean and variance of normal distribution are calibrated to be $2.620 \times 10^{-5.0}$ and $5.623 \times 10^{-5.0}$, respectively. Figure 8 shows the result of stochastic FGM using a Pearson type 3 distribution (Gamma distribution). A Pearson type 3 distribution needs 2 parameters: shape parameter (k_p) and scale parameter (θ). The mean and variance of Pearson type 3 distributions are indirectly determined by $k_p\theta$ and $k_p\theta^2$, respectively. Therefore, it is not simple to calibrate a stochastic FGM. The determined values of k_p and θ for the simulation are $4.780 \times 10^{-1.0}$ and, $5.248 \times 10^{-5.0}$, respectively. Compared to results of the log-normal distribution (Figure 3), FGMs using a normal distribution and a Pearson type 3 distribution overestimate the mean floc size, whereas the overall shape of size distribution is underestimated. Figure 9 represents the results calculated by stochastic FGM using a generalized extreme value (GEV) distribution. A GEV distribution has 3 parameters: mean, variance, and shape parameter (ξ). In this study, the mean, variance, and ξ of the GEV distribution are set to be $1.147 \times 10^{-5.0}$, $1.0 \times 10^{-5.0}$ and 0.5, respectively. In the case of the GEV distribution, the floc size larger than the mean value is overestimated. Although a GEV distribution has one more degree of freedom (ξ) compared to log-normal distribution, no significant advantage is found.

4. Conclusions

This paper presents a stochastic FGM to calculate the size distribution of flocs. A stochastic approach is applied to the breakup process under the assumption that the aggregation process is significantly affected by properties of the sediment and carrier flow. For the stochastic approach, the parameter related to the breakup efficiency (k'_B) is empirically assumed to have a log-normal distribution. From many previous studies under low organic content conditions (e.g., [15,35–37]), it is known that the size distribution of flocs is similar to a log-normal distribution. The capacity and limitations of the developed stochastic models are validated with three experimental results from Spicer *et al.* [23], Biggs and Lant [9], and Burban *et al.* [24]. These three experiments have different experimental conditions such as sediment property, turbulence intensity, and concentration.

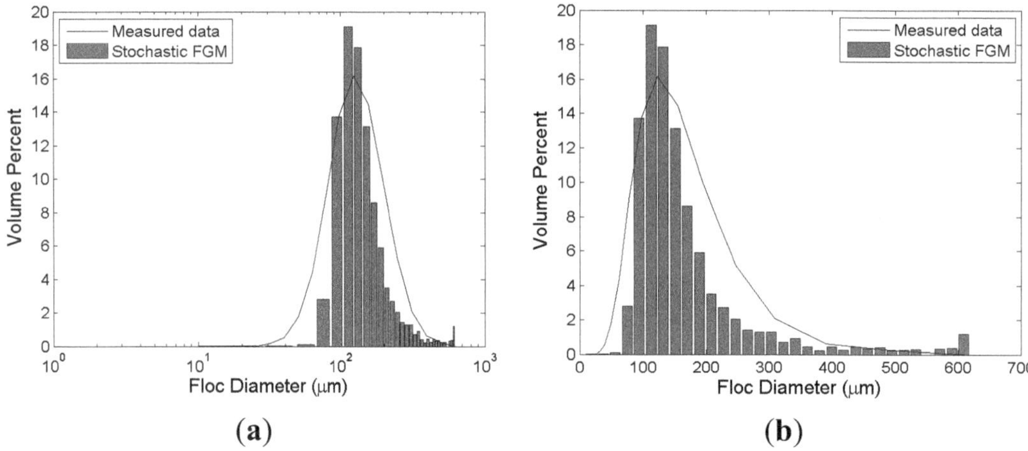

Figure 7. Simulation result of stochastic FGM using a normal distribution (compared to the experimental results of Biggs and Lant [9]). The bars and the solid lines represent the experimental and simulation results, respectively. (**a**) semi-log axis; (**b**) normal axis.

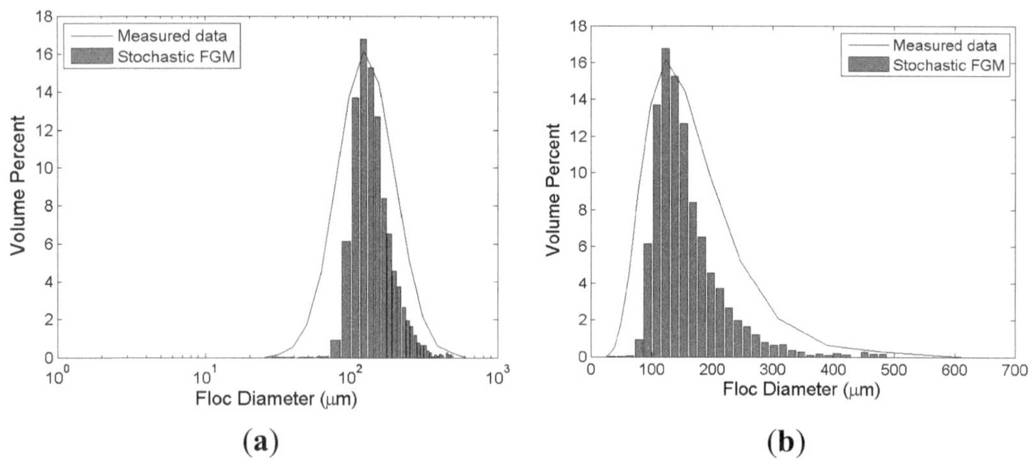

Figure 8. Simulation result of stochastic FGM using a Pearson type 3 distribution (compared to the experimental results of Biggs and Lant [9]). The bars and the solid lines represent the experimental and simulation results, respectively. (**a**) semi-log axis; (**b**) normal (Cartesian) axis.

As examined in Section 3, a stochastic FGM using a log-normal distribution has the capability to replicate the size distribution of flocs to a reasonable extent. Three different probability distributions such as normal, Pearson type 3, and GEV distributions are also tested in this study. Among them, an FGM using a log-normal distribution has the advantage of accuracy and simplicity. When a normal distribution is applied, a stochastic FGM underestimates the distribution of flocs larger than 200 μm, although the variance and mode are calculated in conformity. This finding shows that the positive skewness of k'_B is needed to calculate large flocs reasonably well. The positive skewness of k'_B decreases the breakup possibility for relatively large flocs. Flocs become less dense and more fragile as the size of floc increases. As a result, large flocs have higher possibility of breakup. A stochastic FGM using a Pearson type 3 distribution shows a mean-concentrated distribution of floc size. The skewness of a Pearson type 3 distribution is inversely proportional to the shape parameter $(2/\sqrt{k_p})$ whereas the variance is proportional to the shape parameter $(k_p\ \theta^2)$. Therefore, the variance decreases resulting in the mean-concentrated distribution when the skewness is increased. The result of a GEV distribution is in good agreement with measurement results. Compared to a log-normal distribution, a GEV distribution has one more degree of freedom as mentioned in Section 3. Thus, the calibration of a GEV distribution is not found to be complicated relative to a log-normal distribution.

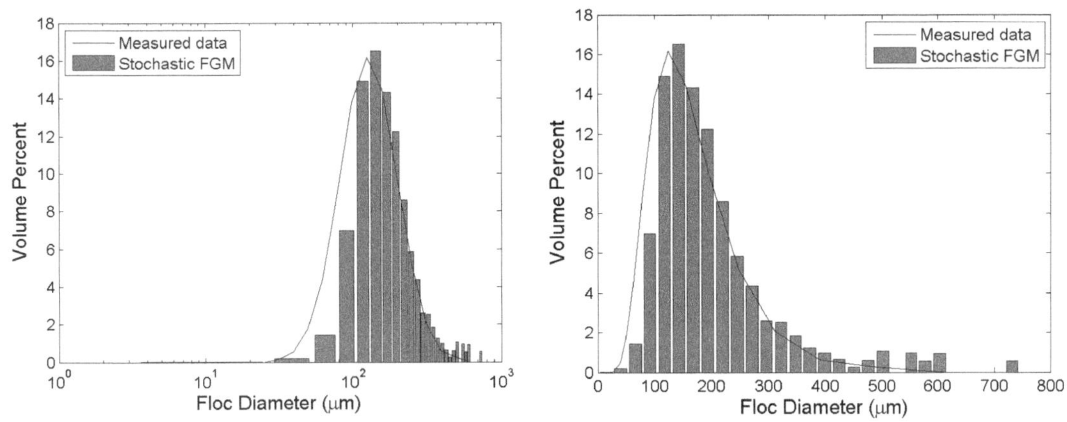

(a) (b)

Figure 9. Simulation result of stochastic FGM using a generalized extreme value distribution (compared to the experimental results of Biggs and Lant [9]). The bars and the solid lines represent the experimental and simulation results, respectively. (**a**) semi-log axis; (**b**) normal (Cartesian) axis.

An FGM has an advantage in the sense that it ensures mass conservation in a closed system and can be easily incorporated into the numerical sediment transport model. The previous FGMs, such as those used by Son and Hsu [7] and Winterwerp [8], calculate the average size of flocs (representative floc size). Therefore, the sediment transport model considering the flocculation process also assumes the suspension of mono-sized sediment (e.g., [25]). When a stochastic FGM discussed in this study is incorporated with the sediment transport model, it has the capability to calculate the size distributions of suspended cohesive sediments (flocs) as well as the temporal and spatial variation of floc concentration. Hence, future incorporation of these results (*i.e.*, stochastic FGMs) to sediment transport models will greatly improve resolution and uncertainty, furthering our knowledge of cohesive sediment transport processes.

In this study, the mean of k'_B (μ_l in Equation (3)) is assumed to be constant following the previous studies (e.g., [1,7,8]). k'_B is a parameter representing the probability of floc breakup. As the size of floc increases, the probability of breakup is also expected to increase because the yield stress of floc usually depends on floc size. Therefore, it is necessary to study the effect of floc size evolution through the flocculation process on calibration of k'_B in the future. It is also notable that the time step size may affect the calculation of floc size distribution. In this study, the time step for the numerical calculation is automatically determined by the Dormand-Prince method. In order to examine the effect of the time step, it is necessary to calculate the stochastic FGM with different fixed time step values.

Acknowledgments

This research was supported by the Basic Science Research Program through the National Research Foundation of Korea (NRF) funded by the Ministry of Education, Science and Technology

(2011-0010749), the research fund of Chungnam National University in 2014, and a project, "*Development of integrated estuarine management system*", funded by the Ministry of Oceans and Fisheries of Korea.

Author Contributions

H. J. Shin and M. Son conceived the experiments; M. Son wrote the paper; H. J. Shin performed the experiments; H. J. Shin and G. Lee analyzed the results.

Conflicts of Interest

The authors declare no conflict of interest.

References

1. Son, M.; Hsu, T.-J. The effect of variable yield strength and variable fractal dimension on flocculation of cohesive sediment. *Water Res.* **2009**, *43*, 3585–3592.
2. Dyer, K.R. Sediment processes in estuaries: Future research requirements. *J. Geophys. Res.* **1989**, *94*, 14327–14339.
3. Dyer, K.R.; Manning, A.J. Observation of the size, settling velocity and effective density of flocs, and their fractal dimensions. *J. Sea Res.* **1999**, *41*, 87–95.
4. Lick, W.; Lick, J.; Ziegler, C.K. Flocculation and its effect on the vertical transport of fine-grained sediments. In *Sed./Water Interactions, Hydrobiologia*; Hart, B.T., Sly, P.G., Eds.; Kluwer Academic Publishers: Dordrecht, The Netherlands, 1992; pp. 1–16.
5. Hunt, J.R. Particle dynamics in seawater: Implications for predicting the fate of discharged particles. *Environ. Sci. Tech.* **1982**, *16*, 303–309.
6. Stolzenbach, K.D.; Elimelich, M. The effect of density on collisions between sinking particles: Implications for particle aggregation in the ocean. *J. Deep Sea Res.* **1994**, I 13, 409–419.
7. Son, M.; Hsu, T.-J. Flocculation model of cohesive sediment using variable fractal dimension. *Environ. Fluid Mech.* **2008**, *8*, 55–71.
8. Winterwerp, J.C. A simple model for turbulence induced flocculation of cohesive sediment. *J. Hydraul. Res.* **1998**, *36*, 309–326.
9. Biggs, C.A.; Lant, P.A. Activated sludge flocculation: On-line determination of floc size and the effect of shear. *Water Res.* **2000**, *34*, 2542–2550.
10. Bouyer, D.; Line, A.; Do-quang, Z. Experimental analysis of floc size distribution under different hydrodynamics in a mixing tank. *AIChE* **2004**, *50*, 2064–2081.
11. Tsai, C.H.; Hwang, S.C. Flocculation of sediment from the Tanshui River estuary. *Mar. Freshw. Res.* **1995**, *46*, 383–392.
12. Tsai, C.H.; Iacobellis, H.S.; Lick, W. Flocculation of fine-grained sediments due to a uniform shear stress. *J. Great Lakes Res.* **1987**, *13*, 135–146.
13. Maggi, F.; Mietta, F.; Winterwer, J.C. Effect of variable fractal dimension on the floc size distribution suspended cohesive sediment. *J. Hydrol.* **2007**, *343*, 43–55.
14. McAnally, W.H.; Mehta, A.J. Aggregation rate of fine sediment. *J. Hydraul. Eng.* **2000**, *126*, 883–892.

15. Verney, R.; Lafite, R.; Brun-Cottan, J.C.; le Hir, P. Behaviour of floc population during a tidal cycle: Laboratory experiments and numerical modeling. *Cont. Shelf Res.* **2011**, doi:10.1016/j.csr.2010.02.005.
16. Lick, W.; Huang, H.; Jepsen, R. Flocculation of fine-grained sediments due to differential settling. *J. Geophys. Res.* **1993**, *98*, 10279–10288.
17. Hounslow, M.J.; Ryall, R.L.; Marshall, V.R. A discretized population balance for nucleation, growth and aggregation. *AIChE* **1988**, *34*, 1821–1832.
18. Maerz, J.; Verney, R.; Wirtz, K.; Feudel, U. Modeling flocculation processes: Intercomparison of a size class-based model and a distribution-based model. *Continental Shelf Res.* **2011**, doi:10.1016/j.csr.2010.05.011.
19. Levich, V.G. *Physicochemical Hydrodynamics*; Prentice Hall: Upper Saddle River, NJ, USA, 1962.
20. Khelifa, A.; Hill, P.S. Models for effective density and settling velocity of flocs. *J. Hydraul. Res.* **2006**, *44*, 390–401.
21. Maggi, F. Variable fractal dimension: A major control for floc structure and flocculation kinematics of suspended cohesive sediment. *J. Geophys. Res.* **2007**, *112*, doi:10.1029/2006JC003951.
22. Sonntag, R.C.; Russel, W.B. Structure and breakup of flocs subjected to fluid stresses. II. Theory. *J. Colloid Interface Sci.* **1987**, *115*, 378–389.
23. Spicer, P.T.; Pratsinis, S.E.; Raper, J.; Amal, R.; Bushell, G.; Meesters, G. Effect of shear schedule on particle size, density, and structure during flocculation in stirred tanks. *Power Tech.* **1998**, *97*, 26–34.
24. Burban, P.Y.; Lick, W.; Lick, J. The flocculation of fine-grained sediments in estuarine waters. *J. Geophys. Res.* **1989**, *94*, 8323–8330.
25. Son, M.; Hsu, T.-J. The effects of flocculation and bed erodibility on modeling cohesive sediment resuspension. *J. Geophys. Res.* **2011**, *116*, doi:10.1029/2010JC006352.
26. Van der Ham, R.; Fontijn, R.H.L.; Kranenburg, C.; Winterwerp, J.C. Turbulent exchange of fine sediments in a tidal channel in the Ems/Dollard estuary. Part I: Turbulence measurement. *Cont. Shelf Res.* **1994**, *21*, 1605–1628.
27. Son, M.; Lee, G.-H. On effects of skewed and asymmetric oscillatory flows on cohesive sediment flux: Numerical study. *Water Resources Res.* **2013**, *49*, 4409–4423, doi:10.1002/wrcr.20365.
28. Man, C.; Tsai, C.W. Stochastic partial differential equation-based model for suspended sediment transport in surface water flow. *J. Eng. Mech.* **2007**, *133*, 422–430.
29. Davie, A.M.; Gaines, J.G. Convergence of numerical schemes for the solution of parabolic stochastic partial differential equations. *Math. Comput.* **1999**, *70*, 121–134.
30. Gyongy, I. Lattice approximation for stochastic quasi-linear parabolic partial differential equations driven by space-time white noise II. *Potential Anal.* **1999**, *11*, 1–37.
31. Yoo, H. Semi-discretization of stochastic partial differential equations on R^1 by a finite-difference method. *Math. Comput.* **2000**, *69*, 653–666.
32. Maggi, F. Stochastic flocculation of cohesive sediment: Analysis of floc mobility within the floc size spectrum. *Water Resour. Res.* **2008**, *44*, doi:10.1023/2007WR006109.
33. McCave, I.N. Size spectra and aggregation of suspended particles in the deep ocean. *Deep Sea Res.* **1984**, *31*, 329–352.
34. Hill, P.S.; Boss, E.; Newgard, J.P.; Law, B.A.; Milligan, T.G. Observations of the sensitivity of beam attenuation to particle size in a coastal bottom boundary layer. *J. Geophys. Res.* **2011**, *116*, C02023.

35. Mikkelsen, O.A.; Milligan, T.G.; Hill, P.S.; Moffatt, D. INSSECT–and instrumented platform for investigating floc properties close to the seabed. *Limnol. Oceanogr. Methods* **2004**, *2*, 226–236.
36. Mikkelsen, O.A.; Hill, P.S.; Milligan, T.G.; Chant, R.J. *In situ* particle size distributions and volume concentrations from a LISST-100 laser particle sizer and a digital floc camera. *Cont. Shelf Res.* **2005**, *25*, 1959–1978.
37. Mikkelsen, O.A.; Hill, P.S.; Milligan, T.G. Single-grain, microfloc and macrofloc volume variations observed with a LISST-100 and a digital floc camera. *J. Sea Res.* **2006**, *55*, 87–102.
38. Syvitski, J.P.M.; Hutton, E.W.H. *In situ* charateristics of suspended particles as determined by the floc camera assembly FCA. *J. Sea Res.* **1996**, *36*, 131–142.
39. Xu, F.; Wang, D.-P.; Riemer, N. Modeling flocculation processes of fine-grained particles using a size-resolved method: Comparison with published laboratory experiments. *Cont. Shelf Res.* **2008**, *28*, 2668–2677.
40. Manning, A.J. Observation of the properties of flocculated cohesive sediment in three western European estuaries. *J. Coastal Res.* **2002**, *SI 41*, 70–81.
41. Manning, A.J.; Dyer, K.R.; Lafite, R.; Mikes, D. Flocculation measured by video based instruments in the Gironde estuary during the European Commission SWAMIEE Project. *J. Coastal Res.* **2004**, *SI 41*, 58–69.
42. Manning, A.J.; Bass, S.J.; Dyer, K.R. Observations of cohesive sediment settling fluxes in high energy tidal estuarine environments. *Mar. Geol.* **2006**, *235*, 177–192.
43. Manning, A.J.; Friend, P.L.; Prowse, N.; Amos, C.L. Estuarine mud flocculation properties determined using an annular mini-flume and the LabSFLOC system. *Cont. Shelf Res.* **2007**, *27*, 1080–1095.
44. Gratiot, N.; Manning, A.J. An experimental investigation of floc characteristics in a diffusive turbulent flow. *J. Coast. Res.* **2004**, *SI 41*, 105–113.
45. Gibbs, R.J. Coagulation rates of clay minerals and natural sediment. *J. Sediment Res.* **1983**, *53*, 1193–1203.
46. Lick, W.; Lick, J. Aggregation and disaggregation of fine-grained sediments. *J. Great Lakes Res.* **1988**, *14*, 514–523.
47. Mietta, F. Evolution of the Floc Size Distribution of Cohesive Sediments. Ph.D. Thesis, Delft University, Delft, The Netherlands, 2010.
48. Maggi, F. Flocculation Dynamics of Cohesive Sediment. Ph.D. Thesis, Delft University, Delft, The Netherlands, 2005.
49. Serra, T.; Casamitjana, X. Effect of the shear and volume fraction on the aggregation and break-up of particle. *Am. Instit. Chem. Eng.* **1998**, *44*, 1724–1730.
50. O'Melia, C.R. Aquasols: The behavior of small particles in aquatic systems. *Env. Sci. Technol.* **1980**, *14*, 1052–1060.
51. Van Leussen, W. Estuarine Macroflocs and Their Role in Fine-Grained Sediment Transport. Ph.D. Dissertation, University of Utrecht, Utrecht, The Netherlands, 1994.

Permissions

All chapters in this book were first published in Water, by MDPI; hereby published with permission under the Creative Commons Attribution License or equivalent. Every chapter published in this book has been scrutinized by our experts. Their significance has been extensively debated. The topics covered herein carry significant findings which will fuel the growth of the discipline. They may even be implemented as practical applications or may be referred to as a beginning point for another development.

The contributors of this book come from diverse backgrounds, making this book a truly international effort. This book will bring forth new frontiers with its revolutionizing research information and detailed analysis of the nascent developments around the world.

We would like to thank all the contributing authors for lending their expertise to make the book truly unique. They have played a crucial role in the development of this book. Without their invaluable contributions this book wouldn't have been possible. They have made vital efforts to compile up to date information on the varied aspects of this subject to make this book a valuable addition to the collection of many professionals and students.

This book was conceptualized with the vision of imparting up-to-date information and advanced data in this field. To ensure the same, a matchless editorial board was set up. Every individual on the board went through rigorous rounds of assessment to prove their worth. After which they invested a large part of their time researching and compiling the most relevant data for our readers.

The editorial board has been involved in producing this book since its inception. They have spent rigorous hours researching and exploring the diverse topics which have resulted in the successful publishing of this book. They have passed on their knowledge of decades through this book. To expedite this challenging task, the publisher supported the team at every step. A small team of assistant editors was also appointed to further simplify the editing procedure and attain best results for the readers.

Apart from the editorial board, the designing team has also invested a significant amount of their time in understanding the subject and creating the most relevant covers. They scrutinized every image to scout for the most suitable representation of the subject and create an appropriate cover for the book.

The publishing team has been an ardent support to the editorial, designing and production team. Their endless efforts to recruit the best for this project, has resulted in the accomplishment of this book. They are a veteran in the field of academics and their pool of knowledge is as vast as their experience in printing. Their expertise and guidance has proved useful at every step. Their uncompromising quality standards have made this book an exceptional effort. Their encouragement from time to time has been an inspiration for everyone.

The publisher and the editorial board hope that this book will prove to be a valuable piece of knowledge for researchers, students, practitioners and scholars across the globe.

List of Contributors

Jui-Yi Ho
Hydrotech Division, Taiwan Typhoon and Flood Research Institute, National Applied Research Laboratories, 11F., No. 97, Sec. 1, Roosevelt Rd., Taipei City 10093, Taiwan

Kwan Tun Lee
Hydrotech Division, Taiwan Typhoon and Flood Research Institute, National Applied Research Laboratories, 11F., No. 97, Sec. 1, Roosevelt Rd., Taipei City 10093, Taiwan
Department of River and Harbor Engineering, National Taiwan Ocean University, No 2, Beining Rd., Keelung City 20224, Taiwan

Yunliang Li
Key Laboratory of Watershed Geographic Sciences, Nanjing Institute of Geography and
Limnology, Chinese Academy of Sciences, 73 East Beijing Road, Nanjing 210008, China

Qi Zhang
Key Laboratory of Watershed Geographic Sciences, Nanjing Institute of Geography and
Limnology, Chinese Academy of Sciences, 73 East Beijing Road, Nanjing 210008, China
Key Laboratory of Poyang Lake Wetland and Watershed Research, Ministry of Education,
Jiangxi Normal University, 99 Ziyang Road, Nanchang 330022, China

Jing Yao
Key Laboratory of Watershed Geographic Sciences, Nanjing Institute of Geography and
Limnology, Chinese Academy of Sciences, 73 East Beijing Road, Nanjing 210008, China

Donatella Caniani
School of Engineering, University of Basilicata, viale dell'Ateneo Lucano n. 10, Potenza 85100, Italy

Donata Serafina Lioi
School of Engineering, University of Basilicata, viale dell'Ateneo Lucano n. 10, Potenza 85100, Italy

Ignazio M. Mancini
School of Engineering, University of Basilicata, viale dell'Ateneo Lucano n. 10, Potenza 85100, Italy

Salvatore Masi
School of Engineering, University of Basilicata, viale dell'Ateneo Lucano n. 10, Potenza 85100, Italy

Lei Huang
Shapotou Desert Research and Experimental Station, Cold and Arid Regions Environmental and Engineering Research Institute, Chinese Academy of Sciences, Lanzhou 730000, China
Key Laboratory of Stress Physiology and Ecology in Cold and Arid Regions Gansu Province, Lanzhou 730000, China

Zhishan Zhang
Shapotou Desert Research and Experimental Station, Cold and Arid Regions Environmental and Engineering Research Institute, Chinese Academy of Sciences, Lanzhou 730000, China
Key Laboratory of Stress Physiology and Ecology in Cold and Arid Regions Gansu Province, Lanzhou 730000, China

Marc Vis
Department of Geography, University of Zurich, Winterthurerstrasse 190, 8057 Zurich, Switzerland

Rodney Knight
U.S. Geological Survey Lower Mississippi—Gulf Water Science Center, 640 Grassmere Park, Suite 100, Nashville, TN 37211, USA

Sandra Pool
Department of Geography, University of Zurich, Winterthurerstrasse 190, 8057 Zurich, Switzerland

William Wolfe
U.S. Geological Survey Lower Mississippi—Gulf Water Science Center, 640 Grassmere Park, Suite 100, Nashville, TN 37211, USA

Jan Seibert
Department of Geography, University of Zurich, Winterthurerstrasse 190, 8057 Zurich, Switzerland
Department of Earth Sciences, Uppsala University, Villavägen 16, 752 36 Uppsala, Sweden

Chungman Moon
Department of Civil and Environmental Engineering, University of Windsor, 401 Sunset Ave.,

Windsor, ON N9B 3P4, Canada
Korea Institute of Energy Research, 152 Gajeong-ro, Yuseong-gu, Daejeon 305343, Korea

Rajesh Singh
Department of Civil and Environmental Engineering, University of Windsor, 401 Sunset Ave., Windsor, ON N9B 3P4, Canada
Central University of Gujarat, School of Environment and Sustainable Development, Sector-30, Gandhinagar 382030, India

Sathyanarayan S. Veeravalli
Department of Civil and Environmental Engineering, University of Windsor, 401 Sunset Ave., Windsor, ON N9B 3P4, Canada

Saravanan R. Shanmugam
Department of Civil and Environmental Engineering, University of Windsor, 401 Sunset Ave., Windsor, ON N9B 3P4, Canada

Subba Rao Chaganti
Great Lakes Institute for Environmental Research, University of Windsor, 401 Sunset Ave., Windsor, ON N9B 3P4, Canada

Jerald A. Lalman
Department of Civil and Environmental Engineering, University of Windsor, 401 Sunset Ave., Windsor, ON N9B 3P4, Canada

Daniel D. Heath
Great Lakes Institute for Environmental Research, University of Windsor, 401 Sunset Ave., Windsor, ON N9B 3P4, Canada
Department of Biological Sciences, University of Windsor, 401 Sunset Ave., Windsor, ON N9B 3P4, Canada

Marco Martello
Department of Agronomy, Food, Natural resources, Animals and Environment, Agripolis,
University of Padova, Viale Dell'Università 16, Legnaro (Padova) 35020, Italy

Nicola Dal Ferro
Department of Agronomy, Food, Natural resources, Animals and Environment, Agripolis,
University of Padova, Viale Dell'Università 16, Legnaro (Padova) 35020, Italy

Lucia Bortolini
Department of Land, Environment, Agriculture and Forestry, Agripolis, University of Padova,
Viale Dell'Università 16, Legnaro (Padova) 35020, Italy

Francesco Morari
Department of Agronomy, Food, Natural resources, Animals and Environment, Agripolis,
University of Padova, Viale Dell'Università 16, Legnaro (Padova) 35020, Italy

Wenjie Chen
School of Civil Engineering and Transportation, South China University of Technology, Guangzhou 510640, China

Chenghao Chen
Hydrology and Water Resources Survey Bureau of Hainan, Hainan 571126, China

Longbing Li
Hydrology and Water Resources Survey Bureau of Hainan, Hainan 571126, China

Litao Xing
Hydrology and Water Resources Survey Bureau of Hainan, Hainan 571126, China

Guoru Huang
School of Civil Engineering and Transportation, South China University of Technology, Guangzhou 510640, China
State Key Laboratory of Subtropical Building Science, South China University of Technology, Guangzhou 510640, China

Chuanhao Wu
School of Civil Engineering and Transportation, South China University of Technology, Guangzhou 510640, China

Rosa-Elena Yaya-Beas
Environmental Engineering Faculty, National University of Engineering, Lima 25, Peru
Sub-Department of Environmental Technology, University of Wageningen, Wageningen 6700 AA,The Netherlands

Christian Ayala-Limaylla
Environmental Engineering Faculty, National University of Engineering, Lima 25, Peru

Katarzyna Kujawa-Roeleveld
Sub-Department of Environmental Technology, University of Wageningen, Wageningen 6700 AA,The Netherlands
Lettinga Associates Foundation (LeAF), P.O. Box 500, Wageningen 6700 AM, The Netherlands

Jules B. van Lier
Section Sanitary Engineering, Department of Water Management, Faculty of Civil Engineering and Geosciences, Delft University of Technology, Delft 2600 GA, The Netherlands
Unesco-IHE Institute for Water Education, Westvest 7, Delft 2611 AX, The Netherlands

Grietje Zeeman
Sub-Department of Environmental Technology, University of Wageningen, Wageningen 6700 AA,The Netherlands
Lettinga Associates Foundation (LeAF), P.O. Box 500, Wageningen 6700 AM, The Netherlands

Jianli Zhang
State Key Laboratory of Simulation and Regulation of Water Cycle in River Basin,
China Institute of Water Resources and Hydropower Research, Beijing 100048, China
Department of Civil and Environmental Engineering (Dept 2470), North Dakota State University, PO Box 6050, Fargo, ND 58108-6050, USA

Xuefeng Chu
Department of Civil and Environmental Engineering (Dept 2470), North Dakota State University, PO Box 6050, Fargo, ND 58108-6050, USA

Xianghu Li
Key Laboratory of Watershed Geographic Sciences, Nanjing Institute of Geography and Limnology, Chinese Academy of Sciences, 73 East Beijing Road, Nanjing 210008, China

Xuchun Ye
School of Geographical Sciences, Southwest University, 2 Tiansheng Road, Chongqing 400715, China

Pengfei Li
Environment Research Institute, Shandong University, Jinan 250100, China

Jian Zhang
School of Environmental Science and Engineering, Shandong University, Jinan 250100, China

Huijun Xie
Environment Research Institute, Shandong University, Jinan 250100, China

Zhen Hu
School of Environmental Science and Engineering, Shandong University, Jinan 250100, China

Haiyan He
School of Environmental Science and Engineering, Shandong University, Jinan 250100, China

Wenxing Wang
Environment Research Institute, Shandong University, Jinan 250100, China

Hyun Jung Shin
Department of Oceanography, Inha University, Incheon 402-751, Korea

Minwoo Son
Department of Civil Engineering, Chungnam National University, Daejeon 305-764, Korea

Guan-hong Lee
Department of Oceanography, Inha University, Incheon 402-751, Korea